G F Walker.

Physico-Chemical Aspects of Food Processing

Physico-Chemical Aspects of Food Processing

Edited by

S.T. Beckett
Nestlé R&D Centre
York

BLACKIE ACADEMIC & PROFESSIONAL
An Imprint of Chapman & Hall

London · Glasgow · Weinheim · New York · Tokyo · Melbourne · Madras

Published by
Blackie Academic & Professional, an imprint of Chapman & Hall,
Wester Cleddens Road, Bishopbriggs, Glasgow G64 2NZ

Chapman & Hall, 2–6 Boundary Row, London SE1 8HN, UK

Blackie Academic & Professional, Wester Cleddens Road, Bishopbriggs, Glasgow G64 2NZ, UK

Chapman & Hall GmbH, Pappelallee 3, 69469 Weinheim, Germany

Chapman & Hall USA, 115 Fifth Avenue, Fourth Floor, New York, NY 10003, USA

Chapman & Hall Japan, ITP-Japan, Kyowa Building, 3F, 2-2-1 Hirakawacho, Chiyoda-ku, Tokyo 102, Japan

DA Book (Aust.) Pty Ltd, 648 Whitehorse Road, Mitcham 3132, Victoria, Australia

Chapman & Hall India, R. Seshadri, 32 Second Main Road, CIT East, Madras 600 035, India

First edition 1995

© 1995 Chapman & Hall

Typeset in 10/12 pt Times by Best-set Typesetter Ltd., Hong Kong
Printed in Great Britain by The University Press, Cambridge

ISBN   0 7514 0240 0

Apart from any fair dealing for the purposes of research or private study, or criticism or review, as permitted under the UK Copyright Designs and Patents Act, 1988, this publication may not be reproduced, stored, or transmitted, in any form or by any means, without the prior permission in writing of the publishers, or in the case of reprographic reproduction only in accordance with the terms of the licences issued by the Copyright Licensing Agency in the UK, or in accordance with the terms of licences issued by the appropriate Reproduction Rights Organization outside the UK. Enquiries concerning reproduction outside the terms stated here should be sent to the publishers at the Glasgow address printed on this page.

The publisher makes no representation, express or implied, with regard to the accuracy of the information contained in this book and cannot accept any legal responsibility or liability for any errors or omissions that may be made.

A catalogue record for this book is available from the British Library
Library of Congress Catalog Card Number: 95-79999

∞ Printed on acid-free text paper, manufactured in accordance with ANSI/NISO Z39.48-1992 (Permanence of Paper)

Preface

Food processing is now the biggest industry in the UK and in many other countries. It is also rapidly changing from what was essentially a craft industry, batch processing relatively small amounts of product, to a very highly automated one with continuously operating high speed production lines. In addition, consumers have developed a greater expectation for consistently high standard products and coupled this with demands for such things as a more natural flavour, lower fat etc. The need for an increased knowledge of the scientific principles behind food processing has never been greater.

Within the industry itself, increased automation, company diversification and amalgamations etc. have meant that those working in it have often to change their field of operation. Whereas twenty years ago, someone starting work in one branch of the food industry could expect, if he or she so desired, to work there all their working lives, this is now seldom the case. This means that a basic knowledge of the principles behind food processing is necessary both for the student at university or college, and for those already in the industry. It is hoped, therefore, that this book will appeal to both, and prove to be a useful reference over a wide range of food processing.

As a physicist working in the food industry, I was very enthusiastic when the publishers suggested producing a book on the physico-chemical aspects of food processing. This was because I felt that the physical side is often neglected, although it is often vital in determining the texture of many products and also plays a prime role in determining the operating performance of much processing plant. There are many books about the chemical changes which take place, and indeed it is not possible to understand food processing without some knowledge of them. In this book, however, the 'chemistry' chapters are restricted to those reactions, such as Maillard, which have a direct impact on the physical properties of food, like colour and texture.

A major concern, when planning the book, was that each chapter would describe water activity, glass transition etc. In order to try to overcome this, the book has deliberately been divided into two sections. The first part describes the most common physico-chemical properties from a scientific angle. Each author has been – and usually still is – involved in research in that particular field, and was asked to write it so as to give the layman a good understanding of the fundamentals behind their topic. This is obviously not an easy task, especially as they were told not to trivialise their subject. The title for each chapter is obviously the same as for many complete text books. Each author has therefore included a comprehensive list of references to enable the reader to study the subject in more detail, should he or she wish to do so.

The second part of the book looks at the changes which occur during food processing from an industrial point of view. Thus, for example, instead of saying what glass transitions are, the authors were asked to say where they occur within their processes and why they are important. The authors were also asked to include other physico-chemical aspects which were specifically applicable to their own particular product. The food industry is so wide that it is impossible to cover every part of it. Ten product areas were chosen as a representative selection and the final chapter added to cover some of the complex situations, where different types of food are combined in a single product. Although not being present as full chapters, other industries, e.g. wine and margarine, are reviewed elsewhere (in this case in the Fermentation and the Fat Eutectics and Crystallisation chapters respectively).

The international importance of food processing is emphasised in the multinational authorship of the book – 5 countries and three continents. In editing the book the enthusiasm of every author for their own topic has been very apparent and I am very grateful to all of them for their contributions.

Finally I would like to thank Nestlé for their permission to edit the book and to write one of the chapters, and those colleagues at the Research and Development Centre York, who aided me with various problems, in particular Rob Keeping and Caryn Douglas for overcoming word processing difficulties and Maxine Nicholson for her invaluable help with the references. I am also grateful to my wife, Dorothy, for her help in correcting texts and indexing and for putting up with me spending so many hours of my free time working on our home computer.

S.T.B.

Contributors

C.W. Bamforth	Director of Research, BRF International, Lyttel Hall, Coopers Hill Road, Nutfield, Redhill, Surrey, RH1 4HY, UK
S.T. Beckett	Nestlé R&D Centre, PO Box 204, York, YO1 1XY, UK
B. Bergenståhl	Institute for Surface Chemistry, Drottning Kristinas väg 45, PO Box 5607, S-114 86, Stockholm, Sweden
J.M.V. Blanshard	Faculty of Agriculture and Food Sciences, Department of Applied Biochemistry and Food Science, University of Nottingham, Sutton Bonington Campus, Loughborough, Leics, LE12 5RD, UK
P. Bowler	RHM Research & Engineering Ltd., High Wycombe, Bucks, UK
C.J.B. Brimelow	Nestlé R&D Centers, USA, Corporate Headquarters, 201 Housatonic Avenue, New Milford, CT 06776, USA
S.M. Clegg	Leatherhead Food RA, Randalls Road, Leatherhead, Surrey, KT22 7RY, UK
P.S. Dimick	The Pennsylvania State University, 111 Borland Laboratory, University Park, PA 16802, USA
M.C. Edwards	Campden and Chorleywood Food Research Association, Chipping Campden, Glos, GL55 6LD, UK
M.F. Eeles	Trebor Bassett Group Ltd., St Peters Street, Maidstone, Kent, ME16 0SP, UK
R.M. Gibson	Nestlé R&D Centre, York, Nestec York Ltd, PO Box 204, Haxby Road, York, YO1 1XY, UK
R.C.E. Guy	Campden & Chorleywood Food Research Association, Chipping Campden, Glos, GL55 6LD, UK
S.J. Haylock	Research Institute, New Zealand Dairy Private Bag 11029, Palmerston North, New Zealand

CONTRIBUTORS

T.M. Herrington — Department of Chemistry, University of Reading, Whiteknights, Reading, RG6 2AD, UK

J.E. Holdsworth — Cream Products Division, New Zealand Dairy Board, PO Box 417, Wellington, New Zealand

J.C. Hoskin — Clemson University Computer Center, Plant and Animal Science Building, 50 New Cherry Road, Clemson, SC 29634-2803, USA

S.J. James — Food Refrigeration & Process Engineering Research Centre, University of Bristol, Churchill Building, Langford, Bristol, BS18 7DY, UK

H.F. Jones — University of Wales Cardiff, Research & Commerical Department, PO Box 923, Cardiff, CF1 3TE, UK

V.Y. Loh — RHM Foods Ltd., Chapel House, PO Box 1445, 69 Alma Road, Windsor, Berks, SL4 3TG, UK

R.A. Marsh — RHM Foods Ltd., Chapel House, PO Box 1445, 69 Alma Road, Windsor, Berks, SL4 3TG, UK

C.A. Moules — CAMTEL Ltd., 5 Carrington House, 37 Upper King Street, Royston, Herts, SG8 9AZ, UK

C.A. Street — Broadoaks, Flaxton, York, YO6 7RJ, UK

G. Talbot — Loders Croklaan Ltd., Walden Court, Parsonage Lane, Bishop's Stortford, Herts, CM23 5DB, UK

F.C. Vernier — Department of Chemistry, University of Reading, Whiteknights, Reading, RG6 2AD, UK

E.J. Windhab — ETH-Zürich, Department of Food Science/Food Engineering, Universitätstrasse 2, CH-8092, Zürich, Switzerland

Contents

1 Vapour pressure and water activity   **1**
   T.M. HERRINGTON and F.C. VERNIER

   1.1 Introduction   1
   1.2 The importance of ERH   1
   1.3 Thermodynamic relationships   3
   1.4 Equilibrium and non-equilibrium systems   9
   1.5 Methods of measurement   11
   1.6 Humectants   13
   1.7 Bound and free water   14
   1.8 Conclusions   15
   References   15

2 The glass transition, its nature and significance in food processing   **17**
   J.M.V. BLANSHARD

   2.1 Nomenclature   17
   2.2 Introduction   18
   2.3 Characteristics of the glassy and rubbery states   18
   2.4 Characteristics of the glass–rubber transition   19
   2.5 Methods of determination of the glass–rubber transition   21
       2.5.1 Calorimetric techniques   21
       2.5.2 Molecular mobility   21
       2.5.3 Mechanical methods   23
       2.5.4 Other methods   24
   2.6 State diagrams   25
   2.7 Molecular mobility and diffusion in the glassy and rubbery states   28
       2.7.1 Molecular mobility   28
       2.7.2 Diffusion   30
       2.7.3 Models of diffusion   30
   2.8 Kinetics   32
   2.9 Plasticization and the glassy state   34
   2.10 More complex biopolymer systems   37
   2.11 The significance of the glass transition in frozen products   39
   2.12 The significance of the glass transition in the storage of foods at ambient temperatures   40
   2.13 Exploitation of the glassy and rubbery states in processing   42
       2.13.1 Understanding extrusion   42
       2.13.2 Understanding the storage stability of cereal products   45
   2.14 Maximizing product quality through the use of state diagrams in dehydration   46
   2.15 Conclusion   46
   References   47

3 Emulsions
B. BERGENSTÅHL
49

3.1 Introduction	49
3.2 The kinetics of instability	49
3.2.1 Sedimentation or creaming	50
3.2.2 Ostwald ripening	50
3.2.3 Coalescence	51
3.2.4 Film rupture	51
3.2.5 Flocculation	53
3.3 Droplet–droplet interactions	55
3.4 The oil–water interface	58
3.4.1 Competitive adsorption	59
3.4.2 Associative adsorption	61
3.4.3 Layered adsorption	61
3.5 Conclusions	62
Acknowledgements	63
References	63

4 Non-enzymatic browning of foods
J.C. HOSKIN and P.S. DIMICK
65

4.1 Introduction	65
4.2 History and name of reaction	65
4.3 Overview of reactions	66
4.4 Initial reactions	67
4.5 Intermediate reactions	68
4.6 Final reactions	73
4.7 Control of Maillard reactions	74
4.8 Other effects of Maillard reaction products	75
4.9 Conclusion	77
References	77

5 Rheology in food processing
E.J. WINDHAB
80

5.1 Introduction	80
5.2 Types of rheological flow	80
5.3 Rheological measurements	82
5.4 Rheological parameters	84
5.5 Food structure and rheological measurements	85
5.5.1 Equilibrium viscosity function, $\eta(\dot\gamma)$	86
5.6 Modelling of viscous behaviour in disperse systems	89
5.6.1 $\tau_{0.1}$ 'structure' model	89
5.6.2 Relative viscosity	91
5.6.3 Structure parameters for the $\tau_{0.1}$ model	92
5.6.4 Structure – viscosity functions	94
5.6.5 Intrinsic viscosity	95
5.7 Time-dependent viscous behaviour	99
5.8 The influence of temperature on rheological properties	102
5.9 Rheology in food processes	103
5.9.1 Relationship of process parameters and rheological behaviour	103
5.9.2 Continuous crystallization of fat systems	105
5.9.3 Emulsifying process for mayonnaise	107
5.9.4 Sensorial texture optimization of continuously processed ketchups	110

5.10	Conclusion	113
	References	115

6 Thickeners, gels and gelling   117
S.M. CLEGG

6.1	Introduction	117
6.2	Thickeners	118
	6.2.1 Intrinsic viscosity and coil dimensions	118
	6.2.2 Concentration dependency of viscosity and coil overlap	120
	6.2.3 Shear-rate dependency of viscosity	123
6.3	Viscoelastic nature of thickeners and gels	126
	6.3.1 Mechanical spectroscopy	127
	6.3.2 Viscoelastic behaviour of random-coil hydrocolloid solutions	127
	6.3.3 Mechanical spectra of hydrocolloid gels	129
6.4	Gels and gelling	130
	6.4.1 Critical gelling concentration and concentration dependency of gel rigidity	131
	6.4.2 Food gel structures	132
	6.4.3 Weak gels	138
6.5	Conclusion	140
	References	140

7 Fat eutectics and crystallisation   142
G. TALBOT

7.1	Glossary of useful terms	142
7.2	Introduction	145
7.3	Triglyceride structure	145
7.4	Molecular packing of triglycerides	147
7.5	Composition and structure of natural fats	149
	7.5.1 Cocoa butter	150
	7.5.2 Milk fat	154
	7.5.3 Palm oil	154
	7.5.4 Palm kernel and coconut oils	156
	7.5.5 Animal fats	156
	7.5.6 Hydrogenated fats	157
7.6	Mixtures of fats and eutectic effects	157
7.7	Blends with β-stable fats	158
	7.7.1 Cocoa butter–milk fat	158
	7.7.2 Cocoa butter–cocoa butter equivalent	159
	7.7.3 Cocoa butter–lauric fats	159
	7.7.4 Cocoa butter–hydrogenated fats	161
7.8	Migration of fats in composite products	162
7.9	Conclusion	166
	References	166

8 Surface effects including interfacial tension, wettability, contact angles and dispersibility   167
C.A. MOULES

8.1	Introduction	167
8.2	Surface/interfacial activity	168

	8.3	Liquid–gas interface		169
		8.3.1 Static/equilibrium measurements		169
		8.3.2 Time-dependent measurements		172
		8.3.3 Dynamic measurements		173
	8.4	Applications of static and dynamic surface tension measurements		175
		8.4.1 Equilibrium tests		175
		8.4.2 Dynamic tests		175
	8.5	Liquid–liquid interface		176
		8.5.1 Static/equilibrium measurements		176
		8.5.2 Dynamic measurements		178
		8.5.3 Applications		178
	8.6	The liquid–solid interface		179
		8.6.1 Contact angle and wetting		179
		8.6.2 Spreading wetting		181
		8.6.3 Adhesional wetting		182
		8.6.4 Immersional wetting		183
	8.7	Theories of wetting		183
	8.8	Other factors affecting wetting		185
	8.9	Measurement of contact angles		185
		8.9.1 Sessile drop method		186
		8.9.2 Wilhelmy technique		187
		8.9.3 Washburn technique		188
	8.10	Dispersion		191
	8.11	Conclusion		192
	References			192

9 Fermentation 193
R.M. GIBSON

9.1	Abbreviations and glossary		193
9.2	Introduction		193
9.3	Microbiology of fermentation		194
	9.3.1 Microbial growth		195
	9.3.2 Carbon source		195
	9.3.3 Nitrogen/sulphur source		195
	9.3.4 Other nutritional requirements		196
	9.3.5 Environmental conditions		196
9.4	Anaerobic fermentation		197
9.5	Characteristics of microbial growth		199
9.6	Secondary metabolite production		200
9.7	The application of fermentation in the beverage industry		202
	9.7.1 Wine		202
	9.7.2 Beer		203
	9.7.3 Vinegar		204
9.8	The application of fermentation in the food industry		205
	9.8.1 Dough fermentation		205
	9.8.2 Fermented milks		206
	9.8.3 Cheese fermentations		207
	9.8.4 Cocoa		208
	9.8.5 Vegetable fermentations		210
	9.8.6 Other food fermentations		210
9.9	Conclusions		211
References			211

10 Change in cell structure
M.C. EDWARDS
212

10.1	Introduction	212
10.2	Plant tissues	212
	10.2.1 Effects of moist heat on plant tissues	214
	10.2.2 Effects of processing on starch	217
	10.2.3 Effects of dry heat on plant tissues	218
	10.2.4 Oil frying of plant tissues	219
	10.2.5 Effects of microwaving on plant-cell structure	219
	10.2.6 Effects of freezing on plant-cell structure	219
10.3	Animal tissues	222
	10.3.1 Effects of freezing on muscle tissue	225
	10.3.2 Effects of heating on muscle tissue	225
	10.3.3 Effects of chemical additives on meat structure	227
	10.3.4 Effects of processing on fish cell structure	228
10.4	Conclusions	230
	References	230

11 Dairy products
J.E. HOLDSWORTH and S.J. HAYLOCK
234

11.1	Introduction	234
11.2	Milks	234
	11.2.1 Pumping	235
	11.2.2 Standardisation	235
	11.2.3 Homogenisation	235
	11.2.4 Heat processing	236
11.3	Butters	236
	11.3.1 Concentration	237
	11.3.2 Crystallisation	237
	11.3.3 Phase inversion	238
	11.3.4 Dispersion	239
	11.3.5 Continuous butter-making	239
	11.3.6 Scraped-surface processing	240
11.4	Creams	242
	11.4.1 Single creams	243
	11.4.2 Whipping creams	243
11.5	Cheese	244
11.6	Cultured products	248
11.7	Ice cream	250
11.8	Powdered consumer products	253
11.9	Conclusion	256
	Acknowledgements	256
	References	256

12 Cereal processing: The baking of bread, cakes and pastries, and pasta production
R.C.E. GUY
258

12.1	Introduction	258
12.2	Bakery products, pastry, cakes and bread	259
	12.2.1 Physical changes during the manufacture of baked products; pastries	259

	12.2.2 Cakes and flour confectionery	263
	12.2.3 Shelf-life of cakes	267
	12.2.4 Bread and morning goods	268
	12.2.5 Wafer products	270
12.3	Pasta products	271
12.4	Colour formation in cereal products	272
	12.4.1 Bakery products, pastries, cakes and bread	272
	12.4.2 Colour in pasta products	274
12.5	Conclusions	274
	References	274

13 Freezing and cooking of meat and fish   276
S.J. JAMES

13.1	Introduction	276
13.2	Freezing	277
	13.2.1 Pre-freezing treatment	277
	13.2.2 The freezing process	280
	13.2.3 During frozen storage	281
13.3	Cooking	283
	13.3.1 Cooking of small meat joints	284
	13.3.2 Cooking of whole poultry	286
	13.3.3 Cooking and cooling of large meat joints	287
13.4	Conclusions	288
	References	289

14 Fruits and vegetables   292
H.F. JONES and S.T. BECKETT

14.1	Introduction	292
14.2	Storage	293
14.3	Preparation	295
	14.3.1 Cleaning	295
	14.3.2 Sorting	296
	14.3.3 Peeling and cutting	297
	14.3.4 Blanching	298
14.4	Preserving	300
	14.4.1 Minimal processing	300
	14.4.2 Bottling and canning	301
	14.4.3 Freezing	302
	14.4.4 Dehydration	304
	14.4.5 Candying of fruit and osmotic dehydration	305
	14.4.6 Irradiation	305
	14.4.7 High pressure	306
14.5	Cooking	306
	14.5.1 Batch cooking	306
	14.5.2 Continuous cooking	308
14.6	Reconstitution	309
	14.6.1 Thawing	309
	14.6.2 Rehydration	310
14.7	Fruit juices	310
	14.7.1 Manufacture	310
	14.7.2 Aseptic packaging	312
14.8	Conclusion	312
	References	313
	Further reading	314

15 Preserves and jellies 315
P. BOWLER, V.Y. LOH and R.A. MARSH

15.1 Introduction 315
15.2 Pectin 316
15.3 Manufacturing processes 318
    15.3.1 Open pan boiling process 318
    15.3.2 Batch vacuum pan boiling process 319
    15.3.3 Continuous vacuum boiling process 320
    15.3.4 Processes for reduced-sugar products 320
    15.3.5 Selection criteria 321
15.4 Set 322
    15.4.1 Measurement of set 322
    15.4.2 The effects of process on set 323
    15.4.3 Raw material effects on set 324
15.5 Effects of pectolytic enzymes 327
15.6 Fruit flotation 328
15.7 High-pressure treatment 329
15.8 Conclusion 330
References 330

16 Sugar confectionery 332
M.F. EELES

16.1 Introduction 332
16.2 Products and processes 332
16.3 High-boiled sweets 334
16.4 Toffee and fudge 338
16.5 Fondant 344
16.6 Hydrocolloid-based sweets 344
16.7 Tableting 345
16.8 Conclusion 345
References 346
Further reading 346

17 Chocolate confectionery 347
S.T. BECKETT

17.1 Introduction 347
17.2 Cocoa beans 347
    17.2.1 Growing 347
    17.2.2 Fermentation and drying 349
    17.2.3 Roasting 350
    17.2.4 Cocoa liquor and powder production 351
17.3 Chocolate manufacture 353
    17.3.1 Grinding 353
    17.3.2 Conching 355
    17.3.3 Factors affecting chocolate viscosity 357
17.4 Chocolate usage 360
    17.4.1 Tempering 360
    17.4.2 Moulding and enrobing 362
    17.4.3 Cooling 363
17.5 Multiple component products 364
    17.5.1 Modified chocolates 364
    17.5.2 Fat and moisture migration 365

	17.6	Conclusion	366
	References		366
18	**Breakfast cereals and snackfoods** R.C.E. GUY	**368**
	18.1	Introduction	368
	18.2	Physico-chemical changes taking place during the technological processes used for breakfast cereals and snackfoods	369
		18.2.1 A general view of the initial changes in mixing and dough formation	369
		18.2.2 Processes involving native grains or grits for breakfast cereals	370
		18.2.3 Processes involving maize grains for tortillas and tacos	374
		18.2.4 Processes from grain involving puffing guns and chambers	376
		18.2.5 Low-moisture extruded products made from flours or grits – directly expanded breakfast cereals and snackfoods	378
		18.2.6 Intermediate-moisture-extruded products – half-products for snackfoods	383
	18.3	Colour and flavour formation	384
	18.4	Conclusions	385
	References		386
19	**Sauces, pickles and condiments** C.J.B. BRIMELOW	**387**
	19.1	Introduction	387
	19.2	History	387
	19.3	Definitions	388
	19.4	Acid preservation and preservative indices	389
	19.5	Vegetable/fruit treatments for sauces and pickles	391
		19.5.1 Fermentation brining	391
		19.5.2 Brining without fermentation	395
		19.5.3 Sugaring/syruping	396
		19.5.4 Concentration	397
		19.5.5 Canning and freezing	399
		19.5.6 Chilling and controlled atmospheres	400
	19.6	Pickle and sauce processing	401
		19.6.1 Processing of sweet pickles, relishes and chutneys	401
		19.6.2 Processing of clear pickles	404
		19.6.3 Processing of thick and thin sauces	405
		19.6.4 Mayonnaises, salad creams, dressings and spreads	408
	19.7	Conclusion	414
	References		415
20	**Beer and cider** C.W. BAMFORTH	**417**
	20.1	Introduction	417
	20.2	Beer foam	417
		20.2.1 Bubble formation	418
		20.2.2 Foam drainage	418
		20.2.3 Bubble coalescence	419
		20.2.4 Bubble disproportionation	420
		20.2.5 Surface tension	420

	20.2.6 Surface viscosity	421
20.3	Flavour	422
	20.3.1 The influence of foam on flavour	422
	20.3.2 Mouthfeel	422
	20.3.3 Flavour partitioning	425
20.4	Colour	425
20.5	The production of beer	426
	20.5.1 Malting	426
	20.5.2 Production of sweet wort	430
	20.5.3 Wort boiling	431
	20.5.4 Beer fermentation	433
	20.5.5 Beer processing and packaging	433
20.6	Cider-making	434
	20.6.1 Milling and pressing of apples	435
	20.6.2 Cider fermentation	435
	20.6.3 Cider processing and packaging	435
20.7	Conclusion	436
Acknowledgements		436
References		436

21 Multi-component foods   440
C.A. STREET

21.1	Introduction	440
21.2	Mathematical modelling of heat and mass transfer	441
21.3	Application of models to models to control and development of processes	444
21.4	Meat pies	448
21.5	Other examples of changes during the processing of multi-component foods	449
	21.5.1 Oxtail soup	449
	21.5.2 Pizza	449
	21.5.3 Chocolate chip cookies	450
	21.5.4 Choc ices	450
21.6	Frozen meals	451
21.7	Conclusion	451
References		452

Index   453

1 Vapour pressure and water activity
T.M. HERRINGTON and F.C. VERNIER

1.1 Introduction

Of the various criteria used to study food stability, undoubtedly the water activity is the most useful for expressing the water requirement for microorganism growth, enzyme activity and chemical spoilage. The amount of loss or gain of water by the foodstuff depends both on the nature and on the concentration of water vapour in the atmosphere. The desorbable or absorbable water present in a foodstuff may be expressed in terms of the equilibrium relative humidity or ERH.

$$\text{ERH} = (p^{\text{equ}}/p^{\text{sat}})_T, \quad p = 1 \text{ atm}$$

where p^{equ} is the partial pressure of water vapour in equilibrium with the sample in air at 1 atm. total pressure and at a temperature T and p^{sat} is the saturation partial pressure of water in air at a total pressure of 1 atm. and temperature T. In other words, this is a measure of the amount of water actually present in the air at equilibrium, divided by the amount which would be present if the air was saturated. T is important because the ERH is temperature-dependent. In the food industry, ERH is normally expressed as a percentage.

A foodstuff in moist air will exchange water until the equilibrium partial pressure at that temperature is equal to the partial pressure of water in the moist air, so that the ERH value is a direct measure of whether moisture will be sorbed or desorbed.

Measurement of ERH or water activity does not give any indication of the state of the water present, e.g. how is it bound to the substrate? To answer this question other techniques have been used, for example differential thermal analysis, dielectric relaxation and nuclear magnetic resonance, although the information is not always easy to analyse.

1.2 The Importance of ERH

Many biological materials in contact with the ambient atmosphere continuously adjust their moisture contents by sorbing or desorbing water. In agriculture and food technology, both the prevention of moisture exchanges (in the packaging and storage of foods) and the controlled removal of water to reduce weight and extend storage life (in dehydration and freeze-drying) are important. The water content of cereals is critical, partly because it is weight that must be paid for when grain is bought or sold. The moisture content may well vary from 8 to 16%, so that the buyer may be purchasing very expensive water! Also, wet grain occupies more shipping and storage space than dry grain; it also flows less readily, an important factor in conveying it from one place to another. Undoubtedly most important of all, the lower the moisture content, the longer grain and other agricultural products will store without deterioration. The rate of growth of insects and mites, as well as of micro-organisms, increases enormously with the moisture content, as does the rate of undesirable chemical changes. The rates of these processes generally increase with increasing temperature, so that more moisture can be tolerated at low temperatures.

The amount of desorbable water at a given temperature does not just depend on the actual concentration of water present in the sample, but it is a sensitive function of the nature and concentration of water-soluble substances present. The amount and rate of loss or gain also depends on the concentration of water vapour present in the atmosphere. This concentration varies over a wide range and it is necessary to have an index of whether the substance will sorb or desorb moisture under given storage conditions.

Microbiologists have long been familiar with the fact that the ability of bacteria to grow on a foodstuff is related to the state of the water present and not just the water content. Scott, while studying the growth of bacteria on beef carcasses in the 1950s, found that the relative humidity of the refrigerated storage chamber had a major influence and that there was a correlation between the water activity of foodstuffs and the ability of micro-organisms to grow on them [1]. Since the work of Scott, the concept of water activity has provided a basis for an increasing understanding of the influence of other factors on microbial growth at low values of the water activity. The water activity can be used to modify the metabolic production or excretion of a micro-organisms, and has been shown by Gervais to be a fundamental parameter in aroma production [2].

It is thus possible to minimize the growth of micro-organisms in food systems by reducing the availability of 'free' water. The water activity is a measure of the availability of water in an equilibrium system. However, in a non-equilibrium system, such as an intermediate moisture food, the water is effectively immobilized by high activation barriers, so that this water is not available to micro-organisms in a period of time commensurate with its shelf-life. Thus there are non-equilibrium ways of 'binding' water in which the water is inaccessible to the micro-organism because of kinetic rather than thermodynamic factors. Part of the art of the food technologist is to create a metastable system by ingenious means to defeat either microbial activity or unacceptable 'mouthfeel'. In toffee biscuits, the transfer of moisture between caramel and wafer is defeated by the low diffusion coefficient of water in a highly viscous medium. Margarine is an example of a system where the art lies in avoiding the true equilibrium state and its accompanying undesirable crystallization processes. The non-equilibrium state may be the result of high activation energies being necessary to enable the water to migrate. These may be produced by the physical state of the system, e.g. its viscosity, or it may result from chemical binding of the water. 'Bound water' and 'hydration' are terms often used to refer to both physically and chemically bound water. During food storage, moisture migration is commonly associated with deleterious changes in food quality. Crackers, potato chips and popcorn, all with initial water activity in the range 0.062–0.074, become unacceptable when this increases to 0.39, 0.51 and 0.49, respectively [3]. The crusts of frozen pizzas and pies can become soggy during frozen storage because of moisture migration from the filling or sauce (see chapter 21).

The effect of the water activity on the stability of the food depends on the humectants used (a humectant is an ingredient which has water-binding properties such as lecithin). The difficulties encountered in preserving foods, when only manipulating the water activity, have led to the development of the 'hurdle technology' [4]. Many processed products of high-moisture content are made stable by combining 'hurdles' such as pH with water activity. However, water activity is also important to food processing; for example, too low a water activity was found to increase the brittleness and lower the cutting resistance of a fruit bar [5]. The water activity of partially desiccated apple is a sensitive measure of the change in textural properties with dehydration [6]. Sauvageot and Blond found that there was a critical value of the water activity above which the crispness of breakfast cereals decreased rapidly [7]. The caking and stickiness of dairy-based food powders occurs when the water activity at the storage temperature is equal to that of the glass transition ([8] and chapter 2).

1.3 Thermodynamic relationships

The amount of water vapour in the atmosphere may be measured by its concentration or partial pressure. The partial pressure, p, of water vapour in a sample of air is defined as:

$$p = yP \tag{1.1}$$

where y is the mole fraction of water in the gaseous phase and P is the total pressure of air plus water vapour. The partial pressure of water vapour in saturated air at pressure P is not, in theory, the same as the vapour pressure of pure liquid water at that temperature. The difference is given by:

$$\ln p_{(P)}^{sat} - \ln p_{(p°)}^° = (P - p°)V°/RT \qquad (1.2)$$

where $p_{(P)}^{sat}$ is the partial pressure of water vapour in saturated air at total pressure P, $p_{(p°)}^°$ is the vapour pressure of water at the same temperature T, $V°$ is the molar volume of liquid water at temperature T and R is the gas constant. It has been assumed in equation (1.2) that the vapour phases are perfect gases; for real gases, p^{sat} and $p°$ should be replaced by the more exact functions p^{*sat} and $p^{*°}$. At ordinary pressures, the quantity on either side of equation (1.2) is negligible. Thus, for all practical purposes, p^{sat} can be replaced by $p°$, the vapour pressure of water at the same temperature, and this pressure dependency of the partial pressure of water vapour can be neglected.

As the amount of water lost from, or taken up by, a water-containing food depends on the nature and concentration of the water-soluble substances present, it is no use attempting to measure the quantity of water actually present in the foodstuff. Instead, we need to know the weight of water desorbed into the vapour phase by the time it reaches an equilibrium. This may be a large quantity, depending on the volume of the vapour phase. A direct measure of the quantity of water in the vapour phase is the equilibrium partial pressure, p^{equ}. Thus the partial pressure, p^{equ}, of the water vapour in equilibrium with the food, is taken as a measure of the concentration of desorbable water present. The rate of loss or gain depends on the value of p^{equ} in relation to the partial pressure of water vapour in the atmosphere. However, p^{equ} is a highly temperature-dependent quantity, so that in practice it is more convenient to find a quantity less dependent on temperature. We shall return to this point later.

In order to compare the effect of different substances on the partial pressure of an aqueous solution, it is helpful to define a reference level, with which the actual observed behaviour can be compared. An ideal binary solution, in which both components are liquids and obey Raoult's Law, is defined as one in which the change in molar Gibbs energy (ΔG) on mixing is given by the equation:

$$\Delta G_m^{Id}/RT = x_1 \ln x_1 + x_2 \ln x_2 \qquad (1.3)$$

where x_1 and x_2 are the mole fractions in the liquid phase of components 1 and 2, respectively. Now, for a real liquid mixture, the chemical potential (μ) of component i at temperature T and total vapour pressure P is given by:

$$\mu_i(T,P) = \mu_i^°(T,P) + RT \ln (p_i^{equ}/p_i^°) \qquad (1.4)$$

where $\mu_i^°(T,P)$ is the chemical potential of pure i at the same temperature T and total pressure P as the solution, P_i^{equ} is the partial pressure of component i in equilibrium with the solution and $p_i^°$ is the vapour pressure of pure i at the same temperature T and pressure P as the solution.

Now if the activity, a_i, of component i is defined by:

$$a_i = p_i^{equ}/p_i^° \qquad (1.5)$$

then:

$$\mu_i = \mu_i^° + RT \ln a_i. \qquad (1.6)$$

(To allow for the fact that the vapour is not an ideal gas, the activity should be more strictly defined as a fugacity ratio.)

The molar Gibbs energy change on mixing for a real binary solution is given by:

$$\Delta G_m = x_1(\mu_1 - \mu_1^°) + x_2(\mu_2 - \mu_2^°) \qquad (1.7)$$

and thus:

$$\Delta G_m/RT = x_1 \ln a_1 + x_2 \ln a_2 \qquad (1.8)$$

It is convenient to define an excess molar Gibbs energy of mixing of a binary solution by:

$$G_m^E = \Delta G_m - \Delta G_m^{Id} \qquad (1.9)$$

and hence:

$$G_m^E/RT = x_1 \ln f_1 + x_2 \ln f_2 \qquad (1.10)$$

where the activity coefficient, f_i, is defined by:

$$f_i = a_i/x_i. \qquad (1.11)$$

Thus the activity, or strictly speaking, the activity coefficient, is a measure of the deviation of component i from 'ideal' behaviour in a liquid mixture. There is nothing prescribed by the laws of thermodynamics in these concepts of 'ideality' and 'activity'; they are merely convenient definitions of a reference level and of departures from it.

Williamson [9] measured the total vapour pressure of mixtures of methanol and water and also the composition of liquid and vapour phases at equilibrium. The partial pressure of each component can then be calculated from equation (1.1) and the activity from equation (1.5). The results are shown in Table 1.1. The activity coefficient reflects the differences in interaction of, say, a molecule of component 1, with its neighbours in solution as compared with its neighbours in the pure liquid. In a molecule of methanol, one hydrogen atom of water has been replaced by a methyl group, and as can be seen from Table 1.1, the activity coefficient of water deviates more from unity, the greater the mole fraction of methanol. Solutions of methanol and water deviate considerably from the concept of ideal behaviour.

If the solution contains a substance with negligibly small vapour pressure

Table 1.1 Activites and activity coefficients of water plus methanol at 35°C (95°F) and 1 atm.

	Water			Methanol	
x_w	a_w	f_w	x_m	a_m	f_m
1.0000	1.0000	1.000	0.0000	0.0000	–
0.9000	0.9378	1.042	0.1000	0.1539	1.539
0.8000	0.8552	1.069	0.2000	0.2854	1.427
0.7000	0.7616	1.088	0.3000	0.3918	1.306
0.6000	0.6762	1.127	0.4000	0.4792	1.198
0.5000	0.5945	1.189	0.5000	0.5580	1.116
0.4000	0.5056	1.264	0.6000	0.6366	1.061
0.3000	0.4002	1.334	0.7000	0.7168	1.024
0.2000	0.2820	1.411	0.8000	0.8040	1.005
0.1000	0.1461	1.461	0.9000	0.8991	0.999
0.0000	0.0000	–	1.0000	1.0000	1.0000

then the definition of water activity based on equation (1.5) is not very helpful. Also substances of negligible vapour pressure are often solids in the pure state. It is also convenient to give the concentration of a solute in terms of its molality, m, which is defined as the number of moles of solute in a kilogram of solvent; the molality of a solute is unchanged on the addition of other solutes.

It is then more convenient to define an activity coefficient, γ_i, by:

$$\mu_i = \mu_i^\ominus + RT \ln \gamma_i m_i, \quad \text{where } \gamma_i \to 1 \text{ as } m_i \to 0 \quad (1.12)$$

where, $\mu_i^\ominus$ is the chemical potential of pure i in a hypothetical liquid state in which $\gamma_i = 1$ at the same temperature and pressure as the solution. The activity, a_i, is defined by:

$$a_i = \gamma_i m_i. \quad (1.13)$$

This definition of activity is convenient for solutes. For an aqueous solution of sucrose, equations (1.5) and (1.6) are used for the water activity and equations (1.12) and (1.13) for the activity of the sucrose. However, how is the activity coefficient of the sucrose determined? For a binary solution, the Gibbs–Duhem equation is:

$$x_1 d\mu_1 + x_2 d\mu_2 = 0. \quad (1.14)$$

Therefore:

$$d \ln a_2 = -(x_1/x_2) d \ln a_1 \quad (1.15)$$

$$= -(1/M_1 m) d \ln a_1 \quad (1.16)$$

where M_1 is the molar mass of solvent in kilograms per mole. Then the activity of solute, γ_2' at molality, m', is given by:

$$\ln \gamma_2' = \Phi - 1 + \int_0^{m'} (\Phi - 1) \mathrm{d} \ln m \tag{1.17}$$

where Φ, called the osmotic coefficient of the solvent, is defined by:

$$\Phi = -\ln a_1 / M_1 m. \tag{1.18}$$

Thus the activity coefficient of the solute can be calculated if that of the solvent is known over the range of integration.

The molecules of water and glycerol both contain hydroxyl groups, so that, in a mixture of the two, any one molecule of water is in an environment only slightly different from that in the pure liquid. In addition, the departure of liquid mixtures from ideal behaviour reflects the difference in the intermolecular forces present in the mixture from that in the unmixed liquids. It might be expected, therefore, that in a mixture of glycerol and water, the water would show only slight deviations from 'ideal' behaviour and the glycerol from 'semi-ideal dilute' behaviour (γ_i equal to unity in equation (1.12)). Values of the activity coefficients of water (solvent) and glycerol (solute) are given in Tables 1.2 and 1.3 [10]; it can be seen that, particularly in dilute solutions, the activity coefficients behave as predicted. Values for aqueous solutions of sucrose are also given in Tables 1.2 and 1.3 [11].

Like glycerol, the sucrose molecule has many hydroxyl groupings and probably intramolecular as well as intermolecular hydrogen bonds are formed; as is evident from the data, the water molecules in the presence of this larger organic molecule show greater deviations from ideal behaviour. It is the activity of the water in a solution which is closely related to the ability of that solution to absorb or desorb water from the ambient atmosphere. Consider the activity, a_w, of the water present in a solution, which from equation (1.5) is defined as:

$$a_w = p^{\mathrm{equ}} / p^\circ \tag{1.19}$$

Table 1.2 Activities and activity coefficients of aqueous solutions of glycerol at 25°C (77°F)

	Water			Glycerol	
x_w	a_w	f_w	Molality (mol kg^{-1})	γ
0.9982	0.9982	1.0000	0.1000	1.003
0.9823	0.9819	0.9996	1.0000	1.027
0.9652	0.9638	0.9985	2.0000	1.051
0.9487	0.9458	0.9969	3.0000	1.071
0.9328	0.9278	0.9947	4.0000	1.091
0.9174	0.9101	0.9920	5.0000	1.108
0.9025	0.8925	0.9889	6.0000	1.125

Table 1.3 Activities and coefficients of aqueous solutions of sucrose at 25°C (77°F)

Water			Sucrose	
x_w	a_w	f_w	Molality (mol kg^{-1})	γ
0.9982	0.9982	1.0000	0.1000	1.017
0.9823	0.9806	0.9983	1.0000	1.188
0.9652	0.9581	0.9926	2.0000	1.442
0.9487	0.9328	0.9832	3.0000	1.751
0.9328	0.9057	0.9709	4.0000	2.101
0.9174	0.8776	0.9567	5.0000	2.481
0.9025	0.8493	0.9411	6.0000	2.878

where p^{equ} is the partial vapour pressure of water in equilibrium with the solution, and p° is the vapour pressure of pure water at the same temperature and pressure as the solution.

Now the temperature dependence of the water activity is given by:

$$(\partial \ln a_w / \partial T)_p = (H_{H_2O} - H^\circ_{H_2O})/R \qquad (1.20)$$

where H_{H_2O} is the partial molar enthalpy of water in the solution and $H^\circ_{H_2O}$ is the molar enthalpy of water at the same temperature T. $H_{H_2O} - H^\circ_{H_2O}$ is called the relative partial molar enthalpy, L_{H_2O}. For a saturated solution of sucrose at room temperature L_{H_2O} (J mol^{-1}) is O(150), which gives $da_w/dT = 10^{-4}$ K^{-1} [11]. For a saturated sodium chloride solution [12] (6.1 mol kg^{-1}), $L_{H_2O}/$(J mol^{-1}) is 0(50); L_{H_2O} has a maximum value of 90 J mol^{-1} for a 4.6 mol kg^{-1} solution. Thus, again the temperature dependence of the water activity is small. Refrigeration is often used as a means of lowering the water activity of a foodstuff, but a decrease in temperature of over 20°C (36°F) was found to decrease the activity of kamaboko and poi, ethnic Hawaiian foods, by <0.1 [13].

As has been pointed out, the activity, a_w, as defined by equation (1.19), is strictly a fugacity ratio determined at a constant overall pressure, equal (the vapour pressure of the solution). For practical purposes, the equilibrium relative humidity or ERH, defined by the equation on p. 1 is used. It is conventional to use ERH $\simeq 100 a_w$. From equation (1.2) the pressure dependence of both p^{sat} and p^{equ} is negligible and p^{sat} may be replaced by p°. The ERH is temperature-dependent, but, as shown by equation (1.20), this can sometimes be ignored; p^{equ} and p^{sat} must, of course, both refer to the same temperature T. In certain circumstances, however, the ERH is highly temperature-dependent. If, for example, a saturated solution is heated in the presence of excess solute, the concentration of solute usually increases and hence the ERH will change with increasing temperature.

1.4 Equilibrium and non-equilibrium systems

In an equilibrium system, the water activity or ERH is a measure of the availability of the water, but in a non-equilibrium system the thermodynamic concept of water activity cannot be defined. With many biological materials, a different value of the equilibrium partial vapour pressure is reached, depending on whether the material is sorbing or desorbing moisture, i.e. hysteresis occurs, and also the ERH is temperature-dependent. Figure 1.1 shows some data for cellulose at 25°C (77°F) and 35°C (95°F) [14]; the ERH increases with increasing temperature. Isotherms for cellulose with adsorbed sucrose (10^{-3} mol of sucrose adsorbed on 1 g of cellulose) are also shown at 25°C (77°F) and illustrate the

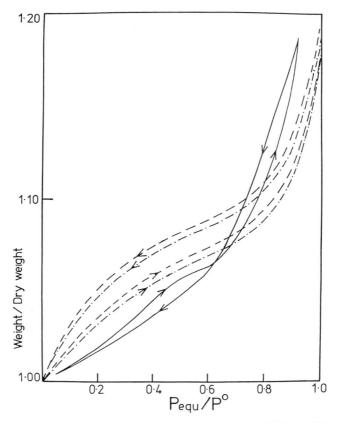

Figure 1.1 Adsorption isotherms of stabilized cellulose. --- stabilized cellulose at 25°C (77°F); –·–·– stabilized cellulose at 35°C (95°F); ——— 10^{-3} mol sucrose adsorbed on 1 g cellulose at 25°C (77°F).

sensitivity of the ERH to the nature of the adsorbate. From its definition, the use of the term 'water activity' is restricted to systems in true thermodynamic equilibrium. The observation of a hysteresis loop raises the question as to whether in one arm of the cycle the system is not in true equilibrium or whether the substrate has been altered by dehydration. Is the movement of water between substrate and atmosphere effectively controlled by kinetic rather than by thermodynamic criteria? In the food industry, dough is a good example of a system in which different 'pseudo-equilibria' are produced by kinetic barriers to free-water diffusion. For a dough of given composition, two entirely different but apparently steady water-vapour pressures may be obtained, depending on whether water is added to the flour before or after the flour has been intimately mixed with the fat (see also chapter 12). Indeed a lot of effort and ingenuity has been concentrated in intermediate moisture food technology in deliberately creating metastable systems so that water in unavailable for use by microorganisms.

A complete analysis of the equilibrium moisture content of a sample at a given temperature is given by its sorption isotherm. The isotherm shows the variation of ERH with change in moisture content of the sample. Most foodstuffs show at least some hysteresis and it must be appreciated that a sample may have an ERH intermediate between that for sorption and desorption, depending on the fluctuations in atmospheric humidity which it has experienced. For many macromolecular materials, the sorption isotherm of the same sample gives a changing series of sigmoid curves. Thus for some materials, particularly those which exhibit marked hysteresis effects, a knowledge of the moisture content does not necessarily give the ERH, even if the sorption isotherm has been determined. Nevertheless, the sorption isotherm of a foodstuff containing water-soluble crystalline compounds, e.g. confectionery products or milk powders, can be useful; the isotherm may show sharp inflexions and discontinuities at certain humidities. The sigmoid isotherm is characteristic of cellulose and of proteins, such as casein, so that it is typical of many foodstuffs. Various models have been proposed for the isotherm, but probably the Brunauer–Emmett–Teller theory (although not strictly applicable to hydrogen-bonded systems) is most useful for food storage and processing. According to this theory, the first portion of the isotherm (Figure 1.2), where the curve is concave to the humidity axis, represents the adsorption of the first layer of water; the second layer of water is adsorbed in the region of inflexion; the final curved portion shows the continued adsorption of additional layers. Thus, the first part of the isotherm represents the forces of interaction between the matrix substrate and water; the greater the displacement towards the moisture content (vertical) axis, the greater the intermolecular forces and the binding energies may indicate chemisorption. In the second and third portions of

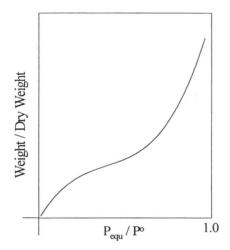

Figure 1.2 Sigmoid isotherm according to the Brunauer–Emmett–Teller theory.

the isotherm, the attractive forces successively approach those present in liquid water.

1.5 Methods of measurement

A complete analysis of the vapour pressure of water in equilibrium with a sample is given by its sorption isotherm. The isotherm shows the variation of equilibrium relative humidity with change in moisture content of the sample. The sorption isotherm can be determined by classical methods in which the water vapour pressure at equilibrium is determined using a manometer. However, equilibrium methods are slow since the system takes time to equilibrate. In dynamic methods, air is repeatedly circulated over the sample in an enclosed system, which hastens the attainment of equilibrium. Non-equilibrium methods using sorption rates have been successfully applied to materials which readily form surface layers of syrup or crystals impeding moisture transfer.

In many instruments, the ERH or water activity is determined in the enclosed space above an equilibrated foodstuff. In the isopiestic technique, the water vapour pressure above the sample is determined at a constant temperature and pressure equal to the equilibrium vapour pressure. Under laboratory conditions, the measurement is carried out in an initially evacuated vessel, against standard salt solutions of known water activity, in a thermostat. Several days are required to reach equilibrium. The method can only readily be used some ±10°C from ambient and is very demanding in experimental skill. The salt solution, usually lithium chloride, because of its high solubility and therefore water activity range, loses or gains moisture until it has the same partial vapour pressure of water as the sample; measurement of conductivity gives the salt concentration at equilibrium. Alternatively, for samples of high water activity, a salt slush is used and weighed before and after equilibration. The method can also be used to bracket a sample by using several salt slushes covering a range of water activities and weighing as a function of time. An ingenious adaptation of the isopiestic method is the Proximity Equilibration Cell, in which the weight gain of a pre-dried filter paper, equilibrated over the sample is determined [15].

Instead of weighing, indirect electrical methods have been used to determine the moisture content using a suitable probe. However, these so-called 'electronic hygrometers' use sensors that have to be frequently recalibrated using standard salt slushes. The method has been adapted for several commercial instruments in which the sample is mounted in a small closed container in air and a suitable probe used to determine the water vapour pressure at equilibrium at a specified temperature. A variety of probes have been used. The electrolytic hygrometers of Dunmore [16] and Spencer-Gregory and Rourke [17] use salt solution as a sensor. The concentration of the salt solution is determined by conductivity. Several modern instruments adopt this approach. The method has been used for foods of widely differing ERH values from chocolate to meat in the range of 30–90%. In capacitance hygrometers the capacitance change of a thin polymer or anodized aluminium film is measured. The film adsorbs water and the resulting change in capacitance is proportional to the equilibrium relative humidity. The method is suitable for ERH values from 1 to 99% over a wide range of temperature. In other commercial instruments the electrical impedance of a hygroscopic liquid or polystyrene wafer is used to determine the moisture content. Again this can be used for a wide range of ERH values. A robust remote sensor recently developed is an analogue transmitter, which uses a robust probe of moisture absorbing polymer overlayed onto an interdigitated gold terminal [18]. At the other end of the range, many commercial instruments are simple adaptations of the traditional hair hygrometer. These use hygroscopic polyamide filaments, whose length is proportional to the humidity. The elongation of the filament is magnified by springs and a direct-dial read out obtained. Commercial instruments are also available based on the dew-point principle, whilst others automatically work out a set of isotherms by changing the temperature and vapour pressure conditions.

Recently, many innovative techniques for determining water activity have been appearing in the literature. At the moment these are only for laboratory use, but it is anticipated that commercial instruments using these methods will be available in the not-too-distant future. Delwiche *et al.* have used a near-infrared reflectance method to study the water activity and moisture content of starch and cellulose samples [19]. Infrared diode laser absorption spectroscopy has been use by Bone *et al.* for the accurate determination of the water vapour concentration in the vapour space above a liquid sample [20]. The method is based on the measurement of the fractional absorption, at line centre, of a single ro-vibrational line.

There is an absence of reliable methods to determine the water activity at temperatures above 100°C (212°F); however, a new manometric, dynamic method suitable for laboratory use has recently been developed [21]. The experimental measurement of water vapour sorption isotherms of mixed solutes is both time-consuming and expensive. Several methods of estimation have been devised for mixtures and have now become more sophisticated with easy access to fast computers [22].

1.6 Humectants

Humectants are added to adjust the water activity of a foodstuff. The choice of these is necessarily limited by the requirement of non-toxicity. The humectant must also be compatible with the food, effective at low concentrations, flavourless at these concentrations and not producing any colour changes. Glycerol (glycerine), sodium chloride, propylene glycol and various sugars, such as fructose, sucrose and corn syrups are common humectants. The growth of micro-organisms for a given water activity may be different for different solutes. For example, Jakobsen and Trolle [23] found that glycerol would support the growth of a given micro-organism, whereas sodium chloride would not.

Many confectionery products contain layers of different sugary material such as nougat, caramel, marshmallow, fudge, fondant cream, chocolate, wafer, biscuit, etc. [Movement of water takes place as each layer, with a different water activity trying to reach equilibrium.] For example, in a wrapped candy prepared using caramel and nougat of the same water activity, there would be no water transfer, but if a layer of marshmallow of higher water activity is added, then both caramel and nougat will take up water from the marshmallow. The caramel will soften and crystallize to a fudge-like consistency and the marshmallow will assume a rubbery consistency [24]. The problem is compounded if the candy is placed on a biscuit base, since biscuit must have a water activity below 0.35 to remain crisp [25,26] (see also chapter 17). Thus the requirement for a quality confectionery product is that each layer has the same water activity. The water activity can be increased, in some of the components, by adding humectants such as glucose syrup, invert sugar or glycerol. The product must still be protected from moisture in the environment by coating with chocolate or enclosing in a water-impermeable packaging material.

1.7 Bound and free water

Although considerations of solute–water interaction in solution are helpful in considering the ERH behaviour of food products such as syrups and fudges, many basic foodstuffs, such as grain, consist of bio-colloidal systems, with, for example, a cellulose system acting merely as a matrix for the sorption and desorption of water. In these systems, is the water held by chemical or physical forces? In a bio-colloidal system, which is made up of various substances and which also possesses an organized structure, there will be many types of water binding. This will range from free water, held in the intergranular spaces and pores of the material by intermolecular forces of the same order of magnitude as hydrogen-bonding energies ($16\,\text{kJ}\,\text{mol}^{-1}$) to chemically bound water forming an integral part of the organic molecule ($400\,\text{kJ}\,\text{mol}^{-1}$). Binding energies intermediate between these two are termed 'chemisorption'.

The attractive forces between molecules comprise several kinds of interaction such as dipole–dipole, dipole–induced dipole and ion–dipole. Water with a large dipole moment for its molecular size plays a prominent part in adsorption phenomena. Starch and protein are the main constituents of many foodstuffs and have many polar sites available for interaction with water. Starch is a high polymer built up from a basic glucose unit into a long or branched chain; hydroxyl groups on the glucose ring, the ring oxygen and the bridge oxygen may all hydrogen bond with water molecules. A protein has a polymeric structure, but here the repeating units are not identical and may be any of the naturally occurring amino acids; a polypeptide backbone is formed with side groups containing, for example, a hydroxyl group in serine, threonine and tryosine, and an amino group in lysine. Water is able to attach itself by hydrogen binding to the hydroxyl and amino groups.

An operational definition of 'bound water' that has been used is that of Kuntz and Kauzmann [27]. Bound water is 'water in the vicinity of macromolecules whose properties differ detectably from those of bulk water in the same system'. However, the disadvantage is that an estimate of the hydration depends very much on the physical property that is used to measure it. The hydration of sugars and polymer molecules is closely connected with the concept of the glass-transition temperature, T_g^* (chapter 2). The connection between the rates of crystallization of boiled sweets and the glass-transition temperature of the sucrose–dextrose–fructose–water system was first discovered by Branfield and Herrington [28]. Since then, a seemingly exponential growth of interest in concepts centred around the glassy state phenomenon and glass-transition temperature in foods has taken place [29]. The glass-transition temperature affects the food quality, both structural and textural, and the long-term storage stability, particularly with respect to micro-organisms. At the glass-transition temperature of a hydrated sugar or protein, the rate of change of viscosity with temperature increases sharply. This in turn affects the water availability, due to the changes in physical properties of the system.

1.8 Conclusions

The concept of water activity applies strictly to an equilibrium situation. For practical purposes, the equilibrium relative humidity (ERH) can be used. The water activity is not necessarily a measure of available water, because the movement of water may be controlled by kinetic factors. Many foodstuffs are in a non-equilibrium situation with respect to a determined water vapour pressure, which may have physical or chemical causes. Food humectants are used to achieve a required water activity, and it is necessary to know their relative rate of approach to their equilibrium values. The concept of 'bound' and 'free' water has been illustrated using the techniques of NMR (nuclear magnetic resonance), dielectric relaxation and DSC (differential scanning calorimetry) to determine the state of the water within a foodstuff. Each method has its own inherent timescale so quantitative agreement of the degree of binding will differ. Measurement of the degree of hydration in an equilibrium situation is necessary to understand the distribution of water between the different phases of a foodstuff in the non-equilibrium situation. However, the primary parameter for food quality is knowledge of the rate of approach to the equilibrium state.

In order to ascertain the best method for determining the water activity of a given product, it is necessary to take account of both the shape of the sorption isotherm, so that it is known in what water-absorption range accurate discrimination is required, and the working temperature. Modern methods of assessing the bound water, together with classical equilibrium methods, are helpful in assessing the shelf-life of a food product.

References

1. Scott, W.J. *Adv Res*, **7** (1957), 83.
2. Gervais, P. *Appl Microbiol Biotechnol*, **33** (1990), 72.
3. Katz, E.E. and Labuza, T.P., *J Food Sci*, **46** (1981), 403.
4. Leistner, L. In *Water Activity: Theory and Applications to Food*, Ed. Rockland, L.B. and Beuchat, L.R. Marcel Dekker, New York (1987), 287.
5. Owen, S.R., Tung, M.A. and Durance, T.D. *J Texture Studies*, **22** (1991), 191.
6. Bourne, M.C. *J Texture Studies*, **17** (1986), 331.
7. Sauvageot, F. and Blond, G. *J Texture Studies*, **22** (1991), 423.
8. Chuy, L.E. and Labuza, T.P. *J Food Science*, **59** (1994), 43.
9. Williamson, A.G., PhD Thesis, University of Reading (1957).

10. Scatchard, G., Hamer, W.J. and Wood, S.E. *J Am Chem Soc*, **69** (1938), 3061.
11. Dunning, W.J., Evans, H.C. and Taylor, M. *J Chem Soc* (1951), 2363.
12. Randall, M. and Bisson, C. *J Am Chem Soc*, **42** (1920), 347.
13. Walter, R.H. and Seeger, S.C. *J Food Protection*, **53** (1990), 72.
14. Dobney, P.W., Cox, C. and Herrington, T.M. Unpublished data.
15. Xavier, S. and Karanth, N.G. *Lett Appl Microbiol*, **15** (1992), 53.
16. Dunmore, F.W. *J Res Nat Bur Stds*, **23** (1939), 701.
17. Spencer-Gregory, H. and Rourke, E. *Hygrometry*, Crosby Lockwood, London (1957).
18. Schultz, G. *Med Electr* (1993), 90.
19. Delwiche, S.R., Pitt, R.E. and Norris, K.H. *Cereal Chem*, **69** (1992), 107.
20. Bone, S.A., Cummins, P.G., Davies, P.B. and Johnson, S.A. *Appl Spectroscopy*, **47** (1993), 834.
21. Bassal, A. *Drying Technol*, **12** (1994), 427.
22. Peleg, M. and Normand, M.D. *Trends Food Sci Technol*, **3** (1992), 157.
23. Jakobsen, M. and Trolle, G. *Nord Vet-Med*, **31** (1979), 206.
24. Morley, G.M. and Groves, R. *Candy Industry*, **152** (1987), 44.
25. Manley, D.J.R. *Technology of Biscuits, Crackers and Cookies*, Ellis Horwood (1983).
26. Cakebread, S.H. *Confect Prod*, **42** (1976), 172.
27. Kuntz, I.D. and Kauzmann, W. *Adv Protein Chem*, **28**, 239.
28. Branfield, A.C. and Herrington, T.M. *J Food Technol*, **19** (1984), 427.
29. Slade, L. and Levine, H. *Water and Food Quality*, Ed. Hardman, T.M. Elsevier Applied Science, London (1989), 37.

2 The glass transition, its nature and significance in food processing
J.M.V. BLANSHARD

2.1 Nomenclature

The following symbols are used in the equations and text of this chapter:

- a_w Water activity (see chapter 1)
- α_p Coefficient of expansion under constant pressure
- a_T Ratio of the relaxation phenomenon at temperature T to the relaxation at the reference temperature T_{ref}
- C'_g Concentration of the unfrozen matrix at the glass-transition temperature T'_g
- C_p Specific heat at constant pressure
- De   Deborah number = λ/θ
- δ Phase angle, where $\tan\delta = E''/E'$
- E' Elastic or storage modulus (equivalent to G', chapter 5)
- E'' Loss modulus (equivalent to G'', chapter 5)
- H Enthalpy – heat content per unit mass
- k Rate of reaction at temperature T
- k_g Rate of reaction at the glass transition
- λ Characteristic relaxation time, i.e. the time for a reaction/change to take place
- η Viscosity
- M_t Mass of water taken up at time t
- θ Characteristic diffusion time, e.g. the time for water to diffuse into a glassy sucrose layer and make it rubbery
- T_g Glass-transition temperature
- T'_g Temperature coordinate of the point of intersection of the extrapolated liquidus curve with the T_g curve in a state diagram. It is the T_g of the maximally freeze concentrated solution
- T_m Melting temperature
- T_E Eutectic temperature, where under equilibrium conditions total solidification would normally be predicted
- T_2 The relaxation time in nuclear magnetic resonance (NMR) measurements which relates to the protons returning to their natural random spin alignment, after having been aligned by the instrument's magnetic field

T_1 The relaxation time in NMR measurements which relates to the time constant with which magnetic spins come to equilibrium with their surroundings

$T_{1\rho}$ A variant of T_1 in the rotating frame, which permits measurements at lower frequencies of molecular motion

2.2 Introduction

Although the glass transition is a comparatively recent subject of study in physical science and even more so in food science, it is a phenomenon which is widely observed in natural systems. For example, the opening of the pods and the scattering of seeds by many plants relies on dehydration from a rubbery to a highly elastic and brittle glassy condition. Likewise, the formation of spiders' webs from a rubbery material that can be spun, to one which produces a robust elastic structure is the consequence of a rubber-to-glass transition. On the domestic scene, the 'ironing' of a cotton sheet relies on the glass → rubber → glass transition of cellulose. There is also strong evidence that the resistance to dehydration damage and the protection against freezing temperatures observed in both animals and plants arises in many cases from transitions in the prevailing physico-chemical system from rubbery to glassy states.

The significance of the glass transition, which has been studied in polymer systems for the past 70 years or so, has been reviewed by Allen [1]. In similar vein, Kauzmann [2] recognized that the behaviour of highly concentrated solid–liquid systems, including low water–sugar systems, can only be comprehended where the importance of the glass-transition temperature (T_g) is fully taken into account. The extrapolation and exploitation of these observations to sugar confectionery systems was made in the mid-1960s by White and Cakebread [3] at Lyons Central Laboratories in London. However, the past ten years have seen a torrent of interest in this area, largely pioneered and promoted by Levine and Slade [4], leading to an exponentially increasing number of publications covering this field.

2.3 Characteristics of the glassy and rubbery states

The glassy state is amorphous in character, i.e. without crystalline order (Figure 2.1). In practice, this is a consequence of the kinetic energy of the system (which is expressed as molecular motion) being removed at a rate in excess of the ability of the molecules to reorganize into the ordered state.

Glassy materials are brittle (a term usually reserved for non-food

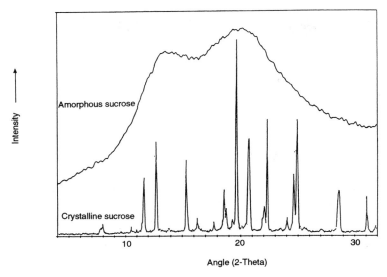

Figure 2.1 X-ray diffractograms of amorphous (broad hump) and crystalline (sharp peaks) sucrose.

materials) or crisp (a term associated with food systems). The latter feature is a reflection of the very high viscosity ($>10^{13.5}$ Pa·s). In addition, a considerable number of glassy systems can be produced in a transparent state. The physical reason for this is not entirely clear, save that there is an absence of ordered regions which might otherwise scatter the incident light beams. Perhaps most importantly from the physical chemist's point of view, the glassy state is characterized by molecular immobility whereas there is significant molecular motion in the rubbery state. In a pure, glassy polymer system, for example, it is believed that the only motion that occurs is that of small side-chains; any movement of the polymer 'backbone' is much slower. Where substantial motions of more than ten adjacent units within a polymer chain occur, then we can infer that we have moved from the glassy to the rubbery state.

2.4 Characteristics of the glass–rubber transition

In contrast to the transitions associated with ice crystallization (a first-order transition), the glass transition is termed a second-order transition. Such a transition has the following characteristics:

- There is no discontinuity in volume (V) and enthalpy (H) at the transition from highly viscous liquid to glass.

- In contrast to crystallization which is characterized by spikes in C_p and α_p' no such counterpart is observed at the glass transition.
- The absolute value of the glass-transition temperature (T_g) is a function of the cooling rate.

From a thermodynamic point of view, Figure 2.2 shows the important parameters. On heating a glass from low temperature there is an increase in the slope at the glass transition (T_g; Figure 2.2a) which suggests that the internal volume of the system is increasing more rapidly in the rubbery state with temperature, than was the case in the glassy state. Figure 2.2(b) illustrates the volume compressibility versus temperature

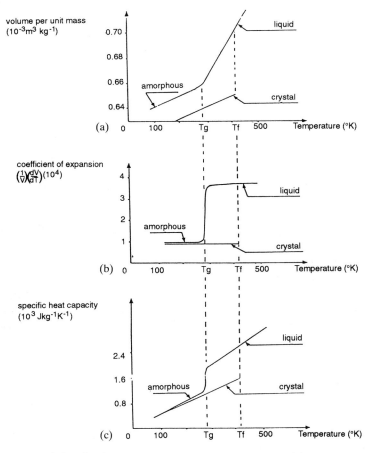

Figure 2.2 Variation in three parameters around the glass-transition temperature, T_g. Changes in (a) the volume per unit mass, (b) the coefficient of expansion and (c) the specific heat for glucose, in both amorphous and crystalline states.

and dramatically shows the substantial change at T_g. A plot of the specific heat (dH/dT) versus the temperature again shows the increased gradient at the glass transition. It would appear therefore that the volume change evident in Figure 2.2(a) is associated with an increase in energy of the system and we may reasonably conclude that we are observing an increased energy of the molecules, increased molecular mobility and consequently an internal expansion of the system.

2.5 Methods of determination of the glass–rubber transition

The methods of determination of the glass-transition temperature are the outcome of the phenomena that we have already described. They may be listed under a number of headings.

2.5.1 Calorimetric techniques

These are a consequence of the change in specific heat at the glass transition. The use of differential scanning calorimetry (DSC, which works by noting the change in temperature with time of the substance relative to a standard as it is heated or cooled), to perform this experiment is widely reported and extensively used. For small, simple molecules, e.g. sugars the results are usually excellent. However, with increasing molecular weight and structural complexity, the observation becomes more problematical. In a homogeneous polymer such as amylopectin, the T_g can be observed with little difficulty, but in a protein where different regions of the chain may have different amino acid compositions, then, according to one explanation, it is not surprising that each will have distinctive 'micro' T_g values and the overall T_g response will be smeared over an extended temperature range and be difficult to observe. The technique has also been used extensively for monitoring T_g at sub-zero temperatures in aqueous systems. This has proved very valuable, but particular attention needs to be paid to extracting such parameters as T'_g (see section 2.6) as the situation is usually confused by the recrystallization or melting of ice crystals in the regions between T_g and T_m [5,6].

2.5.2 Molecular mobility

A range of techniques is available to perform this type of measurement, e.g. dielectric techniques and those associated with nuclear magnetic resonance (NMR) relaxation. The determination of the NMR T_2 relaxation time has proved very valuable, but it is important to realize that this defines the 'rigid lattice limit' (RLL) which is not necessarily the same as T_g [7,8]. Figure 2.3(a) illustrates a typical T_2 decay curve from a pulse

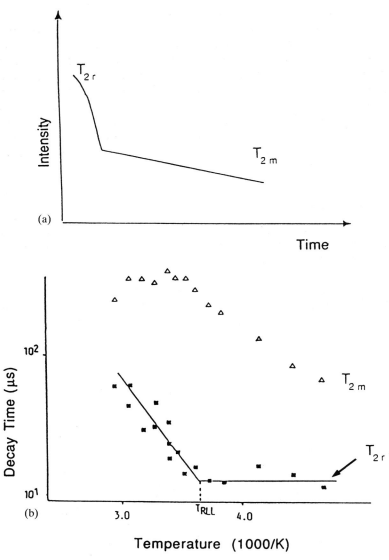

Figure 2.3 (a) Typical two-component decay after application of a 90° pulse to an amylopectin–water sample in the glassy state; (b) typical values of T_{2R} and T_{2M} as a function of inverse temperature in the region of T_g. (R = rigid; M = mobile.)

NMR experiment. The existence of two T_2 components is immediately evident, arising from 'rigid' and 'mobile' protons. The results for the same experiment conducted at different temperatures spanning the RLL are shown in Figure 2.3(b).

The great virtue of the NMR technique is that it can give not only information about whole molecules, but also about the mobilization of individual groups or regions within such a structure. Such evidence usually arises from high-resolution spectra and this may involve more sophisticated techniques using, for example, magic angle spinning (MAS). However, it is worth noting that direct measurements of the spectra of, for example, amylopectin, over a range of temperatures clearly show that the 83 ppm peak associated with the amorphous amylopectin disappears as the temperature is raised, or the water content increased (Figure 2.4), both of which, under appropriate conditions, promote passage from the glassy to the rubbery states.

2.5.3 Mechanical methods

At the very simplest, it is possible to monitor the change from the brittle to the rubbery state using a three-point bend test, in which the characteristic brittle fracture becomes a non-linear yielding process after the transition. However, more elegant experiments which yield further information, can be performed using dynamic mechanical thermal analysis (DMTA) [7,9]. Figure 2.5 shows a typical trace which emerges from

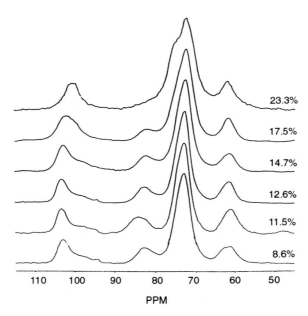

Figure 2.4 ^{13}C CP-MAS NMR spectra of amylopectin at room temperature at various moisture contents.

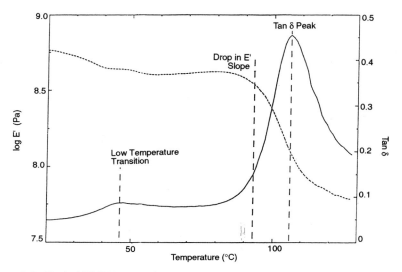

Figure 2.5 Typical DMTA trace of an amylopectin–water system showing the E' and $\tan \delta$ (E''/E') traces.

DMTA, a process which scans with an increasing temperature ramp the loss modulus and the elastic modulus and coincidentally computes $\tan \delta$. A change from the glassy to the rubbery state is attended by a fall in the elastic modulus (the ability of a system to store energy), a modest peak in the loss modulus and a sharp peak in the derived $\tan \delta$. This major peak is termed the α-relaxation and is associated with mobilization of the polymer backbone. Not infrequently, minor peaks are observed at lower temperatures and are termed β- and γ-relaxation processes, usually reflecting mobilization of side-chains.

The critical question, of course, is where does the glass transition occur on such a trace and what relation does this measurement bear to those from other techniques? The usual experience is that the RLL of NMR in amylopectin, for example, coincides with the incipient fall in the elastic modulus. It also appears that the consumer perceives significant changes in texture at that point. The mid-point of the DSC change in specific heat coincides with the maximum in the loss modulus and these appear in the mid-point region of the fall in elastic modulus. These values are usually slightly below the temperature of the $\tan \delta$ maximum.

2.5.4 Other methods

A number of other approaches can be used to provide information. These include measurements of volume and spectroscopic techniques, e.g. infrared, electron spin resonance and dielectric spectroscopies.

2.6 State diagrams

The use of phase diagrams to record the *equilibrium* relationships between different components and phases is well known and has provided considerable insight into manufacturing processes in many industries including the food industry. Such diagrams may include information on solubilities, melting and boiling points. However, as already mentioned, the glass-transition temperature, T_g, does not reflect a true first-order transition with a rigorously reproducible value in a system with a given composition. In fact, its value is dependent on both the previous history of the system and the rate of heating when DSC is used. To include such information on a phase diagram would make the latter a misnomer. Phase diagrams, however, which are supplemented with such non-equilibrium data as T_g behaviour are now widely called 'state diagrams' and have proved valuable in accounting for the behaviour of food materials in the glassy and rubbery states.

Figure 2.6 records the state diagram of sucrose. The liquidus and solidus curves represent the ice melting point and the sucrose solubility curves when plotted in the temperature–concentration plane. T_E is the

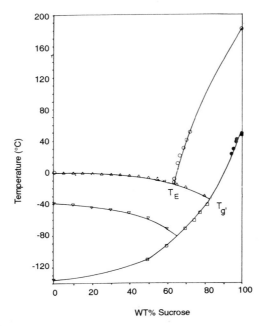

Figure 2.6 State diagram of sucrose–water system. ○ = Young and Jones; △ = Muhr; □ = Luyet and Rasmussen; ▽ = MacKenzie; ● = Gough.

eutectic temperature where, under equilibrium conditions, total solidification of the system might be anticipated. However, crystallization of sucrose from a concentrated solution at sub-ambient temperatures is extremely sluggish and therefore, in practice, with continued cooling, further ice crystallization and consequent concentration of the sucrose solution occurs along the pathway between T_E and T'_g making further crystallization of the sucrose even more unlikely. Ultimately, the system will attain the position where the remaining sucrose solution at T'_g has a concentration C'_g. At this point, the concentrated sugar solution transforms into a glass. The actual anatomy of the system consists of ice crystals embedded in a sucrose–water glass, composition C'_g. It will be evident from Figure 2.6 that T'_g and C'_g values are the coordinates of the point of intersection of the equilibrium liquidus curve with the kinetically controlled glass-transition curve on the supplemented phase (i.e. state) diagram.

A variety of approaches have been used to determine T_m, T_g, C'_g and T'_g. The exact values of C'_g and T'_g have been the subject of considerable debate, largely through different approaches to the interpretation of DSC thermograms. In view of the widespread use of DSC, it is desirable to outline the problem and what is believed to be the correct solution.

Figure 2.7 shows DSC thermograms from frozen malto-oligomer solu-

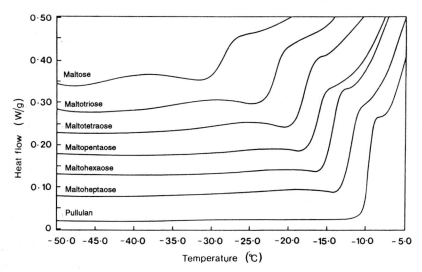

Figure 2.7 DSC thermal transitions for 30% frozen solutions of malto-oligomers and pullulan as a function of increasing molecular weight. (Reproduced with permission.)

tions together with the polysaccharide pullulan [8]. Inspection of the maltose thermogram shows that a double transition is evident before the ice dissolution endotherm. The series also demonstrates that with increasing molecular weight (maltose to pullulan):

- there is a decrease in the change in heat capacity with the initial transition;
- there is an increase in the temperatures at which the transitions occur;
- there is a decrease in the temperature differential between the two transitions.

Experimental and modelling studies for sucrose by Ablett *et al.* [5,6] have suggested that the initial transition was the glass transition and that the larger, second transition (clearly seen in Figure 2.7 for maltose) is not a glass transition at all, but is due to a change in heat capacity resulting from the delayed initiation of ice dissolution. Recent studies with modulated DSC [10] give additional support to this view.

It is, however, interesting that the two transitions come closer together and ultimately merge with increasing molecular weight. The fact that coalescence occurs and that the observed phenomena are demonstrated by the much larger change in heat capacity (associated with the delayed onset of ice dissolution in contrast to the much smaller heat-capacity changes associated with the glass transition underlying this large transition), suggests that it is appropriate to use this transition to characterize the T'_g temperature of the large molecular-weight biopolymers.

ESR [11], relaxation methods [12] and NMR [8,13] have all been used to provide information on the glass-transition temperature behaviour of sugar systems.

T_m, the melting temperature, can be determined directly from the DSC thermogram, but this becomes more difficult at concentrations >60% sucrose. Simatos and Blond [14] have reported the use of UNIQUAC and UNIFAC computer programs to calculate the T_m. The method relies on deriving the activity coefficient from the excess Gibbs function by estimating the interactions between molecules. Thereafter T_m can be calculated by the use of the appropriate equation.

The really significant feature of a state diagram is that it identifies the glassy regions where molecular motion is minimal and the rubbery state where molecular motion is more pronounced. Figure 2.8 specifies regions where various time-dependent changes may occur in the rubbery state. Convention has it that the kinetics of the glassy state, in so far as they occur, are Arrhenius in type, while those in the 'rubbery' regions obey the Williams, Landel and Ferry (WLF) equation and/or the related Vogel, Tamman and Fulcher (VTF) equation (see section 2.7). The more distant the system conditions are from the glass transition, the more rapid is the

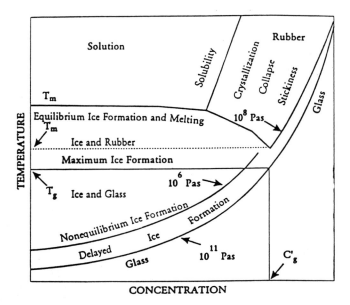

Figure 2.8 Schematic state diagram showing processes which may occur in the rubbery state of carbohydrates, some of which may result in the formation of an amorphous state. Typical processes affected by the physical state are shown (e.g. stickiness). Maximum ice formation only occurs in the region above T'_g but below the equilibrium ice melting temperature of ice in contact with the maximally freeze-concentrated unfrozen matrix. The concentration of the unfrozen matrix (C'_g) is equal to that of solution with $T_g = T'_g$. Reproduced with permission of *Food Technology*.

kinetic process. The diagram also shows a variety of important food quality-modifying events including crystallization of water and solutes, collapse, stickiness and phenomena associated with storage stability in the frozen state.

2.7 Molecular mobility and diffusion in the glassy and rubbery states

A basic assumption in the study of the glassy state is that macromolecules are largely immobile, at least in terms of their backbones, but that some motion of side-chains and small plasticizer molecules is feasible. Much greater motion is, however, possible in the rubbery state.

2.7.1 Molecular mobility

The mobilization of the polymer backbone at T_g can simply be demonstrated through pulse NMR studies and analysis of the resultant multiexponential decay curve. On resolving into two components, the fast relaxing protons associated with the polymer backbone show a sharp increase in T_2 at the so-called 'rigid lattice limit' (see Figure 2.3b) which approximates to the glass transition [7]. Similarly, Hemminga et al. [11] have shown that electron spin resonance spectroscopy (ESR) is effective in detecting changes in the mobility of macromolecules (around T_g) in a polymer matrix. ESR also provided information on the mobility of included smaller molecules and the existence of cavities in the lattice. As already indicated (section 2.5.2) high-resolution solid-state NMR can also be used to demonstrate molecular mobilization by providing evidence of the development of new molecular structures (Figure 2.4). At a more sophisticated level, it can monitor changes in the T_1, $T_{1\rho}$ and cross-polarization behaviour of specific nuclei, which thereby report on the mobility of distinct chemical groups, or even whole molecules, as a system is transformed from a glass to a rubber.

The existence of mobile side-chains in preference to the backbone may also be detected by NMR procedures, but frequently give evidence of their presence as a β-relaxation in mechanical spectroscopy, e.g. DMTA (section 2.5.3). The possible significance of this in amylopectin for textural changes and methods of manipulating it, by modifying temperature or water content, have already been explored by various groups [7,15,16].

The mobility of water molecules within the glassy and rubbery states was confirmed by the T_2 studies of Kalichevsky et al. [7]. It is noticeable in Figure 2.3(b) that there is a continuous but no sharp change while passing through the T_g of the polymer matrix. Studies of the rotational and translational diffusion of water in a 26% w/w H_2O–ovalbumin system suggest that there is only a modest reduction in these parameters (two orders of magnitude less than pure bulk water, viz. $2.3 \times 10^{-11} \, m^2 s^{-1}$) with a self-diffusion coefficient of $6 \times 10^{-11} \, m^2 s^{-1}$ at $-5°C$ (23°F) [8] for the system. The direct determination of diffusion coefficients by drying, reported in the classical studies of Fish [17] on potato likewise give no suggestion of a sharp change in gradient at values close to the T_g.

Lüdemann and his coworkers have published a series of papers reporting studies of the rotational mobility of sucrose and water molecules in concentrated (≤90% sucrose), rubbery solutions [18], the molecular mobility of the sucrose by ^{13}C relaxation [19], the molecular mobility of the water molecules by ^2H-NMR relaxation [13], and the concentration and temperature dependence of the self-diffusion of sucrose and water molecules in these systems [20]. MacNaughtan [21] has also recorded changes in the mobility of specific groups of even whole sugar rings in the concentration range (90–100% sugar) using NMR T_1, variable contact time and $T_{1\rho}$ techniques. As might be expected, the T_1 of exocyclic groups (—CH_2OH) and the T_1 of whole sugar molecules, exhibit substantial changes over the glass transition.

2.7.2 Diffusion

Extensive measurements have been made by pulse field gradient (PFG) techniques of the self-diffusion coefficient of H_2O molecules in polymers in the rubbery state (a measure of the rotational and translational mobility of molecules in the absence of net transport), but the technique becomes more problematical in the glassy region due to the short T_2,H_2O values. Magnetic resonance imaging (MRI, the technique used in medical body scanners) shows considerable potential, either again in measuring the distribution of self-diffusion coefficients in a system, or in determining the translational (mutual) diffusion coefficient by monitoring the change in water distribution in real time. The translational diffusion of fluorine-labelled compounds (e.g. sugars) can, if required, be followed by using an ^{19}F-probe.

2.7.3 Models of diffusion

The diffusion process is of enormous significance in food processing and preservation. In some cases (e.g. dehydration, rehydration, conditioning of wheat grains, pickling or preparation of intermediate moisture foods), the intention is to maximize the diffusion of the critical components. In other instances (such as flavour retention, prevention of oxidative rancidity, extending shelf-life of products with more than one component of different water activities, a_w), the aim is to minimize diffusion. In practice, relating the process of diffusion to changes in molecular structure, and the dynamics that are involved, leads to a far from simple situation as will be evident from the following brief overview.

A very common situation is where water is being transported through a rubbery to a glassy material [22]. In such conditions, the diffusion that is observed in glassy biopolymers often deviates from the prediction of Fick's Law due, it has been suggested, to the finite rate at which a polymer structure rearranges to accommodate water molecules. Two broad types of behaviour have been identified which actually are the extremes of a continuum. In so-called Case I (Fickian) diffusion, the mobility of the water is much less than the segmental relaxation rate. In Case II diffusion, water permeation is much higher than the segmental relaxation rate and the rate-determining step is the relaxation occurring at the sharp boundary between swollen and the largely unplasticized polymer. The continuum reflects different ratios of the relative rates of penetrant mobility and biopolymer relaxation. The abnormal diffusion has characteristics intermediate between the Case I and Case II models.

The distinction between these two types has been placed on a quantitative basis by Vrentas *et al.* [23,24] who have introduced a diffusional Deborah number (De) defined as the ratio of a characteristic relaxation time λ to a characteristic diffusion time, θ, i.e. De = λ/θ. In Case I diffusion, $\lambda \gg \theta$ and therefore De $\gg$ 1. In Case II diffusion, $\lambda \ll \theta$, which means that changes in the macromolecular structure occur instantaneously with respect to the time of diffusion. Figure 2.9 shows regions of differences in a polymer–penetrant system as a function of penetrant concentration and temperature.

If we consider water transport in a plane sheet, where M_t is the mass of water taken up at time t, and M_∞ that taken up as time approaches infinity, M_t/M_∞ can be related to a selection of variables, e.g. the diffusion coefficient D, and the relaxation constant, k_0, the initial water content C_0, the thickness of the plane sheet l and the time t, depending on whether Case I or Case II diffusion dominates.

Case I assumes the form:

$$M_t/M_\infty = 4(D_t/\pi l^2)^{1/2} \tag{2.1}$$

while in Case II:

$$M_t/M_\infty = 2k_0 t/C_0 l. \tag{2.2}$$

Peppas and Brannon-Peppas [22] have, on the basis of the above considerations, proposed a simple semi-empirical equation of the form:

$$M_t/M_\infty = k_1 t^{1/2} + k_2 t \tag{2.3}$$

which can be generalized as:

$$M_t/M_\infty = kt^n \tag{2.4}$$

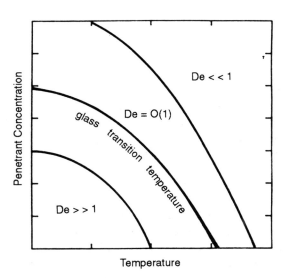

Figure 2.9 Regions of penetrant transport in a polymer–penetrant system as a function of penetrant concentration and temperature. (Reproduced with permission of Food Engineering.)

where k is a constant incorporating characteristics of the macromolecular network system and the penetrant, while n, the exponent, points to the mechanism of transport. Case I (Fickian) and Case II transport have values of n, respectively, of 1 and 0.5. The constant $k = 4(D/\pi l^2)^{1/2}$ for Fickian diffusion while $k = 2k_0/C_0 l$ for Case II diffusion.

2.8 Kinetics

A significant understanding of kinetic processes has been gained by recognizing that other physical parameters which reflect molecular mobility (e.g. viscosity, relaxation processes and translational diffusion) provide valuable insights and also a mathematical formulation which can be transferred to the physical or chemical processes responsible for changes in foodstuffs.

In some liquids [12], the increase in viscosity (η) with decrease in temperature takes an Arrhenius form, i.e.

$$\eta = A \exp(E/RT). \tag{2.5}$$

A plot of $\log \eta$ vs. $1/T$ or T_g/T (when comparing different liquids) shows a linear dependence (Figure 2.10). The network of molecules must be disrupted for flow to occur and the disruption process has a constant

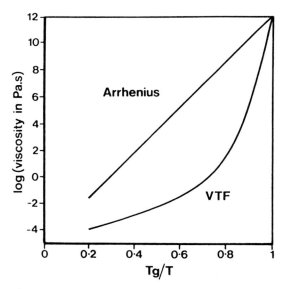

Figure 2.10 Typical plots of $\ln \eta$ versus T_g/T for liquids showing Arrhenius and VTF behaviour. (Reproduced with permission.)

activation energy. These materials are known as strong liquids and exhibit a small ΔC_p at T_g. More commonly, liquids exhibit non-Arrhenius behaviour with no constant activation energy for viscous flow, which indicates that the mechanism of flow changes with temperature. The viscosity of such liquids is described by the semi-empirical Vogel, Tamman and Fulcher (VTF) equation:

$$\eta = \eta_0 \exp[DT_0/(T - T_0)] \quad (2.6)$$

where D, η_0 and T_0 are constants [25]. Such materials are termed 'fragile liquids' and display a relatively large ΔC_p at T_g. T_0 is the temperature of viscosity divergence which lies below T_g by an amount which depends on the parameter D. Liquids with structures that resist thermal degradation ('strong' liquids) have large D parameters; those with small D are termed 'fragile'.

An alternative approach to modelling the temperature dependence of dielectric relaxation (which provides details of molecular motion) and mechanical behaviour within the rubbery state was formulated by Williams et al. [26] and is known as the Williams, Landel and Ferry (WLF) equation taking the form:

$$a_T = \tau(T)/\tau(T_{ref}) = \exp[C_1(T - T_{ref})/\{C_2 + (T - T_{ref})\}] \quad (2.7)$$

where C_1 and C_2 were system-dependent coefficients and a_T was defined as the ratio of the relaxation phenomenon at T to the relaxation at the reference temperature, T_{ref} (i.e. η/η_{ref} for viscosity). It has also become common practice to use average, 'universal' values of the coefficients C_1 and C_2, namely $C_1 = 17.44$ and $C_2 = 51.6$. There is now strong evidence that the thoughtless use of these values is misguided and erroneous (see Peleg [27]). Where diffusion is dependent on the free volume, the rate of reactions can be expressed using the WLF equation in the form

$$k_{ref}/k = \exp[C_1(T - T_{ref})/\{C_2 + (T - T_{ref})\}]. \quad (2.8)$$

The equivalence of the VTF and WLF equations can readily be demonstrated, if we express both in terms of relaxation times:

WLF:

$$\log \tau(T)/\tau(T_g) = C_1(T - T_g)/\{C_2 + (T - T_g)\} \quad (2.9)$$

VTF:

$$\tau(T)/\tau_0(T_0) = \exp[B/(T - T_0)]. \quad (2.10)$$

Letting $C_2 = T_g - T_0$ and $2.3 C_1 C_2 = B$ in equation (2.10), then the equivalence can be demonstrated [25].

The VTF equation is more useful than the WLF form in non-polymeric liquids, such as sugars, as a consequence of the pre-exponential factor,

η_0, which factors out having a universal value of approximately 10^{-3} Pa·s. In contrast, in polymers, the VTF pre-exponent is extremely non-universal and depends on molecular weight, which is one reason why the WLF equation is frequently preferred by food scientists.

Much use has been made of the fact that on applying the 'universal' values of C_1 and C_2 in the WLF equation, the rate of increase over $(T - T_g = 10°C\ (18°F))$ is much greater if WLF kinetics prevail (as is supposed to be the case in the rubbery state) rather than the Arrhenius-type kinetics. To make such a wide-ranging assumption at this stage cannot be justified (Table 2.1; see also section 2.11).

2.9 Plasticization and the glassy state

If we examine the typical results that are obtained as we monitor the change in the glass-transition temperature of amylopectin, and some other food biopolymers, as we increase the water content, then we shall observe behaviour as represented in Figure 2.11. It will be noted that increasing the water content produces a pronounced fall in the glass transition, particularly at low water contents. This process is known as 'plasticization' and is widely employed in synthetic polymers. In biopolymers which are of pre-eminent importance in food systems, the plasticizer *par excellence* is water. No other conventional food component matches water for its effectiveness in this role. Alternatively, the data may be presented by plotting the glass transition against the a_w associated with these water contents. This modifies the appearance of the plot (see Figure 2.12), but the differences are sustained and clearly it suggests that, in a mixed system, it is possible to have two components at a given temperature and water activity with one in the glassy state and one in the rubbery state. Such effects have been demonstrated in practice, for example with gluten and casein.

Once we have observed such behaviour, it is obviously important to try and define the behaviour more rigorously in terms of physico-chemical

Table 2.1 Rate increases predicted using the WLF equation with average values for C_1 and C_2, and observed by Karmas *et al.* [36] for non-enzymic browning within a lactose–CMC–trehalose–xylose–lysine model system at initial $a_w = 0.12$

Temperature range	Relative rate increase predicted	Relative rate increase observed
$T_g \rightarrow T_g + 10°C\ (18°F)$	678	21.8
$T_g + 10°C\ (18°F) \rightarrow T_g + 20°C\ (36°F)$	110	7.39
$T_g + 20°C\ (36°F) \rightarrow T_g + 30°C\ (54°F)$	3.5	4.06
$T_g + 30°C\ (54°F) \rightarrow T_g + 40°C\ (72°F)$	1.58	2.82

THE GLASS TRANSITION

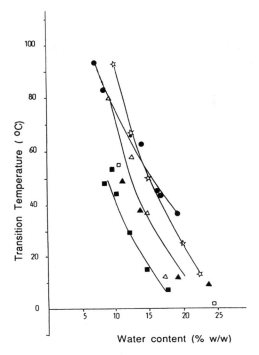

Figure 2.11 Plot of T_g versus water content for various food biopolymers.

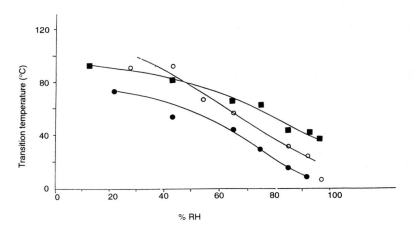

Figure 2.12 Plot of T_g versus a_w for selected food biopolymers.

concepts. Early attempts were made to employ free volume theory which had proved satisfactory in synthetic polymers. This did not prove to be appropriate for biopolymer materials and no adjusting of parameters could make it so. Fortunately, an alternative approach based on equilibrium thermodynamics had been devised by Couchman and Karasz in 1978 [28] and relied on the following assumptions:

- Continuity of the configurational entropy at T_g.
- Intimate miscibility of the components.
- The absence of crystallinity or cross-linking.

A classical form which assumes $\ln T_g$ can be modified, so long as we assume that ΔC_p is proportional to $1/T_g$ and takes a linear character which is more easy to manipulate. Hence:

$$\ln T_g = \frac{W_1 C_{p1} \ln T_{g1} + W_2 C_{p2} \ln T_{g2}}{W_1 \Delta C_{p1} + W_2 \Delta C_{p2}} \quad (2.11)$$

becomes

$$T_g = \frac{W_1 \Delta C_{p1} T_{g1} + W_2 \Delta C_{p2} T_{g2}}{W_1 \Delta C_{p1} + W_2 \Delta C_{p2}} \quad (2.12)$$

A quite empirical approach is that based on the Gordon and Taylor equation [29]:

$$T_g = \frac{W_1 T_{g1} + K W_2 T_{g2}}{W_1 + K W_2} \quad (2.13)$$

which, if we assume that:

$$K = \Delta C_{p2}/\Delta C_{p1} \quad (2.14)$$

takes the form of the Couchman–Karasz equation. The great virtue of this latter equation is that if the data can be fitted to it, and we know the two T_g values and one ΔC_p, then it is possible to calculate the other ΔC_p. The major hindrance, however, with it is that only a two-component system can be used. In contrast, the Couchman-Karasz equation can be extended to three components.

It is quite possible to apply this to amylopectin and to fit the equation to the actual experimental results. The results fit well in terms of values for the T_g of water and amylopectin in the pure states, and the ΔC_p of the amylopectin, but there has been much more discussion about the ΔC_p of the water. This point has been reviewed in some detail by Kalichevsky et al. [28].

2.10 More complex biopolymer systems

A simple system like amylopectin, though instructive, does not reflect the complexity found in food materials, where several different macromolecules are admixed with small molecules such as sugars and lipids.

A relatively simple system such as amylopectin:fructose (4:1) is instructive in that predictions about the ultimate T_g can be made from the Couchman–Karasz equation and DMTA studies suggest the system is fully miscible. If, however, higher concentrations of fructose (amylopectin:fructose 2:1) are examined by DMTA, the rather simple trace which had been reported earlier (Figure 2.5), is replaced by a more complex trace where two clear peaks emerge in the tan δ trace (Figure 2.13). It is not unreasonable to imagine that this is a consequence of phase separation. Figure 2.14 contains the fitted line for pure amylopectin,

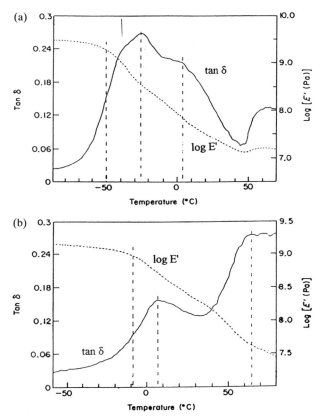

Figure 2.13 DMTA traces for amylopectin:fructose (2:1) systems after equilibration to (a) 85% and (b) 43% relative humidity.

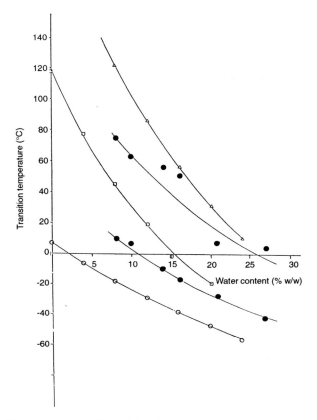

Figure 2.14 Theoretical results for individual components and a fully miscible mixture of amylopectin:fructose (2:1) system compared with the experimental, phase-separated behaviour. △ = Theoretical amylopectin; □ = theoretical mixture (method 1); ○ = theoretical fructose; ● = tan δ peaks.

the fitted line for pure fructose of varying water contents and a fitted line for a 2:1 amylopectin:fructose mixture, assuming total miscibility. When the actual experimental results are plotted, they in fact provide a series of points which suggest that we have two phases (tan δ traces), one of which is amylopectin-rich, but is plasticized by both the fructose and the water, while the other is fructose-rich and water but is antiplasticized by the amylopectin. The exact nature of these phases requires further examination.

It should be emphasized, however, that the situation that is observed in biopolymers generally is far from simple. For example, the work that has been done thus far on gluten suggests that, whereas sugars in small concentration may have some plasticizing effect, as that level is increased there is a distinct antiplasticizing effect. It is possible this may arise from unequal partitioning of the water. Similarly, the effect of simple lipids has been explored. In some cases, the effect is zero due, presumably, to total nonmiscibility. On the other hand, where materials have a hydrophobicity/hydrophilicity of the macromolecule which approximates to that of a plasticizing molecule, as appears to be the case with hydroxycaproic acid and gluten, then there is a substantial plasticizing effect [31,32].

A number of examples will now be used to illustrate the significance of glass transitions in product stability and processing.

2.11 The significance of the glass transition in frozen products

Reid et al. [33] suggest that the best way to describe frozen foods is that of an unfrozen matrix surrounding ice crystals. Other crystals and entities may also be present, but usually a proportion of the residue will occur as a non-crystalline, unfrozen matrix. This may transform into the glassy state, with reduced mobility of the glassy state. If on warming the material, the glass transforms into the rubbery state, both molecular mobility and diffusion increase.

Simatos and Blond [14] have examined the relevance of the WLF equation as a means of describing the effect of storage temperature on the rate of deterioration of frozen-food products. The rate constant was assumed to be proportional to the inverse of viscosity and the numerical constants C_1 and C_2 were taken to be 17.44 and 51.6, respectively. Figure 2.15 records the results of their analyses.

The curve for the WLF equation is compared with rates of deterioration for several frozen-food products. The T'_g value was taken as $-35°C$ ($-31°F$) for egg yolk and $-40°C$ ($-40°F$) for peas and beef. It will be immediately evident that the curves corresponding to the food products are not as steep as the WLF master curve. Simatos and Blond [34] suggested that a possible explanation for these effects was that, at temperatures above T'_g, the melting of ice had two consequences: it both decreased the viscosity, which adds to the WLF effect on the viscosity of the freeze-concentrated phase, and diluted reactants which could partially offset the effect of viscosity on reaction kinetics. The authors also suggested that values of C_1 and C_2 relevant to the system should be used, and that a superior model could be developed by using the T_g value appropriate to the concentration of the freeze-concentrated phase at temperature T as the reference temperature (T_{ref}) in equation (2.8), instead of T'_g, which corresponds to the maximum freeze concentration.

Reid et al. [33] have explored the viscosity of the unfrozen phase in equilibrium with ice as a function of temperature for both sucrose and maltodextrin systems. Their results suggest that the use of the T_g of the

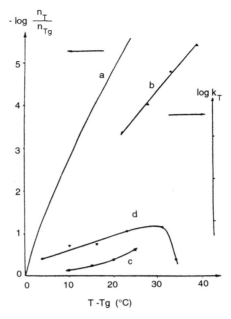

Figure 2.15 Application of the WLF equation and 'universal constants' to model reactions in the frozen state. Curve (a) is a plot of the WLF equation ($\log(\eta_T/\eta_{T_g})$ equivalent to $\log(k_{T_g}/k_T)$); other curves represent the rate constant ($\log k_T$) for several processes of deterioration during storage in the frozen state; (b) decrease in ascorbic acid in peas; (c) ice crystal growth in meat; and (d) increase in apparent viscosity for egg yolk. After Simatos and Blond [31]. (Reproduced with permission.)

glass of the composition of the unfrozen medium, at the temperature of the viscosity measurement, gives the optimum fit, and is slightly better than when using T'_g. On the other hand, Kerr et al. [35] have shown a good correlation between T'_g and enzyme reaction rates between T'_g and T_m.

2.12 The significance of the glass transition in the storage of foods at ambient temperatures

Although water activity has proved a valuable concept in understanding the stability and/or deterioration of foods in the dry or semi-dry state, from time to time its conclusions can be erroneous. There is much in favour of setting T_g and the temperature differential $(T - T_g)$ centre stage, which should then model the mobility of the biomolecules themselves. Already reference has been made to the strong evidence that passage from the glassy to the rubbery state is accompanied by an increase in molecular mobility. Such mobility can express itself in a variety of ways, for example the presence of the physical processes of molecular translation or reorganization, the initiation of chemical reactions, sometimes beneficial, other time deleterious and in certain situations by microbiological activity.

The application of these ideas to non-enzymic browning is an obvious area of study (see also chapter 4). Karmas et al. [36] have established that non-enzymic browning in a carbohydrate model system (lactose–amioca–lysine) was very slow below the system T_g. Depending on the composition and moisture content of the carbohydrate system, the rate of non-enzymic browning increased substantially at some 20–75°C (36–135°F) above the T_g. For systems at a_w = 0.12 and 0.33, dramatic increases in the rate of browning occurred at 50°C and 75°C (90°F and 135°F) above T_g, respectively. Diffusion would have been expected to substantially increase at T_g, but this delayed increase until a much higher temperature may well reflect the increase in free volume that is required before movement of the larger reactant molecules can take place.

The rubbery state is metastable, which means that various changes to the physical state of the system can occur. For example, the sucrose which is initially glassy in an egg white–sucrose meringue may crystallize leading to a deterioration in texture and the release of saturated solution of sucrose. Likewise, the collapse of matrix porosity in gels can lead to surprising changes in permeability – a possible explanation for the reduction in limonene oxidation of orange oil by encapsulating it in maltodextrin M100 at a_w 0.56 [37].

Nelson and Labuza [38] have examined the relative merits of the Arrhenius and WLF equations for modelling reactions in the rubbery state. In some instances, for example Karmas et al. [35], an Arrhenius plot of non-enzymic browning (optical density at 420 nm) within a lactose–CMC–trehalose–xylose–lysine model system at an initial a_w = 0.12 yielded a reasonably straight line, but with an inflection adjacent to the T_g of the system (50°C, 122°F). Even though there was a 40°C (72°F) range of temperature in the rubbery state, the Arrhenius plot was quite linear.

Karmas et al. [36] also compared the predicted increase in rate of the above non-enzymic browning reaction using the WLF equation and the conventional values for C_1 and C_2 with experimental values. There were substantial discrepancies (Table 2.1).

The WLF equation has been used by, for example, Roos and Karel [39] to fit their data for the rates of crystallization of lactose. The applicability of the equation is usually acknowledged if the equation fits the experimental data fairly well. The use of the 'universal' coefficient gave a good fit, but equally, so did other values of the coefficients. The values appropriate for the given system can be found by rearranging the WLF equation into the form:

$$[\log k_g/k]^{-1} = C_2/C_1(T - T_g) - 1/C_1 \qquad (2.15)$$

which in a plot of $[\log k_g/k]^{-1}$ vs. $1/(T - T_g)$ yields the value of $-C_2/C_1$ as the gradient and $1/C_1$ as the intercept. Difficulties in determining the k_g at T_g can be overcome by measuring k at T_{ref} ($T_{ref} > T_g$) and transforming the value by the method of Peleg [27]. Application of this approach to the results of Karmas yields values of C_1 and C_2 of 18.0 and 180.2, respectively using 55°C (131°F) as T_{ref} (Figure 2.16). The fit is very good ($r^2 = 0.99$) but the values are very different from those generally quoted in food science literature.

2.13 Exploitation of the glassy and rubbery states in processing

2.13.1 Understanding extrusion

The process of extrusion of starch involves the exposure of starch granules to a water content and a temperature which result in gelatinization while passing along the extrusion barrel. At water contents approximately <30% w/w, gelatinization will only occur at temperatures >100°C (>212°F), a consequence of which is that after emerging from the die, the gelatinized extrudate expands, largely due to instantaneous vaporization of water. At higher water contents, the extrudate shrinks on cooling. This shrinkage process is also evident in starch–sugar extrudates. Studies have established that the starch–water system is fully gelatinized and the material behaves as a viscous fluid. On cooling, the product is in the glassy state.

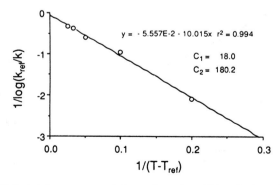

Figure 2.16 WLF plot for non-enzymic browning rates within a lactose–CMC–trehalose–xylose–lysine model system at initial $a_w = 0.12$ [36]. $C_1 = 18.0$; $C_2 = 180.2$. Slope: $y = -5.557E-2 - 10.015x$; $r^2 = 0.994$.

Fan et al. [40] have been able to model the expansion of a water vapour-filled cell as an approximation to this system. They took into account heat transfer and moisture losses, and mathematically treated the rheological behaviour of the rubbery, viscous mass during expansion by either a power-law model or by the WLF equation, using each in its appropriate domain. The water vapour pressure inside the cell (P_b), and the ambient pressure (P_a) were also taken into account. The effect of water and sugars on T_g starch was modelled using the data of Kalichevsky et al. [7] fitted to the Couchman–Karasz equation. Figure 2.17 shows the changes in various parameters that occur during expansion while Figure 2.18 records the effect of varying water contents upon the expansion process. Increasing the water content enhances initial expansion but exacerbates shrinkage, the ultimate effect being a reduced final expansion. The explanation can be seen in the state diagram (Figure 2.19). As the moisture content increases, the trajectory followed by a given sample in the temperature–moisture content plane will spend more time traversing the shrinkage domain the higher the moisture content, which will favour more extensive shrinkage. Experience has shown that at temperatures $<T_g + 30$, the system is so mechanically viscous that neither expansion nor shrinkage will occur. The T_c curve shows the situation where the inner and outer pressures are the same, i.e. $P_a - P_b = 0$.

The effects of sugar in depressing the ultimate degree of expansion can be accounted for by, on the one hand, a reduction in the extent of starch

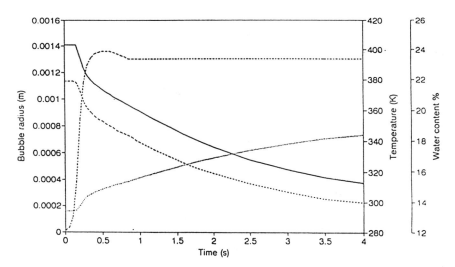

Figure 2.17 Progressive changes in bubble radius, water content, temperatures (T_m and T_g) within an extrudate after emerging from the extruder die. $C_0 = 22\%$ w.b.; $T_0 = 130°C$ (266°F).

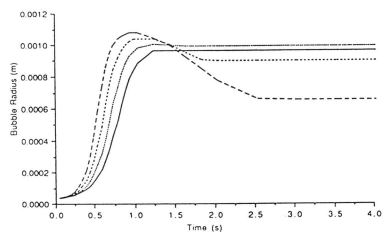

Figure 2.18 Effect of initial moisture contents on bubble growth and shrinkage ($T_0 = 120°C$ (248°F)).

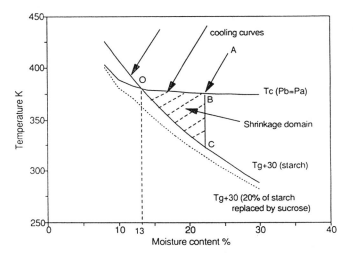

Figure 2.19 State diagram of the shrinkage domain in starch–water and starch–sugar–water systems.

gelatinization (starch 'conversion') and, on the other, by its plasticizing effect resulting in the lowering of the T_g and thereby extending the trajectory of the rubbery extrudate across the shrinkage domain, promoting more extensive shrinkage (Figure 2.19).

2.13.2 Understanding the storage stability of cereal products

Many products are produced from wheat flours and none is simpler than bread. A widely recognized problem is that of staling of bread, which for many years has been known to involve starch recrystallization. Staling of bread with a moisture content of approximately 45% w/w H_2O is at a maximum about 4°C (39°F). It is possible to simulate these general effects by examining the recrystallization of a starch gel. Marsh and Blanshard [41] have reported a study of the rates of crystallization as determined by wide angle X-ray diffraction. The results were modelled according to the theory of crystal growth of polymers developed by Lauritzen and Hofmann which shows the variation of log[crystal growth rate] with temperature as an inverted parabola (Figure 2.20). The growth rate falls to zero as cooling approaches T_g (−75°C, −103°F) and also on heating to T_m (65°C, 149°F), the melting temperature. A cursory observation of the 50% wheat starch gel shows that the maximum rate of recrystallization lies between 0 and 5°C (32–41°F) and that the rate processes at ambient temperature have (as experimentally observed) a negative temperature gradient. At significantly lower water contents (or through the substitution of water by sugar), T_g and T_m are raised, and the temperature at which the rate of recrystallization is at a maximum, is increased. If that effect is sufficiently pronounced, the gradient at ambient temperature becomes positive – a fact which is observed for cakes (in contrast to bread).

Mapping out the regions where the A and B starch polymorphs (two different crystalline states) are stable on the starch state diagram makes it

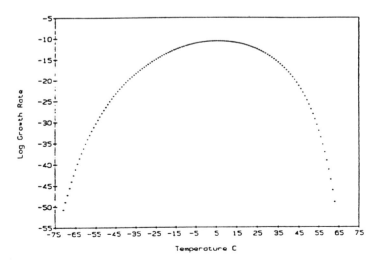

Figure 2.20 Log[crystal growth rate] versus temperature for a 50% w/w starch–water gel.

possible to follow the baking process of bread and the various parboiling processes of rice and to understand the nature of the product formed [42]. Slade and Levine have used the state diagram of sucrose to plot changes during the baking of cookies and to determine conditions of optimum storage stability [43].

2.14 Maximizing product quality through the use of state diagrams in dehydration

The process of dehydration almost always involves the transformation of a material from the rubbery to the glassy state. Frequently dehydration is performed with little understanding of the physico-chemical states and processes and not surprisingly is attended by deterioration in quality or biological activity. If we seek to establish a strategy for the process which minimizes deteriorative reactions, we may well attempt to do this by processing along a reaction pathway within the rubbery state but close to the T_g line. Regrettably, at any significant water content, this will involve sub-ambient temperatures with a corresponding reduction in dehydration rate.

A recent patent [44] which illustrates how proximity to the glassy state can be exploited, has described methods for the dehydration of enzymes that are normally unstable on storage. Two principal approaches have been described. The first relies on the production of a 'dough' of glassy sucrose with a $T_g < 20°C$ (68°F). The second uses Ficoll, a copolymer of sucrose and epichlorhydrin ($T_g = 97°C$, 206°F). In some instances, the Ficoll is used as a 'dough', in others as a relatively dilute solution (ca. 2.5%) with a similar amount of the material to be preserved. Drying takes place frequently under reduced pressure (80% to atmospheric) but always in the rubbery state. Temperatures commonly used are 20–40°C (68–104°F). The level of activity retained after the process and on storage is excellent.

2.15 Conclusion

There is no doubt that the perspectives and new understanding of food processing, storage and stability provided by a recognition of the existence and characteristics of the glassy and rubbery states have been one of the most stimulating and pervasive developments in food research in the past half century. However, much further work is required at both fundamental and applied levels. Continuing progress should assist in providing an important component in a general understanding of the physical chemistry of food systems.

References

1. Allen, G. A history of the glassy state. In *The Glassy State in Foods*, Eds Blanshard, J.M.V. and Lillford, P.J. Nottingham University Press, Loughborough (1993), 1–12.
2. Kauzmann, W. The nature of the glassy state and the behaviour of liquids at low temperatures. *Chemical Reviews*, **43** (1948), 219–256.
3. White, G.W. and Cakebread, S.H. The glassy state in certain sugar-containing food products. *Journal of Food Technology*, **1** (1966), 73.
4. Levine H. and Slade, L. Glass transitions in foods. In *Physical Chemistry of Foods*, Eds Schwartzberg, H.G. and Hartel, R.W. Marcel Dekker, New York (1992), 83–221.
5. Ablett, S., Izzard, M.J. and Lillford, P.J. Differential scanning calorimetric study of frozen sucrose and glycerol solutions. *Journal of the Chemical Society Faraday Transactions*, **88** (1992), 789–794.
6. Ablett, S., Clark, A.H., Izzard, M.J. and Lillford, P.J. Modelling of heat capacity–temperature data for sucrose–water systems. *Journal of the Chemical Society Faraday Transactions*, **88** (1992), 795–802.
7. Kalichevsky, M.T., Jaroszkiewicz, E.M., Ablett, S., Blanshard, J.M.V. and Lillford, P.J. The glass transition of amylopectin measured by DSC, DMTA and NMR. *Carbohydrate Polymers*, **18** (1992), 77–88.
8. Ablett, S., Darke, A.H., Izzard, M.J. and Lillford, P.J. Studies of the glass transition in malto-oligomers. In *The Glassy State in Foods*, Eds Blanshard, J.M.V. and Lillford, P.J. Nottingham University Press, Loughborough (1993), 1–12.
9. Kalichevsky, M.T., Blanshard, J.M.V. and Marsh, R.D.L. Applications of mechanical spectroscopy to the study of glassy polymers and related systems. In *The Glassy State in Foods*, Eds Blanshard, J.M.V. and Lillford, P.J. Nottingham University Press, Loughborough (1993), 133–156.
10. Izzard, M.T. (1995), Personal communication.
11. Hemminga, M.A., Roozen, M.J.G.W. and Walstra, P. Molecular motions and the glassy state. In *The Glassy State in Foods*, Eds Blanshard, J.M.V. and Lillford, P.J. Nottingham University Press, Loughborough (1993), 157–171.
12. Noel, T.R., Ring, S.G. and Whittam, M.A. Dielectric relaxations of small carbohydrate molecules in the liquid and glassy states. *Journal of Physical Chemistry*, **96** (1992), 6662–5567.
13. Girlich, D. and Lüdemann, H.-D. Molecular mobility of the water molecules in aqueous sucrose solutions studied by ^{2}H-NMR relaxation. *Zeitschrift für Naturforschung*, **49c** (1994), 250–257.
14. Simatos, D. and Blond, G. Some aspects of the glass transition in frozen food systems. In *The Glassy State in Foods*, Eds Blanshard, J.M.V. and Lillford, P.J. Nottingham University Press, Loughborough (1993), 395–415.
15. Shogren, R.L. Effect of moisture content on the melting and subsequent physical aging of corn starch. *Carbohydrate Polymers*, **19** (1992), 83–90.
16. Livings, S.J., Donald, A.M. and Smith, A.C. Ageing in confectionery wafers. In *The Glassy State in Foods*, Eds Blanshard, J.M.V. and Lillford, P.J. Nottingham University Press, Loughborough (1993), 507–511.
17. Fish, B.P. Diffusion and equilibrium properties of water in starch DSIR. *Food Investigation Technical Paper No. 5*, London, HMSO (1957).
18. Karger, N. and Lüdemann, H.-D. Temperature dependence of the rotational mobility of the sugar and water molecules in concentrated aqueous trehalose and sucrose solutions. *Zeitschrift für Naturforschung*, **46c** (1991), 313–317.
19. Girlich, D. and Lüdemann, H.-D. Molecular mobility of sucrose in aqueous solution studies by ^{13}C NMR relaxation. *Zeitschrift für Naturforschung*, **48c** (1993), 407–413.
20. Girlich, D., Lüdemann, H.-D., Buttersach, C. and Buchholz, K.C.T. Dependence of the self-diffusion in concentrated aqueous sucrose solutions. *Zeitschrift für Naturforschung*, **49c** (1994), 258–264.
21. MacNaughtan, W. (1995), Personal communication.
22. Peppas, N.A. and Brannon-Peppas, L. Water diffusion and sorption in amorphous macromolecular systems and foods. *Journal of Food Engineering*, **22** (1994), 189–210.
23. Vrentas, J.S., Jarzebeski, C.M. and Duda, J.L. *AIChEJ*, **21** (1975), 894.

24. Vrentas, J.S. and Duda, J.L. *Journal of Polymer Science, Polymer Physics*, **15** (1977), 441.
25. Angell, C.A., Bressel, R.D., Green, J.L., Kanno, H., Orguni, M. and Sare, E.J. Liquid fragility and the glass transition in water and aqueous solutions. *Journal of Food Engineering*, **22** (1994), 115–142.
26. Williams, M.L., Landel, R.F. and Ferry, J.D. The temperature dependence of relaxation mechanisms in amorphous polymers and other glass-forming liquids. *Journal of Chemical Engineering*, **77** (1955), 3301–3307.
27. Peleg, M. On the use of the WLF model in polymers and foods. *CRC Critical Reviews in Food Science*, **32** (1992), 59–66.
28. Couchman, P.R. and Karasz, F.W. A classical thermodynamic discussion of the effect of composition on glass-transitions. *Macromolecules*, **11** (1978), 117–119.
29. Gordon, M. and Taylor, J.S. Ideal copolymers and the second-order transitions of synthetic rubbers. I. Non-crystalline copolymers. *Journal of Applied Chemistry*, **2** (1952), 493–500.
30. Kalichevsky, M.T., Jaroszkiewicz, E.M. and Blanshard, J.M.V. A study of the glass transition of amylopectin–sugar mixtures. *Polymer*, **34** (1992), 346–358 [1993].
31. Kalichevsky, M.T., Jaroszkiewicz, E.M. and Blanshard, J.M.V. The glass transition of gluten. 1. Gluten and gluten–sugar mixtures. *International Journal of Biological Macromolecules*, **14** (1992), 257–266.
32. Kalichevsky, M.T., Jaroszkiewicz, E.M. and Blanshard, J.M.V. The glass transition of gluten. 2. The effects of lipids and emulsifiers. *Internation Journal of Biological Macromolecules*, **14** (1992), 267–273.
33. Reid, D.S., Kerr, W. and Hsu, J. The glass transition in the freezing process. *Journal of Food Engineering*, **22** (1994), 483–494.
34. Simatos, D. and Blond, G. DSC studies and stability of frozen foods. In *Water Relationships in Foods*, Eds Levine, H. and Slade, L. Plenum Press, New York (1991), 139–155.
35. Kerr, W., Liui, M.H., Chen, H. and Reid, D.S. Chemical reaction kinetics in relation to glass transition temperatures in frozen polymer solutions. *Journal of the Science of Food and Agriculture*, **61** (1993), 51–61.
36. Karmas, R., Buera, M.P. and Karel, M. Effect of glass transition on rates of non-enzymatic browning in food systems. *Journal of Agricultural Chemistry*, **40** (1992), 873–879.
37. Ma, Y., Reineccius, G.A., Labuza, T.P. and Nelson, K.A. The solubility of spray-dried micro-capsules as a function of glass transition temperature. Presented at the IFT Annual Meeting, New Orleans, LA, USA (1992).
38. Nelson, K.A. and Labuza, T.P. Water activity and food polymer science: implications of state on Arrhenius and WLF models in predicting shelf life. *Journal of Food Engineering*, **22** (1994), 271–289.
39. Roos, Y. and Karel, M. Crystallization of amorphous lactose. *Journal of Food Science*, **57** (1992), 775–777.
40. Fan, J.-T., Mitchell, J.R. and Blanshard, J.M.V. A computer simulation of the dynamics of bubble growth and shrinkage during extrudate expansion. *Journal of Food Engineering*, **23** (1994), 337–456.
41. Marsh, R.D.L. and Blanshard, J.M.V. The application of polymer crystal growth theory to the kinetics of formation of the B-amylose polymorph in a 50% wheat starch gel. *Carbohydrate Polymers*, **9** (1988), 301–317.
42. Ong, M.H. and Blanshard, J.M.V. The significance of the amorphous–crystalline transition in the parboiling process of rice and its relation to the formation of the amylose–lipid complex and the recrystallization (retrogradation) of starch. *Proceedings of the IFST Annual Conference*, London (1994).
43. Slade, L. and Levine, H. Beyond water activity: recent advances based on an alternative approach to the assessment of food quality and safety. *Critical Reviews in Food Science and Nutrition*, **30** (1991), 115–360.
44. Franks, F. and Hatley, R.H.M. Storage of materials. US Patent 5 098 893 (1992).

3 Emulsions
B. BERGENSTÅHL

3.1 Introduction

Emulsions, a mixture of two immiscible liquids, are by definition unstable. Therefore in trying to develop a 'stable emulsion', scientist are limited to controlling the kinetics of the processes that lead to the breakdown of emulsions. The technologist has two main tools available for this purpose: the use of mechanical devices to disperse the system and the addition of stabilising chemical additives (low-molecular-weight emulsifiers and polymers) to keep it dispersed.

Destabilisation involves aggregation, droplet fusion and settling processes. The emulsion would disappear extremely rapid if every droplet–droplet collision event resulted in fusion. Repulsive droplet–droplet interactions slow down this process, and the interparticle interactions are in fact mainly determined by the properties of the emulsion droplet surface. In food emulsions, this droplet surface is coated by various surface-active molecules of, in most cases, biological origin. However, to approach the ultimate goal of food emulsion technology, which is to permit predictions of the properties and stability of real foods, we have to consider the adsorption onto emulsion droplets of a number of different surface-active molecules from the surrounding solution.

This chapter discusses the kinetics of the instability processes, how the emulsifying additives give rise to stabilising forces and how the emulsifiers in complex technical systems form the oil–water interface of the emulsion droplets.

3.2 The kinetics of instability

Emulsion instability is a process which involves several different mechanisms contributing to the transformation of a uniformly dispersed emulsion into several totally separated phases.

- Creaming (settling). This is a separation caused by the upwards (or sometimes downwards) motion of emulsion droplets due to a density difference between the droplets and the surrounding medium.
- Ostwald ripening. This is a diffusion transport of the dispersed phase in small droplets into larger ones.

- Coalescence. This is a process in which droplets fuse together and lose their identity.
- Flocculation. This is an aggregation of droplets due to collisions.

All four of these mechanisms are present in an emulsion system and may influence each other during its breakdown.

The droplet size is the key parameter determining the destabilisation kinetics of an emulsion. Large droplets are prone to sedimentation and coalescence, whereas finely dispersed emulsions are more sensitive to flocculation and Ostwald ripening [1]. Coalescence is usually regarded as being the final result of flocculation.

3.2.1 Sedimentation or creaming

The sedimentation of particles in a gravitational field is described by the Stokes law:

$$v_{Stokes} = d^2 g \Delta\rho / 18\eta \qquad (3.1)$$

were v_{Stokes} is the settling rate of the particles, η the viscosity, g the gravitational constant, $\Delta\rho$ the density difference between the phases and d the droplet diameter.

However, the Stokes law assumes that the particles are alone and that the settling is not influenced by the presence of other particles. The reduction of the settling due to high concentrations is described by the term 'hindered settling' and is of importance as soon as the particle concentration is above a few percent. The settling is reduced to almost zero at a dispersed phase of about 40% [2,3].

3.2.2 Ostwald ripening

Inside a droplet, there is a positive pressure (the Laplace pressure) caused by the surface tension. When liquid from the interior of the droplet escapes and becomes dissolved in the continuous phase, energy is released which increases the solubility. The increased solubility around the smaller droplets leads to a transport of material from the smaller droplets toward the larger ones, the driving force being proportional to $1/r$. The mass transport from smaller drops to larger ones is also proportional to the specific interfacial area (scaling with $1/r$) and the reciprocal distance between the droplets (also scaling to $1/r$). Hence, the overall kinetic of the ripening process is proportional to the r^{-3} [4].

The solubility of the dispersed phase is the main rate-determining parameter of relevance to the Ostwald ripening process. Triglycerides have very low solubility in water, which makes the Ostwald ripening negligible. The solubility of water in triglycerides is more significant, which makes the Ostwald process more important in oil continuous emulsions. However, the aqueous phase in water–oil emulsions, such as margarine or spreads, also contains salt with a low solubility. This causes an osmotic counterpressure, blocking the progress of the ripening process. The ripening phenomena are much more pronounced in systems where a pure phase is in equilibrium with the bulk, typically crystal dispersions (ice or sucrose crystals in water and fat crystals in a semi-solid fat) where this process will determine the particle growth rate.

3.2.3 Coalescence

Coalescence is a process in which two droplets combine to form a single droplet. The coalescence process, viewed experimentally, follows the following basic steps:

1. The concentrated emulsion (or the cream layer of a more dilute emulsion) slowly changes into a more dense cream through a consolidation process (i.e. instead of being uniformly distributed, the droplets tend to come together in clumps). The frequency of droplet–droplet contacts is increased during this consolidation process.
2. Drainage of the thin film from between the droplets. The rate of the drainage process determines how rapidly the critical breakdown thickness of the film is reached.
3. Rupture of the film. This is a stochastic (random) process, the probability of which is determined by the thickness of the film. This probability would be expected to become significant at a critical thickness (this process is described in more detail in section 3.2.4).
4. Merging of the droplets. If the viscosity is reasonably low, this is an irreversible and rapid process. However, emulsions with a highly viscous (e.g. partly crystalline) oil phase merge very slowly (partial coalescence [5]) and may even be disrupted again if the fusion process is interrupted (for example, due to dilution).

Such complex conditions are very difficult to describe and predict and there is no theory which completely covers the area. Table 3.1 summarises both observational and theoretical evidence about the different conditions which contribute to the various aspects of coalescence.

3.2.4 Film rupture

A liquid film is a metastable aggregate which suddenly may disappear through rupture, similar to a soap bubble. The rupture can be viewed as a process induced by the self-propagation of a disturbance [6]. The disturbance may be externally induced (shear, crystals, gradients) or induced by surface waves. The process can also be viewed as a sudden blast

Table 3.1 Principal effects of increasing different variables on the various steps involved in a phase separation. The qualitative influence on the stability is indicated

Variable increment	Creaming	Consolidation	Drainage	Rupture
Droplet viscosity	0	0	0	+
Continuous phase viscosity	+	0	+	0
Droplet size	−	+	+	−
Volume fraction of dispersed phase	+	−	−	0
Density difference	−	−	−	0
Surface tension	0	0	−	+
Surface viscosity	0	0	0	+
Surface elasticity	0	0	+	+
Depth of attractive minima	0	+	−	0
Repulsive interaction	0	−	0	+
Adsorbed layer thickness	0	−	+	+
Solubility of emulsifier in the dispersed phase	0	0	−	−
Hydrophilic–lipophilic balance of the emulsifier, oil–water emulsion	0	−	+	+

caused by a protrusion (or 'bridge') over the film. In this case, the protruding intermediate form of the film can be recognised as a transition state for the process. The energetically unfavourable transition state may then be viewed as being equivalent to the activation energy in an Arrhenius treatment of the process.

One proposed transition is a solubilisation of the emulsifier at the interface of the two phases [13]. The energy barrier should then be dependent on the ratio between the solubilities of the emulsifier in the two phases. This is closely related to empirically based emulsion rules (used for the selection of emulsifiers) in Table 3.2.

Other proposed transition-state theories involve compression modules [14] in that they propose that the probability of instability is related to disturbances at the interface. Others have proposed the formation of a bridge of the dispersed phase, which makes the energy of the transition state directly proportional to the interfacial tension [15]. The 'formation of a bridge' hypothesis is supported by the observed reduction in stability at low interfacial tensions (emulsions based on low-molecular-weight surfactants, giving an oil–water interfacial tension of 1–5 mN/m, are usually less stable than emulsions based on surface active polymers generating an oil–water interfacial tension of about 10–20 mN/m (see also chapter 8)). There has also been a discussion around the possibility that the emulsifier forms a liquid crystalline bridge and that a phase transition in this bridge (from lamellar into cubic phase) induces the phase transition [16].

Table 3.2 Empirically based guidelines for selecting emulsifiers

Rule	Statement	Reference
Bancroft rule	The phase in which the emulsifier is most soluble will be the continuous phase. *Example*: Oil-soluble technical lecithin gives good oil continuous emulsions while water-dispersible hydrolysed lecithins are usable in water continuous emulsions.	[7]
The hydrophilic–lipophilic balance (HLB) concept	An emulsifier with an HLB value <7 forms an oil continuous emulsion, whereas an emulsifier with an HLB number >7 forms water continuous emulsion. *Example*: The HLB numbers of numerous emulsifiers have been listed and can be used as a guideline for selection and for mixing of emulsifiers. The HLB number can be calculated from the chemical formula of the emulsifier [8].	[8,9]
The phase-inversion temperature (PIT)	Below the PIT, an oil-in-water emulsion is formed and above the PIT a water-in-oil emulsion. *Example*: The PIT concept is useful when strongly temperature-dependent emulsifiers are discussed (e.g. ethoxylated surfactants). Other emulsifiers (anionic, cationic, lecithins, monoglycerides) usually do not display any clear phase-inversion temperature.	[10]
The phase diagram concept	The most stable emulsion is formed when the emulsifier is present in a lamellar liquid crystalline or a semi-crystalline α-form.	[11,12]
	Example: Monoglyceride-stabilised emulsions are stable as long as the emulsifier is present in the semi-crystalline α-form but separates rapidly if the emulsifier is crystallising in the stable β-form.	[12]

3.2.5 Flocculation

Flocculation is the formation of permanent flocs through interparticle collisions. The collision mechanisms may be:

- Brownian motion (perikinetic flocculation).
- Sedimentation motion (gravity-induced flocculation).
- Motion in a shear field (orthokinetic flocculation).

The rate of loss of particles can be obtained from the rate of collisions (if we assume that there is a certain probability that a collision leads to permanent adhesion). In Table 3.3, the collision frequency is given as a function of the number concentration of droplets in the emulsion. The rate of the instability processes (breakdown of the emulsion) is also related to the increase of the particle radius with time, assuming that all flocculated particles coalesce. This way to describe the emulsion corresponds much more closely to the type of measurements that can be

Table 3.3 A comparison of different flocculation-rate mechanisms in emulsions. The instability is expressed in terms of the change in number concentration (dN/dt) of the increase in particle radius (dr/dt), or as the particle radius as a function of time

dN/dt	dr/dt*	$r(t)$
Brownian flocculation [17]		
$\frac{dN[t]}{dt} = -\frac{4k_B T}{3\eta} N[t]^2$	$\frac{dr}{dt} = \frac{2}{3} \cdot \frac{k_B T}{\eta} \cdot \frac{\phi}{\pi r^2} \cdot \frac{1}{w_B}$	$r = \left(r_0^3 + \frac{\phi 2 k_B T}{9\pi \eta w_B} \cdot t\right)^{1/3}$
Shear-induced flocculation: laminar flow [17]		
$\frac{dN}{dt} = -\frac{16}{3} N^2 S r^3$	$\frac{dr}{dt} = \frac{4\phi r S}{3\pi} \cdot \frac{1}{w_s}$	$r = r_0 \exp\left(\frac{4\phi S t}{3\pi} \cdot \frac{1}{w_s}\right)$
Shear-induced flocculation: turbulent flow [18]		
$\frac{dN}{dt}\Big|_{Turb.} = -6\pi\beta\sqrt{\frac{\varepsilon_{intensity}}{\eta}} r^3 N^2$	$\frac{dr}{dt} = \frac{3}{2}\beta\sqrt{\frac{\varepsilon_{int.}}{\eta}} \phi r \cdot \frac{1}{w_T}$	$r = r_0 \exp\left(\frac{3}{2}\beta\sqrt{\frac{\varepsilon_{int.}}{\eta}} \phi t \cdot \frac{1}{w_T}\right)$
Gravity-induced flocculation [19]		
$\frac{dN}{dt} = -\frac{2\pi g \Delta\rho}{9\eta} \lambda' N_1 N_2$	$\frac{dr}{dt} = \frac{\Delta\rho g \phi}{18\eta} \cdot \frac{\lambda'}{w_g} \cdot r^2$	$r = \dfrac{1}{\left(\dfrac{1}{r_0} - \dfrac{\Delta\rho g \phi}{18\eta} \cdot \dfrac{\lambda'}{w_g} \cdot t\right)}$

*The flocculation rate in terms of particle concentration has been transformed to growth in particle size by using the chain rule:

$$\frac{dN}{dt} = \frac{dN}{dr} \cdot \frac{dr}{dt}$$

and by treating the particle concentration as a function of the particle radius:

$$N = \frac{3\phi}{4\pi r^3} \rightarrow \frac{dN}{dr} = -\frac{9\phi}{4\pi r^4}.$$

N = Particle concentration, t = time, k_B = Boltzmann constant, T = absolute temperature, η = viscosity, ϕ = volume fraction dispersed phase, w_B = stability factor for the Brownian flocculation according to Fuchs [20], w_s = a stability factor for shear-induced flocculation, w_T = a stability factor for flocculation in a turbulent flow, S = the shear rate, β = a constant describing the relation between the diffusive flux in a turbulent flow field and the liquid velocity, w_g = a stability factor for gravity-induced flocculation,

$$\lambda' = \left(\frac{r_1}{r_2} + 1\right)^2 \left(\left(\frac{r_1}{r_2}\right)^2 - 1\right)$$

for two classes of flocculating particles, $\Delta\rho$ = density difference, g = gravity, r_0 the radius at time zero.

performed (e.g. particle-size analyses using laser diffraction) and enables the stability to be followed experimentally in an unstable emulsion system.

An interesting observation from Table 3.3 is that the Brownian flocculation (in terms of dr/dt) rapidly declines with increasing radius. The sedimentation-induced flocculation has a weak dependence upon the particle radius and the shear-induced flocculation increases with increasing particle radius. This gives the characteristic property of a shear destabilisation, i.e. an extended lag phase (long initial slow growth) with an apparently stable system followed by a rapid destabilisation.

So far we have considered all drops as being compact droplets. However, under most conditions, the coalescence is a relatively slow process and the flocculated aggregates consist of several droplets loosely aggregated. During the flocculation, the average diameter, the effective density of the aggregates and the apparent volume fraction are changed, which alters the kinetic characteristics of the system.

The flocculation is more or less a stochastic process that results in a disordered structure. The size and density of the flocs can be described as fractal objects [21]. The packing density in the flocs is characterised by the fractal dimension d_f which is defined by:

$$R/r \propto N_p^{1/d_f} \tag{3.2}$$

where R is the apparent radius of the flocs, r the radius of the primary particles, N_p the number of particles in the floc and d_f the fractal dimensionality.

Fractal flocs have the property that they become less dense as they increase in size. In addition, the lower the dimensionality, the more space filling the flocculated structure will be. One result will be that the apparent volume of the dispersed phase will increase during a flocculation process and finally fill the complete volume of the dispersion. This will result in a gelation and total block of simple settling or flocculation, and limit further ageing of the system to slow consolidation and coalescence processes.

3.3 Droplet–droplet interactions

The above description of the flocculation includes interparticle interactions, which protect the particles from fusing after a collision event, in the form of a stability factor, w, where $1/w$ is the probability for attachment after a collision event. The stability factor is a function of the ratio between energy of the collision event and that of the repulsive barrier.

The interactions forming the repulsive barrier are (for a modern extensive review see Israelachvili [22]):

- Van der Waals attraction
- Hydrophobic attraction
- Electrostatic repulsion
- Steric repulsion
- Hydration force

Forces between molecules and particles caused by permanent and induced dipoles are collectively known as van der Waals forces [23]. The van der

Waals force between the particles, droplets or bubbles dispersed in a media is always attractive.

The hydrophobic interaction [24] between macroscopic surfaces has been experimentally demonstrated, though the origin is still unclear. The greatest long-range interaction is observed between very hydrophobic particles. Its role between particles covered by emulsifier is strongly reduced.

When a charged surface is immersed in a liquid, an excess of counterions accumulates close to the surface. If two identically charged surfaces approach each other in an electrolyte solution, a surplus of counterions accumulates between the surfaces and causes a repulsive interaction, usually termed a 'diffuse double-layer force' or 'electrostatic repulsion' [25]. The range of the repulsion is dependent on the thickness of the diffuse double layer, determined by the electrolyte concentration in the aqueous phase. In electrolyte solutions rich in counterions (for monovalent counterions >100 mM and for divalent counterions >25 mM), the strength of the electrostatic repulsion is greatly reduced.

The presence of polymers in the continuous phase and/or on the emulsion droplet surface affects the forces between the droplets. The magnitude and type of interactions are determined by whether they are adsorbing at the surface irreversibly or reversibly or whether they do not adsorb at all (Table 3.4).

A dense layer of irreversibly adsorbed polymers gives rise to steric repulsion between droplets. A surface with adsorbed polymers has an excess of polymer segments close to the surface. If two identical surfaces with adsorbed polymer approach each other in a solvent, a surplus of polymer segments accumulates between the surfaces and causes a repulsive interaction, usually termed 'steric repulsion' [26]. The range of the repulsion is dependent on the thickness of the adsorbed layer, which is itself determined by the molecular weight of the adsorbed molecule.

Table 3.4 The influence of polymers (e.g. hydrocolloids) on food emulsions

Effects	Property	Influence on the inter-droplet interactions	Influence on emulsions (with small droplets)
Influences on the interactions	The polymers well adsorbed	Steric repulsion	Stable emulsion
	Incomplete or reversible adsorbed layer	Bridging	Flocculation and rapid creaming
	Non-adsorbing polymer	Depletion attraction	Flocculation and rapid creaming
Influences on the solvent	Increased viscosity	No effect	Slower flocculation and creaming

A dilute or incomplete layer of adsorbed polymers might give rise to bridging attraction between droplets. If two surfaces with an incomplete coverage of polymer approach each other in a solvent, polymer molecules may attach to both surfaces and form a bridge between them. The bridging causes an attractive interaction, the range of which is dependent on the molecular weight of the adsorbed molecule.

Most systems with adsorbing polymers change from bridging to sterically stabilised when the concentration of polymer is increased. The bridging is favoured by a very strong affinity between the interface and the adsorbing polymers (for instance due to an oppositely charged surface and polymer).

Non-adsorbed polymers give rise to depletion attraction between droplets. A surface in equilibrium with a solution of a non-adsorbing polymer is surrounded by a zone depleted of polymer segments close to the surface (this is due to packing restrictions close to the interface). If two identically surfaces depleted of polymer segments approach each other in a solvent, a depletion zone grows between the surfaces and causes an attractive interaction. The range of the attraction is dependent on the size of the molecule and the concentration. At very low concentration the forces are small, at intermediate concentration the force is pronounced and at high polymer concentration the depletion zone is strongly compressed [27,28].

The removal of water molecules, that solvate polar groups at an interface, gives rise to a repulsive force, usually termed a 'hydration force' [29]. Although this concept has been traditionally accepted, it is intuitive in character. The range and magnitude of this force have been quantified experimentally, but the theoretical explanation is still rather controversial [30]. This repulsive force was introduced to explain the forces observed between stacked bilayers in a lamellar phase (for instance, formed by lecithin and water). The forces were determined by lowering the osmotic pressure of the interlayer water. This was first demonstrated in a series of measurements by Rand, Parsegian and co-workers [29,31] who, for a range of different phospholipid systems, measured the osmotic pressure in relation to pure water as a function of bilayer separation. They observed a strong force with a range of a few nanometres. The distance and salt-dependence of the force was such that it could not be an electrostatic double-layer force. Instead, it was assumed that a change in the lecithin–water interaction occurred as the interlamellar separation changed, and that this gave rise to a repulsive force. The hydration force in this experiment displays an exponential decay usually expressed as [29]:

$$\Pi_{\text{Hydration}} = \Pi_0 e^{-d_w/\lambda} \qquad (3.3)$$

where Π_0 the repulsive pressure, λ is the decay constant and d_w is the water layer thickness.

A typical example is the repulsive interaction operating between layers of monoglycerides illustrated in Figure 3.1 [32]. However, it should be pointed out that the relationship is empirical and is a description of the results, rather than a theoretical representation of the repulsive pressure [29].

The hydration force has been measured for a range of important food emulsifiers (Table 3.5).

3.4 The oil–water interface

In section 3.3, the different classes of surface forces that act between the surfaces within an emulsion were discussed. Their type and magnitude depend on the composition of the surface. Food emulsions are complex mixtures. They usually contain both low-molecular-weight surface-active lipids and a wide range of more or less surface-active proteins and polysaccharides. The actual chemical composition of the emulsion droplet surface is then the key factor which determines most of the surface interactions.

In systems containing several surface-active components, three types of adsorbed layers can be identified, based upon how the layers are formed. In reality, the differences between the three adsorption structures discussed below are not sharp, but this simplified description can provide a basis for evaluating the properties for complex systems.

- Competitive adsorption. A monolayer containing one predominant type of molecule at the interface. This builds up through competition

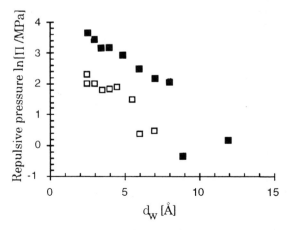

Figure 3.1 The hydration repulsion between bilayers of monopalmitin in the liquid crystalline lamellar phase and the crystalline α-gel states. ■ = Lamellar phase, 60°C (140°F); □ = gel phase, 23°C (73°F). From Pezron *et al.* [32].

Table 3.5 The hydration force measured in osmotic stress experiments. The results are described according to the empirical model $\Pi_{Hydration} = \Pi_0 e^{-(d_w/\lambda)}$

Substance	Temperature (°C)	Temperature (°F)	λ (nm)	Π_0 (MPa)	d_w (Å)	Reference
Phosphatidylcholine						
from egg	25	77	2.6	580	25	[31]
soybean	22	72	2.8	470	20	–[4]
Phosphatidylethanolamine						
from egg	25	77	2.1	3700	19	[31]
Monoglycerides						
monoolein	23	73	2.0	140	9	[32]
monopalmitin, lamellar	60	140	2.4	110	16	[32]
monopalmitin, gel	23	73	2.4	30	7	[32]
monocaprylin[1,3]	20	68	1.3	360	21	[33]
Polyoxyethylene surfactants						
C12EO3[1]	5	41	3.0	88	–[2]	[34]
	15	59	3.1	77	–[2]	[34]
	25	77	3.3	73	–[2]	[34]
	35	95	3.4	45	–[2]	[34]
C12EO4[1]	25	77	4.0	54	–[2]	[34]

Notation: d_w = Thickness of the aqueous layer, Π_0 = the repulsive pressure. 1. Recalculated from data in the reference. 2. The upper limits of the lamellar phase are in equilibrium with other aggregated phases. 3. Using the initial slope, the slope of the long-range repulsion is weaker. 4. Unpublished results obtained by the author.

with other less surface-active components, which it may replace at the interface.
- Associative adsorption. An adsorbed layer containing a mixture of several different surface-active components is formed.
- Layer adsorption. One component adsorbs on top of an other.

3.4.1 Competitive adsorption

In a system with several surface-active components, a homogeneous monolayer is formed by the most surface-active component.

When competitive adsorption of macromolecular emulsifiers is discussed, we first have to consider that polymers are flexible molecules anchored to the surface at several points. A desorption of the molecule demands that all contact points are separated simultaneously. This makes the desorption on dilution of firmly adsorbed material an unlikely event for polymers. Once adsorbed onto a surface, a polymer may undergo slow conformational changes causing changes in the adsorbed amount and in surface properties. This is observed as an ageing effect. The ageing causes the interfacial tension to decrease slowly over a period of several hours and the surface rheological properties to change slowly [35]. If a second component is present, the changes may allow the second component to adsorb and segment by segment to replace the first component at the interface. In the presence of a competing more 'surface-active' polymer, even polymer adsorption may then be reversible [36].

A special case is the observed sequence of adsorbed proteins to a glass surface in contact with blood plasma (Table 3.6). When the surface is placed in contact with blood, the small proteins adsorb first, as they have a higher molar concentration and a more rapid diffusion. As the process proceeds, the composition at the surface continues to change. The smaller molecules are replaced by larger molecules until the largest and most adaptable molecules dominate the surface (the Vroman series [37]).

The competitive adsorption between various food proteins has been investigated by Dickinson and co-workers [38–42]. Their results are summarised in Table 3.7. The table shows that β-casein displays the highest surface activity and is able to displace a significant amount of

Table 3.6 The order of adsorption to a glass surface in contact with blood plasma [37]

Stage	Protein	Concentration (mg/ml)	Molecular weight
1	Albumin	40	69 000
2	IgG	12.1	160 000
3	Fibronectin		
4	Fibrinogen	3	340 000
5	High-molecular-weight kininogen		

Table 3.7 Competitive displacement of food proteins in emulsions

Protein	Displaced protein	Change in total load (%)	Fraction of displaced protein (%)	Reference
β-Casein	α_{s1}-casein	+46	40	[38]
	β-lactoglobulin	+13	13	[39]
	α-lactalbumin	+27	60	[40]
	phosvitin	+88	76	[41]
β-Lactoglobulin	α-lactalbumin	+12	4	[42]
	phosvitin	+13	58	[41]
α-Lactalbumin	β-lactoglobulin	+6	0	[42]

All the experiments were performed by Dickinson and co-workers [38–42]. The general procedure is as follows: A 20% tetradecane emulsion was produced with 0.5% of the 'displaced protein', the emulsion was washed and 0.5% of the competing 'protein' was added. The protein load was then measured. The 'fraction of displaced protein' was determined with respect to the 'displaced protein' load before the competing protein was added.

proteins from the interface (α_{s1}-casein, α-lactalbumin and phosvitin). However, it is also observed that all proteins do not display this sensitivity to the presence of casein. For instance, an adsorbed layer of β-lactoglobulin is reduced only by about 10% when casein is added and the new layer consists of about 25% casein and 75% lactoglobulin.

It is also observed that large flexible proteins are able to dominate the interface over smaller and/or globular proteins, as in the Vroman series. Other observations of competitive adsorption with food or 'food-like' polymers have been reported. For instance, albumin has been observed to displace dextran from silver iodide particles [43], and casein displaces gelatin from the interface between water and hexadecane [44]. From these observations it is possible to generalise the order of surface activity of common food ingredients as follows: lipid emulsifiers > large flexible hydrophobic proteins (e.g. β-casein) > small hydrophilic proteins (e.g. phosphovitin, lactalbumin) > gelatin ≃ surface active polysaccharides (e.g. gum arabic) > starch.

3.4.2 Associative adsorption

In associative adsorption, a mixed surface is formed. The properties displayed by the surface are then an average of those of the emulsifiers present.

A typical associative system may be a sorbitan ester and ethoxylated sorbitan ester. The sorbitan ester acts as a spacer between the bulky ethoxylated groups, which decreases the lateral head-group repulsion within the layer and reduces the surface energy. This increases the adsorption and enhances the surface activity [45]. This is also a necessary prerequisite of the commonly made assumption that an average hydrophilic–lipophilic balance value should describe the properties of an emulsifier blend [13].

In the case of associative adsorption, both components are expected to be present at the surface. If this situation is to be stable, the adsorption of the second component should either be enhanced by the presence of the first component, or not influenced by it. The total amount of adsorbed material should be greater than or equal to the sum of the two components.

3.4.3 Layered adsorption

Adsorption in layers is possible when different classes of surface-active components are present in a mixture. The two components must be very different in character to give a structure with a layered character rather than a mixed layer.

The second component adsorbs to a particle displaying the characteristic properties of the primary adsorbing emulsifier. This usually means a more

Table 3.8 The relationship between the adsorption sbucture and the component determining the interactions in solutions containing several surface-active components

Adsorption	Component determining the interactions
Competitive	The most hydrophobic
Associative	An average
Layers	The most hydrophilic

hydrophilic surface than the clean oil–water interface, and can be expected to reduce the amount adsorbed. However in some cases, the presence of certain groups increases the adsorption of specific substances. For instance, it has been observed that the presence of phosphatidic acid in a mixed phospholipid interface enhances the adsorption of apolipoprotein B [46], and the presence of a bile acid in an emulsion emulsified with lecithin and cholesterol strongly enhances the adsorption of chitosan [47].

A mixture of several surface-active components adsorbs differently depending on the type of adsorbed layer formed (Table 3.8). This determines the resulting types of interactions and hence the stability properties.

3.5 Conclusions

Emulsions are thermodynamically unstable systems. The instability processes within these systems, which control technically important properties such as the shelf-life, texture and taste release, are very complex. No universal theory describing all aspects of the processes exists and empirical development will continue to be the basis for all formulation work with food emulsions.

From various theories, we may obtain an approximate understanding of the role of fundamental parameters such as particle size, rheology, composition, temperature, etc. This knowledge enables us to use modern measuring techniques as efficient and powerful development tools in the field of food emulsions.

An understanding of the functional role of low-molecular-weight emulsifiers, proteins and polysaccharides in emulsions acts as a guideline for formulation work and is of crucial importance when interpreting experimental results. An understanding of the various types of adsorption structures, that are developed in multicomponent mixtures, might help in rationalising the sometimes confusing results which are obtained with real foods.

Acknowledgements

Grateful thanks are due to Mr Norman Burns for comments on the draft manuscript.

References

1. Bergenståhl, B. and Claesson, P. In *Food Emulsions*, Eds Larsson, K. and Friberg, S., Marcel Dekker, New York (1990).
2. Walstra, P. and Oortwijn, H. *Neth. Milk Dairy J.*, **29** (1975), 263.
3. Buscall, R., Goodwin, J.W., Ottewill, R.H. and Tadros, Th.F. *J Colloid Interface Sci*, **85** (1982), 78–86.
4. Kabalnov, A.S., Pertzov, A.V. and Shchukin, E.D. *J. Colloid Interface Sci.*, **118**, (1987), 590–597.
5. Boode, K. Partial Coalescence in Oil in Water Emulsions. Dissertation, Wageningen, Netherlands (1992).
6. Vrij, A. *Disc. Faraday Soc.*, **42** (1966), 23–33.
7. Bancroft, W.D. *J. Phys. Chem.*, **17** (1913), 501.
8. Davies, J.T. *Proc. 2nd Intern. Congr. Surf. Activity*, London, **1** (1957), 426.
9. Griffin, W.C. *J. Soc. Cosm. Chem.* (1949), 311–326.
10. Shinoda, K. and Saito, H. *J. Colloid Interface Sci.*, **30** (1968), 258–263.
11. Friberg, S. and Wilton, I. *Am. Parf. Cosm.*, **85** (1970), 27–30.
12. Wilton, I. and Friberg, S. *J. Am. Oil Chemists Soc.*, **48** (1971), 771–774.
13. Davies, J.T. and Rideal, E.K. *Interfacial Phenomena*, Academic Press, London (1963).
14. MacRitchie, F. *J. Colloid Interface Sci.*, **56** (1977), 53–56.
15. Walstra, P. In *Research in Food Science and Nutrition*, Vol. 5, Eds McLoughlin J.V. and McKenna, B.M. Boole Press, Dublin (1984).
16. Siegel, D.P. *Chem. Phys. Lip.*, **42** (1986), 279–301.
17. von Smoluchowskij, M. *Phys. Z.*, **17** (1916), 593.
18. Levich, V.G. *Physico-chemical Hydrodynamics*, Prentice-Hall, Englewood Cliffs, NJ (1962).
19. Reddy, S.R., Melik, D.H. and Fogler, H.S. *J. Colloid Interface Sci.*, **82** (1981), 116–127.
20. Fuchs, Z. *Physik*, **89** (1934), 736.
21. Bremer, L., Fractal Aggregation in Relation to Formation and Properties of Particle Gels. Dissertation, Wageningen, Netherlands (1992).
22. Israelachvilli, J. *Intermolecular and Surface Forces*, Academic Press, London (1992).
23. Hamaker, H.C. *Physica*, **4** (1937), 1058.
24. Israelachvili, J.N. and Pashley, R.P. *Nature*, **300** (1982), 341.
25. Verwey, E.J.N. and Overbeek, J.Th.G. *Theory of the Stability of Lyophobic Colloids*, Elsevier, Amsterdam (1948).
26. Napper, D.H. *Polymeric Stabilisation of Colloidal Dispersions*, Academic Press, London (1983).
27. Sperry, P. *J. Colloid Interface Sci.*, **99** (1984), 97.
28. Fleer, G.J., Scheutjens, J.H.M.H. and Vincent, B. In *Polymer Adsorption and Dispersion Stability ACS Symposium Series 240*, Eds Goddard, E.D. and Vincent, B. ACS (1984).
29. Parsegian, V.A., Fuller, N. and Rand, R.P. *Proc. Natl. Acad. Sci. USA*, **76** (1979), 2750.
30. Israelachvili, J. and Wennerström, H. *Langmuir*, **6** (1990), 873–876.
31. Lis, L.J., McAlister, M., Fuller, N., Rand, R.P. and Parsegian, V.A. *Biophys. J.*, **37** (1982), 657.
32. Pezron, I., Pezron, E., Bergenståhl, B. and Claesson, P. *J. Phys. Chem.*, **94** (1990), 8255–8261.
33. McIntosh, T.J., Magid, A.D. and Simon, S.D. *Biophys J.*, **55** (1989), 897.

34. Lyle, I.G. and Tiddy, G.J.T. *Chem. Phys. Lett.*, **124** (1986), 432–436.
35. Dickinson, E., Murray, A., Murray, B.S. and Stainsby, G. In *Food Emulsions and Foams*, Ed. Dickinson, E. Royal Society of Chemistry, London, **86** (1987).
36. Dickinson, E. *Food Hydrocolloids*, **1** (1986), 3.
37. Vroman, L. and Adams, A.L. *J. Colloid Interface Sci.*, **111** (1986), 391–402.
38. Dickinson, E., Rolfe, S.E., Dalgleish, S.E. *Food Hydrocolloids*, **2** (1988), 397.
39. Dalgleish, S.E., Euston, S.E., Hunt, J.A. and Dickinson, E. In *Food Polymer and Gels*, Ed. Dickinson, E. Royal Society of Chemistry, Cambridge (1990), 487.
40. Dickinson, E. *ACS Symp. Ser.*, **448** (1991) 114.
41. Dickinson, E., Hunt, J.A. and Dalgleish, D.G. *Food Hydrocolloids*, **4** (1991), 403–414.
42. Dickinson, E., Rolfe, S.E. and Dalgleish D.G. *Food Hydrocolloids*, **3** (1989), 193–203.
43. Matuszewska, B., Norde, W. and Lyklema, J. *J. Colloid Interface Sci.*, **84** (1982), 403.
44. Dickinson, E., Pogson, D.J., Robson, E.W. and Stainsby, G. *Colloids Surfaces*, **14** (1985), 135.
45. Boyd, J.V., Krog, N. and Sherman, P. *Theory and Practice of Emulsion Technology*, Ed. Smith, A.L. Academic Press, London (1976).
46. Bergenståhl, B, Fäldt, P. and Malmsten, M. In *Food Macromolecules and Colloids*, Eds Dickinson, E. and Lorient, D. Royal Society of Chemistry, in press (1995).
47. Fäldt, P., Bergenståhl, B. and Classon, P. *Colloid and Surfaces*, **71** (1993), 1877–1895.

4 Non-enzymatic browning of foods
J.C. HOSKIN and P.S. DIMICK

4.1 Introduction

To the uninitiated, the browning of foods could simply be dismissed as the degree of colour developed during the toasting of bread. Excess colour change coupled with development of burnt flavours might define the extent of interest in the subject by the consumer. The food scientist realizes that Maillard or non-enzymatic browning of food cannot be understood as a simple colour change. Food components may, or may not, change colour with the application of heat. Normally, colour modifications are accompanied by flavour changes, often desirable, such as that specifically associated with heating of cocoa beans, coffee beans, nuts, meats, bakery goods, or the roasting of foods in general. Beyond visual and flavour (taste and aroma) changes, the food professional also notes relevant changes in food structure, nutritional value, production of other desirable and/or undesirable compounds and most important, the need to control these heat-induced changes. The purpose of this chapter is to review the subject of Maillard reactions as an overall concept. Specifics of all the chemistry involved are beyond the scope of this limited text. Reactions will be presented simply with initial and final compounds. Interested readers should refer to the cited literature for an in-depth explanation of reactions.

4.2 History and name of reaction

The vast majority of food coloration changes can be divided into two general types of reactions described as either enzymatic or non-enzymatic. More specific types of browning, such as caramelization of sugar or ascorbic acid browning, may also contribute to colour changes, but typically to a considerably lesser extent. Natural browning of fresh fruit or vegetables occurs via enzymatic mechanisms. Non-enzymatic browning is caused primarily by artificial heating of foods with accompanying colour and flavour change in possibly, otherwise, stable foods. It is the application of artificial heat that greatly differentiates food materials available for human consumption from that available for all other animal species and bypasses some evolutionary plant protective mechanisms [1]. Depending on the food in question, both enzymatic and non-enzymatic browning can potentially occur at the same time, although this is rather unlikely.

The Maillard reaction is named after Louis-Camille Maillard (1878–1936) and, by referring to it as 'my reaction', he may have initiated its use [2]. The term first occurred in print in 1950. Certainly, having conducted fundamental and ground-breaking work, including eight papers on the sugar–amino acid reactions, it is appropriate to honour him by using his name, although non-enzymatic browning is certainly more descriptive and that phrase, along with carbonyl–amine reaction, continue to be used interchangeably. Interest in Maillard reactions has continued to expand and is reflected in numerous books, symposia and review papers [3–10].

4.3 Overview of reactions

An appreciation of the the fundamental concept of non-enzymatic browning is necessary before studying it in more detail. Initial food components, amino acids and sugars provide the starting compounds that participate in the non-oxidative, non-enzymatic browning reactions (Figure 4.1). These reactants participate in a variety of condensation (or addition) reactions that may be initially reversible and produce condensation products. Unless the conditions are mild regarding time and temperature of thermal exposure and the components relatively unreactive, these components will continue to interact and produce various rearrangement products, some of which have similar reactivity as the initial food components. Continuation of such reactions result in compounds that contribute aroma, taste and colour, such as in the toasting of bread or the roasted flavour and colour developed by numerous foods.

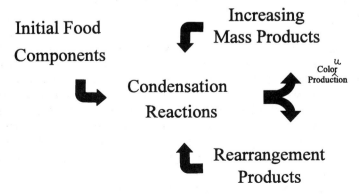

Figure 4.1 Maillard browning overview.

With excessive heating, either higher temperature, longer time or both, non-enzymatic browning reaction products begin to produce condensation products of increasing mass and complexity, which convey extreme dark coloration and the acrid and disagreeable flavour referred to as 'burnt'. The food preparer can control the time and temperature of heating, and the food processor can also obtain increased control over the nature of the food components by appropriate formulation. Conceptually, one can understand the mass balance shifting from the rearrangement reactions producing desirable components to the undesirable build-up of increasingly larger compounds. The reactions may in part be the same but as more and larger components participate in condensing reactions, the reaction products become larger, generally less desirable and exceedingly difficult to identify or chemically characterize. It should be noted that because of the complexity involved, most of what is known about non-enzymatic browning is derived from model systems. The classic work, which explains non-enzymatic browning, was written by Hodge [11].

4.4 Initial reactions

The presence of available free amino acids and/or peptides and reducing sugars in most foods has led to the studies by researchers into these components with respect to non-enzymatic browning reactions. Because non-enzymatic browning reactions are also known as carbonyl–amine reactions, a more accurate definition of participatory compounds includes any carbonyl or amine, regardless of source. The concept of initial reactions is misleading because it suggests that such reactions are required for further reactions and that only sugars and amino acids, the most likely reactants, participate. They should, instead, be considered a high probability mechanism for the introduction of carbonyls and amines to enter the Maillard reactions. Otherwise, the food scientist may not consider non-enzymatic browning as a possible mechanism for colour or flavour development in an apparently unaffected food system, such as degrading fats and oils, as with the discoloration known as 'rusting' of fish [12].

The initial reactions are, however, the most understood, because the participating compounds are known or can be presumed (Figure 4.2). Characterizing aspects of this stage are limited to possible slight colour changes, and are unlikely to include flavour development or physical and structural change and most have at least a degree of reversibility. The limited number of reactions are documented and named in the literature and are less complex than subsequent reactions. With high concentrations of glucose available in nature and the increased reactivity of aldoses, considerations of initial reactions normally focus on glucose. This involves addition of an amine to aldose and reversible formation of a glucosylamine

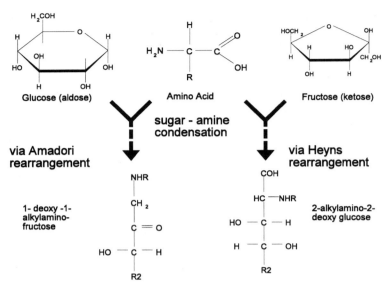

Figure 4.2 Initial Maillard browning reactions.

(or fructosylamine from a ketose). Amadori rearrangement of the N-substituted glycosylamine results in a 1-amino 1-deoxy-2-ketose (1-deoxy-1-alkylaminofructose). A ketose sugar, fructose and amino acid condenses to an N-substituted fructosylamine forming 2-amino-2-deoxy-1-aldose via the Heyns rearrangement. The majority of ensuing reactions are in reality similar to the initial reactions in that carbonyls and amine compounds continue to interact. Fission may produce both carbonyls and amines, which, because of continued rearrangement, give rise to unexpected compounds. Mechanisms, which involve the same reactants but bypass either Anadori or Heyns rearrangement are also known.

4.5 Intermediate reactions

Intermediate reactions involve the above-mentioned compounds and could also be considered as a natural extension of the initial reactions, because they include analogous carbonyl–amine condensation reactions. Additionally, a complex mixture of nitrogen-containing and non-nitrogen compounds are produced, which in turn, could interact in similarly analogous carbonyl-amine reactions. In general terms, this might suggest that the same reactions and components are produced in any food system, but this would be misleading, because the reactions are largely controlled by the nature of the reactants. Differing amounts and types of naturally occurring proteins, free amino acids, sugars, other components, and the amount or type of heating involved, dictate that a roasted potato cannot taste like roasted coffee beans, a roasted pork loin or a loaf of bread.

Intermediate reactions are more complicated in scope, pathways, reactants and consequences (Figure 4.3). Although often presented as separate modules or pathways, a myriad of possible interactions between unstable intermediates from other 'pathways' remain to be considered. In general, colour begins to develop and numerous types of compounds may be formed, which contribute in a desirable, or undesirable, manner to the natural food flavour. Possible compounds, which may be formed, include various substituted acids, carbonyls, cyclopentenes, furans, imidizoles, oxazoles, pyrazines, pyridines, pyrones, pyrroles, sulphur compounds and thiazoles. If unspecified, compounds or groups may be referred to as 'non-enzymatic browning products' and have also been called Maillard browning products. These reactions also incorporate sugars and compounds formed from the breakdown of initial reactions, which inherently allow interaction of unreacted components from other sources. Structural changes are primarily related to the heating and accompanying loss of moisture, although changes to proteins and sugars can affect physico-chemical properties. The reactions begin with the changes to Amadori rearrangement and/or Heyns rearrangement compounds.

In an acidic environment, further degradation of Amadori rearrangement products favours production of furfurals. Under acidic conditions,

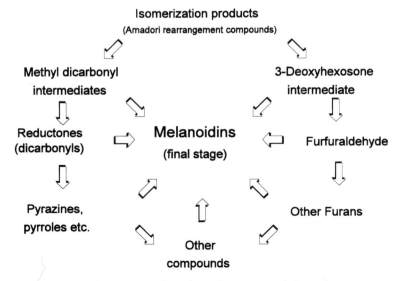

Figure 4.3 Intermediate stage of non-enzymatic browning.

formation of 5-hydroxymethyl-2-furfural from 1-amino-1-deoxy-2-ketose is favoured (Figure 4.4). Although formation of furfurals may be spectrophotometrically monitored, they are not initially important with respect to food colour, because furfurals absorb primarily in the ultraviolet and not the visible range. With further interactions, furfural intermediates and final reactants will contribute to visible changes. Thus, the roasting of acidic cocoa beans and cheese, a fermented milk product, will produce visible colour changes.

Under neutral or basic conditions, a variety of reactions may cause a change in Amadori rearrangement compounds, which eventually are important for the development of visible spectrum absorption or colour. Continued heating of Amadori rearrangement compounds can cause direct fragmentation of the amino compound producing a variety of low-molecular-weight compounds, such as carbonyls and amines, which generally have little direct effect on colour or flavour, but may further participate in carbonyl–amine reactions. Amadori rearrangement compounds may also lose water and form reductones and then dehydroreductones or osones (dioxo or dicarbonyl compounds; Figure 4.5). These compounds retain the amino component and further interact in carbonyl–amine condensation reactions. Fission of deoxyosones or direct fragmentation of the sugars in a high heat environment can produce carbonyls including pyruvaldehyde (pungent stinging odour), pyruvic acid (slightly caramel-like) and diacetyl (butter-like), which can also interact as substrates for further reactions (Figure 4.6). The potential for the production

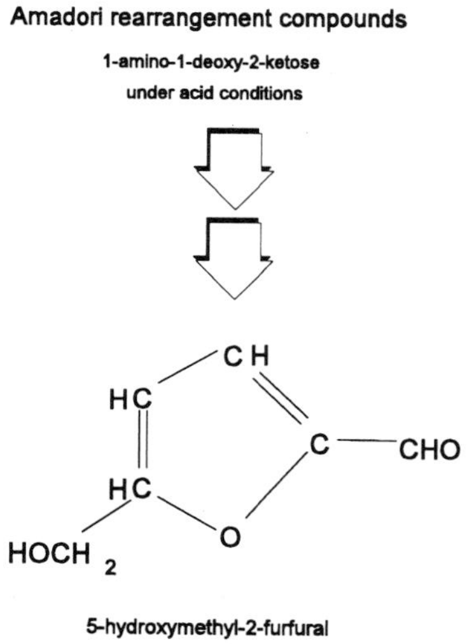

Figure 4.4 Formation of 5-hydroxymethyl-2-furfural.

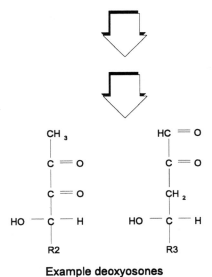

Figure 4.5 Example deoxyosones produced from Amadori rearrangement products.

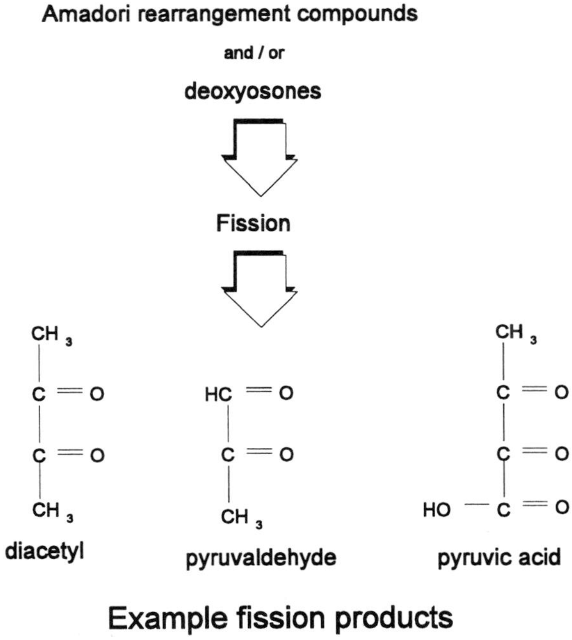

Figure 4.6 Example Amadori rearrangement fission products.

of numerous compounds can be understood when considering that simply heating glucose (250–500°C, 480–930°F) has produced 96 compounds [13].

Regardless of source, when dicarbonyls are produced, they can interact with amino acids via Strecker degradation [14]. The resulting aldehyde derivative of the amino acid and amino carbonyl may contribute to flavour and colour, but can also further interact (Figure 4.7). Evaluation of simple systems provides an indication of much greater complexity. A mixture of leucine and glucose when heated was described as 'sweet chocolate' at 100°C (212°F) and 'moderate breadcrust' at a 20°C (36°F) higher temperature [6].

Numerous heterocyclic flavour compounds including oxazoles, thiazoles, quinoxalines, cyclopentapyrazines and pyrazines have been isolated from roasted foods and contribute to the roasted flavour [15,16] (Figure 4.8). Each type of compound has a proposed pathway of formation that incorporates non-enzymatic browning reaction products. Interaction between alanine and two diacetyl molecules was proposed to form 2,3,5-trimethyloxazole [17]. Although the thiazole component has been reported as a breakdown product of thiamine, a condensation aminecarbonyl reaction also has been suggested [18,19]. A mechanism for the formation of cyclopentapyrazines, such as 3,5-dimethyl-6,7-dihydro-5H-cyclopentapyrazine was proposed [20]. Pyrroles could be formed by addition of amines to pentose or the reaction of a furan with amino acids [21,22].

Perhaps the most desirable compounds produced via the intermediate reactions are the pyrazines. They have been isolated from virtually all foods we associate with desirable roasted flavour: coffee, cocoa, chocolate, roasted nuts (almonds, filberts, macadamia nuts, peanuts, pecans),

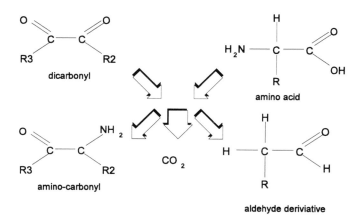

Figure 4.7 Strecker degradation of an amino acid.

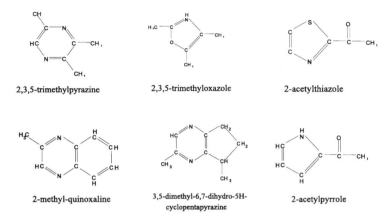

Figure 4.8 Heterocyclic compounds formed via amino-carbonyl condensation.

and various cooked meats (boiled beef, beef fat, fried veal) with more than ten identified in many foods [23,24]. Somewhat misleading is their presence in raw vegetable foods: asparagus, cabbage, carrot, chard, cucumber, leeks and lettuce. The latter are usually methoxy-derivatives of pyrazines, and not generally found in roasted foods, therefore their source is suggestive of a biochemical formation pathway. The proposed general pathway for pyrazines has been the condensation of amino–carbonyls (Figure 4.9), possibly produced via Strecker degradation or other condensation reactions involving various dicarbonyls with additional nitrogen [25,26]. This led to a study of the components needed for the production of identified pyrazines in some food systems, assuming the carbonyl–amine condensation reactions were the method of formation [20,27]. However, the complexity of some pyrazines and the other compounds suggested an additional formation pathway. This was confirmed when a free-radical pyrazine intermediate was identified, which would lead to the production of additional pyrazines [28].

4.6 Final reactions

The final reactions are so named because unlike the previous reactions, they are not reversible. In excess, these are generally undesirable and produce the typical brown or darker coloration and eventual burnt taste and aroma. Excessive moisture loss can also result in unacceptable dry and hard texture. Further reactions, although they may occur, only add to the large number of high-molecular-weight compounds formed. These compounds, melanoidins, are non-linear compounds, which may be

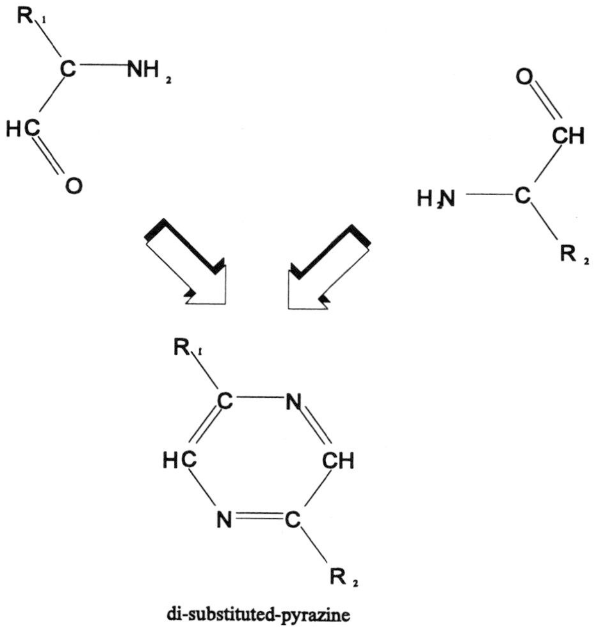

Figure 4.9 Pyrazine formation via condensation reaction.

beyond the current limits of analysis. Perhaps not surprising, their formation may involve all the preceding compounds in carbonyl–amine condensations and other reactions partially determined by the growing three-dimensional structure formed.

4.7 Control of Maillard reactions

Non-enzymatic browning reactions need to be controlled for a variety of reasons. During a heating process, loss of control by either the food preparer or processor results in the undesirable consequences of excess browning, which may render the food product visually undesirable, unpalatable, or less nutritious. Ambient temperature browning of stored products, such as milk powder provide another challenge. Understanding the process involved and utilizing control processes often requires balancing of parameters. Too little inhibition may allow continued browning, while excess inhibition may limit desirable colour or flavour change and result in reduced sensory attributes. For example, a processor may want to increase product flow through an oven, and therefore need to increase the speed of product browning to maintain sensory quality with greater throughput.

Control of non-enzymatic browning reactions has been accomplished in the manner similar to controlling other reactions; control of the type of reactants, the concentrations of the reactants and control of the time and temperature of the reaction. Generally, lower molecular-weight compounds or those less hindered by their structure (steric hindrance) are more reactive: pentoses > hexoses > reducing disaccharides. Reducing capability also influences reactivity, thus, replacing glucose with sucrose in food formulations would reduce browning, while the opposite would increase browning. Because sulphite and its derivatives can combine with the carboxyl group and block the carbonyl–amine condensation reactions, it has been used as an inhibitor. Blocking the carbonyl group during either the initial or intermediate reaction may have the same effect and, therefore, does not of necessity need to block the sugar. The time and temperature of processing, such as blanching of canned foods or pasteurization of milk, are carefully controlled to ensure they meet their primary goal, but also to maintain product quality by limiting non-enzymatic browning. Browning of soups is initially limited by the low concentrations of reactants but as reactants concentrate, non-enzymatic browning becomes the limiting factor in vacuum pan concentration, as further heat would produce the undesirable compounds. Non-heating processes, such as filtration or freeze-drying, could then be utilized. Reduction of water activity also limits the possibility of non-enzymatic browning (chapter 1).

As with the carbonyl group, reactivity of the amine moiety would be moderated by steric hindrance and available amino groups: amino acids > peptides > proteins. The presence of an additional amino group, ε-amino, makes lysine particularly reactive among amino acids [29]. As noted above, other amino byproducts of non-enzymatic browning with low molecular weight will also be very reactive. Careful food formulation also allows the food scientist to control non-enzymatic browning from the amino side of the reaction. Reduction of lysine, replacing milk protein with soy protein, for example, and choosing food components with minimal free amino acids, will therefore reduce the initial potential for non-enzymatic browning. Conversely, if rapid browning is desired, the opposite should increase the rate of non-enzymatic browning.

4.8 Other effects of Maillard reaction products

It should be apparent that participation in non-enzymatic browning reactions by amino acids eliminates any subsequent nutritional activity unless, as in the initial reactions, the reaction reverses, generating the free amino acid or peptide. Generally, production of colour and flavour changes reduces the nutrient concentration of amino acids in a given food system, although it possibly makes the food more appealing to the visual and taste senses. When proteins and peptides participate, condenstation reactions may interfere with enzymatic protein hydrolyses, reducing digestibility and rendering some portions unavailable as a nutrient source, such as with interaction of lysine. Additionally, ascorbic acid may also be destroyed [30].

Degradation of meat flavour, after cooking, short-term storage followed by reheating, is known as 'warmed-over flavour', although it may also occur in minced raw meat. Catalytic oxidation of unsaturated fatty acids by iron has been identified as the cause. The change is characterized by a reduction in desirable meaty flavour and an increase in rancid, metallic and off-aroma. Because of the volume of precooked or convenience fast food meat dishes and dinners, the need to control this off-flavour has been highlighted [31]. Anti-oxidation properties of non-enzymatic browning products were first described in overcooked meat and the participation of beef diffusates in browning have been demonstrated [32,33]. Temperature is a critical factor and must be about 100°C (212°F) to produce inhibitory effects with temperatures of the order of 70°C (160°F) actually accelerating warmed-over flavour [34]. It has been subsequently shown that Amadori rearrangement products conveyed the anti-oxidant properties of free-radical inactivation and decreasing hydroperoxide formation [35]. Compounds such as maltol and dihydroxyacetone have been identified as non-enzymatic browning products with anti-oxidative effects [36,37]. The use of non-enzymatic browning products as anti-oxidants is discussed in several papers on the topic [38–41].

The short- and long-term effects of non-enzymatic browning products on health have been a concern beginning with Maillard's investigations. This concern reflects not only interactions between food materials and consumers, but also the non-enzymatic browning reactions known to occur *in vivo*, including participation in the ageing process [42]. As food materials, they have been divided into four categories, beginning with the loss of free amino acids as suggested above. Reduced digestibility of proteins and a decrease in biological value of the protein beyond that which would be accounted for by destruction of amino acids, are also nutritional concerns. The last and least well-described category involves other undesirable physiological changes which occur with the ingestion of moderate amounts of heavily browned foods. Possible toxicological and mutagenic aspects of non-enzymatic browning products continue to be investigated. Although outside the limits of this chapter, undue concern about this topic would seem currently unfounded.

Although much research has centred on possible negative aspects of non-enzymatic browning, such as the anti-nutrient or mutagenic properties, positive attributes beyond sensory changes are also noted. The compound deoxyfructoserotonin and related compounds have been produced *in vivo* via non-enzymatic browning reactions and have a variety of beneficial properties including inhibition of *Mycobacterium leprae* [43]. Other studies have noted that non-enzymatic browning products affect the growth of micro-organisms, which may ultimately result in undesirable food functionality and structural changes [44–47].

4.9 Conclusion

Non-enzymatic browning plays an important role in much of food processing. It can add to our enjoyment of certain foods in providing attractive flavours and colours. On the other hand, some browning reactions produce very undesirable attributes in food products and may even make them inedible. It is very important, therefore, for food scientists and technologists to understand these reactions as far as possible. This will then help them to optimize their product recipes with respect to the results of non-enzymatic browning.

References

1. De Bry, Luck. Anthropological implication of the Maillard reaction: an insight. *5th International Symposium on the Maillard Reaction, Minneapolis, MN* (1993).
2. Kawamura, S. Seventy years of the Maillard reaction. In *The Maillard Reaction in Foods and Nutrition*, Eds Waller, G.R. and Feather, M.S. American Chemical Society Symposium Series No. 215, American Chemical Society, Washington, DC (1983), 1–18.
3. Davidek, J., Velisek, J. and Pokorny, J. (Eds) Chemical changes during food processing, *Developments in Food Science 21*, Elsevier, Amsterdam (1990).
4. Finot, P.A., Aeschbacher, H.U., Hurrell, R.F. and Liardon, R. *The Maillard Reaction in Food Processing, Human Nutrition and Physiology*, Birkhauser Verlag (1990).
5. Labuza, T.P. Nutrient losses during drying and storage of nonenzymatic browning. In *Physical Chemistry of Foods*, Eds Schwartzberg H.G. and Hartel R.W., Marcel Dekker, New York (1972), 595–649.
6. Mabrouk, A. Flavor of browning reactions products. In *Food Taste Chemistry*, Ed. Boudreau, J.C. American Chemical Society Symposium Series No. 115, American Chemical Society Washington, DC (1979), 205–245.
7. Nursten, H.R. Aroma compounds from the Maillard reaction. In *Developments in Food Flavours*, Eds Birch, G.G. and Lindley, M.G. Elsevier Science, London (1986), 173–190.
8. O'Brian, J. and Morrissey, P.A. Nutritional and toxicological aspects of the Maillard reaction in foods. *CRC Critical Reviews in Food Science and Nutrition*, **3** (1989), 211–248.
9. Parliment, T.H., McGorrin, R.J. and Ho, C.T. (Eds) *Thermal Generation of Aromas*. American Chemical Society, Washington, DC (1989).
10. Waller, G.R. and Feather, M.S. (Eds) *The Maillard Reaction in Foods and Nutrition*, American Chemical Society ACS Symposium Series No. 215, American Chemical Society Washington, DC (1983).
11. Hodge, J.E. The chemistry of browning in model systems, *Journal of Agricultural Food Chemistry*, **1** (1953), 928–943.
12. Pokorny, J., Tax, P.C. and Gunnysack, G. Non-enzymatic browning. 4. Browning reaction of 2-furfuraldehyde with protein. *Zeitschrift für Lebensmittel-Untersuchung und Forschung*, **151** (1973), 36–40.

13. Fagerson, I.S. Thermal degradation of carbohydrates: A review. *Journal of Agricultural Food Chemistry*, **17** (1969), 747–750.
14. Schonberg, A. and Moubacher, R. The Strecker degradation of amino acids. *Chemical Review*, **50** (1952), 261–277.
15. Hoskin. J.C. and Dimick, P.S. Role of nonenzymatic browning during the processing of chocolate–A review. *Process Biochemistry*, **19**(3) (1984), 92–104.
16. Hoskin, J.C. and Dimick, P.S. Role of sulfur compounds in the development of chocolate flavor–A review. *Process Biochemistry*, **19**(4) (1984), 150–156.
17. Rizzi, G.P. The formation of tetramethylpyrazine and 2-isopropyl-4,5-dimethyl-3-oxazoline in the Strecker degradation of DL-valine with 2,3-butanedione. *Journal of Organic Chemistry* (1969), 2002–2004.
18. Mulders, E.J. Volatile components from the non-enzymatic browning reaction of the cysteine/cystine–ribose system. *Zeitschrift für Lebensmittel-Untersuchung und Forschung*, **152** (1973), 193–201.
19. Obermeyer, H.G. and Chen, L. Biological estimation of the thiazole and pyrimidine moieties of vitamin B_1. *Journal of Biolgical Chemistry*, **159** (1945), 117–122.
20. Shibamoto, T., Akiyama, T., Sakaguchi, M., Enomoto, Y. and Massuda, H. A. study of pyrazines formation. *Journal of Agricultural Food Chemistry*, **27** (1979), 1027–1031.
21. Kato, H. and Fujimaki, M. Formation of N-substituted pyrrole-2-aldehydes in the browning reaction between D-xylose and amino compounds. *Journal of Food Science*, **33** (1968), 445–449.
22. Rizzi, G.P. Formation of N-alkyl-2-acylpyrroles and aliphatic aldimine in model non-enzymatic browning reaction. *Journal of Agricultural Food Chemistry*, **22** (1974), 279–282.
23. Maga, J.A. Pyrazines in foods: an update. *Critical Reviews of Food Technology*, **15** (1982), 1–48.
24. Maga, J.A. and Sizer, C.E. Pyrazines in foods. *Critical Reviews of Food Technology*, **4** (1973), 39–115.
25. Newell, J.A., Mason, M.E. and Matlock, R.S. Precursors of typical and atypical roasted peanut flavor. *Food Chemistry*, **15** (1967), 767–772.
26. Rizzi, G.P. A mechanistic study of alkylpyrazine formation in model systems. *Journal of Agricultural Food Chemistry*, **20** (1972), 1081–1085.
27. Hoskin, J.C. The Nature of Flavor Changes During Conching in Chocolate Manufacture, PhD thesis, The Pennsylvania State University (1982).
28. Namiki, M. and Hayashi, T. A new mechanism of the Maillard reaction involving sugar fragmentation and free radical formation. In *The Maillard Reaction in Foods and Nutrition*, Eds Waller, G.R. and Feather, M.S. American Chemical Society Symposium Series No. 215, American Chemical Society, Washington, DC (1983), 21–46.
29. Wolf, J.C., Thompson, D.R. and Reineccius, G.A. Initial losses of available lysine in model systems. *Journal Food Science*, **42** (1977), 1540–1544.
30. Archer, M.C. and Tannenbaum, S.R. Vitamins. In *Nutritional and Safety Aspects of Food Processing*, Ed. Tannenbaum, S.R. Marcel Dekker, New York (1979).
31. Love, J.D. and Pearson, A.M. Lipid oxidation in meat and meat products – A review, *Journal of the American Oil Chemists Society*, **48** (1971), 547–549.
32. Zipser, M.W. and Watts, B.M. The inhibition of warmed-over flavor in cooked meats, *Journal of Food Science*, **38** (1961), 398–403.
33. Wasserman, A.E. and Spienlli, A.M. Sugar-amine interaction in the diffusate of water soluble extract of beef and model systems, *Journal of Food Science*, **35** (1970), 328–332.
34. Porter, W.L. Recent trends in food application of antioxidants. In *Autoxidation in Food and Biological Systems*, Eds Simic, M.G. and Karel, M. Plenum Press, New York (1980), 295–365.
35. Eichner, K. Antioxidant effects of Maillard reaction intermediates. In *Autoxidation in Food and Biological Systems*, Eds Simic, M.G. and Karel, M. Plenum Press, New York (1980), 367–385.
36. Sato, K. and Herring, H.K. The chemistry of warm-over flavor in cooked meats. *Proceedings Annual Reciprocal Meat Conference*, **26** (1973), 64–75.
37. Shin-Lee, S.Y. Warmed-Over Flavor and its Prevention by Maillard Reaction Products, PhD thesis, University of Columbia (1988).

38. Pearson, A.M. and Gray, J.I. Mechanism responsible for warmed-over flavor in cooked meat. In *The Maillard Reaction in Foods and Nutrition*, Eds Waller, G.R. and Feather, M.S. American Chemical Society Symposium Series No. 215, American Chemical Society, Washington, DC (1983), 285–300.
39. Pearson, A.M. Love, J.D. and Shorland, F.B 'Warmed-over' flavor in meat, poultry and fish, *Advances in Food Research*, **23** (1977), 1–74.
40. Bailey, M.E. Inhibition of warmed-over flavor, with emphasis on Maillard reaction products, *Food Technology*, **42**(6) (1988), 123–126.
41. Bailey, M.E., Shin-Lee, S.Y., Dupay, H.P. St., Angelo, A.J. and Vercellotti, J.R. Inhibition of warmed-over flavor by Maillard reaction products. In *Warmed-over Flavor of Meat*, Eds St. Angelo, A.J. and Bailey, M.E. Academic Press, Orlando, FL (1987), 237–266.
42. Monnier, V.M. and Cerami, A. Non-enzymatic glycosylation and browning of proteins *in vivo*. In *The Maillard Reaction in Foods and Nutrition*, Eds Waller, G.R. and Feather, M.S. American Chemical Society Symposium Series No. 215, American Chemical Society, Washington, DC (1983), 431–449.
43. Mester, L., Szabados, L. and Mester, M. Maillard reactions of therapeutic interest. In *The Maillard Reaction in Foods and Nutrition*, Eds Waller, G.R. and Feather, M.S. American Chemical Society Symposium Series No. 215, American Chemical Society, Washington, DC (1983), 451–463.
44. Dworschak, E. Non-enzymatic browning and its effect on protein nutrition. *CRC Critical Reiews in Food Science and Nutrition*, **13** (1980), 1–40.
45. Einarsson, H. The mode of action of antibacterial Maillard reaction products. In *The Maillard Reaction in Food Processing, Human Nutrition and Physiology*, Eds Finot, P.A., Aeschbacher, H.U., Hurrell, R.F. and Liardon, R. Birkhauser Verlag (1990), 215–220.
46. De Lara, R.C. and Gilliland, S.E. Growth inhibition of microorganisms in refrigerated milk by added Maillard reaction products, *Journal of Food Protection*, **48** (1985) 138–141.
47. Einarsson, H. and Eriksson, C. The antibacterial effects of Maillard reaction products and sorbic acid at different pH levels and temperatures. In *The Maillard Reaction in Food Processing, Human Nutrition and Physiology*, Eds Finot, P.A., Aeschbacher, H.U., Hurrell, R.F. and Liardon, R. Birkhauser Verlag (1990) 227–232.

5 Rheology in food processing
E.J. WINDHAB

5.1 Introduction

Rheology is the study of the deformation and flow of materials under well-defined conditions. Factors such as the moving force, for example pumping speed, and the surface of pipes, influence the way in which materials flow, whereas pressure and temperature will affect the deformation behaviour. Rheology uses laboratory measurements to predict and explain some of the flow, deformation and textural changes which take place during processing. This involves both steady, changing and unstable flow conditions.

Rheometric flow is normally classified into two types, namely shear and elongational. Shear flow can be visualized by placing the material between two parallel plates (see Figure 5.1), which are moving at a constant velocity relative to each other. The material will move according to its internal frictional resistance to motion, a parameter called shear viscosity, η_s. Particles will rotate in shear flow. In elongational flow, the material is stretched, and the particles will not rotate. The measured material parameter is the elongational viscosity, η_e.

The simplest measurements to make are normally one-directional (uniaxial) shear tests, although two-dimensional tests are possible, as are one- and two-directional elongational techniques. In most food processing, the flow is very complex, but generally the uniaxial measurements can satisfactorily estimate or predict actual process conditions.

Rheology looks at the flow–deformation properties of materials at both the bulk scale and at the molecular/particulate level. In doing so, it is a powerful tool for investigating how industrial processes influence food structures during their manufacture, and perhaps even more importantly, how they affect the food's sensory and functional properties. This can be taken even further in that rheology can help with the development of new products with predetermined textures.

5.2 Types of rheological flow

This section represents an introduction to rheology.

In Figure 5.1, a material is shown as being under shear between the two surfaces, which are moving at velocities v_1 and v_2 at a distance h

RHEOLOGY IN FOOD PROCESSING

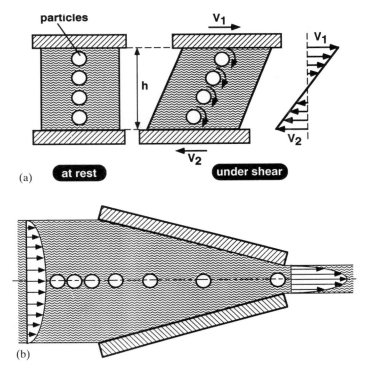

Figure 5.1 Representation of simple shear and elongational flow.

apart. The velocity gradient $(v_1 - v_2)/h$ is measured in s^{-1} (reciprocal seconds) and is known as the 'shear rate' ($\dot{\gamma}$). If each surface has an area A, then the frictional stress (τ, sometimes called 'shear stress') at the surface, due to the movement is given by:

$$\tau = \eta(v_1 - v_2)/h \quad \text{or} \quad \tau = \eta\dot{\gamma} \quad (5.1)$$

where η is the shear viscosity.

Consequently, it is the ratio of shear stress to shear rate ($\tau/\dot{\gamma}$) which enables the rheologist to learn something about the viscous behaviour of a liquid. At any one point this is known as the 'apparent viscosity'. Newton suggested that this value was constant for all shear rates, and although this is approximately true for some fluids like water, most food substances are what is called 'non-Newtonian' in behaviour. Figure 5.2 [1] shows some of the different types of flow behaviour which can occur when the apparent viscosity is determined at different shear rates.

When a stress is applied to some materials they may remain virtually solid until a critical stress is reached, after which they suddenly behave as liquids. This critical stress is termed the 'yield value', τ_0.

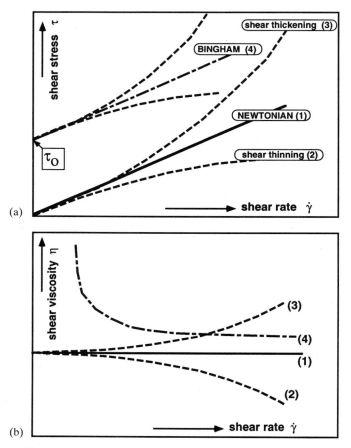

Figure 5.2 Comparison of Newtonian and non-Newtonian plots [1].

5.3 Rheological measurements

Gaps between rotating surfaces, for example concentric cylinders, are used to generate shear flow. The speed of rotation is related to the shear rate and the force between them to the shear stress. Typical rotating measuring systems are shown in Figure 5.3 together with some capillary systems which can also be used to carry out shear-flow measurements.

Viscometers may operate by varying the shear rate, for example by increasing or decreasing the rate of rotation or oscillation of the moving surface. Alternatively, the stress can be controlled, in which case a known force is applied and the degree of movement (deformation) measured.

RHEOLOGY IN FOOD PROCESSING 83

(a)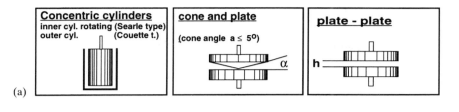

(b)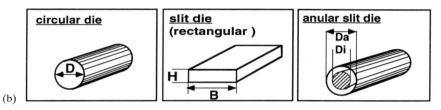

Figure 5.3 Rheometric gap geometries. (a) Rotational rheometers. (b) Capillary rheometers.

Table 5.1 Typical rheological tests

Control	Small deformation		Large deformation	
	Static	Dynamic	Static	Dynamic
Stress controlled (τ)	–	Small τ-amplitude shear oscillation test	Creep shear test (τ = const.)	Creep relaxation test Large amplitude oscillation test
Deformation controlled (γ)	–	Small γ-amplitude shear oscillation test	Step shear/ relation test (γ = const.)	Relaxation test Large amplitude oscillation test
		Wave propagation (kHz–GHz)		Steady flow (eq Maxwell orthog. rheometer)

The tests are normally carried out at either very small deformations (often to determine the 'unchanged' structure within a material) or at large deformation (e.g. at high shear rates) to determine what happens under processing flow conditions. In the former, even fluids with complex viscoelastic behaviour behave like 'simple' Newtonian fluids. Once the fluid is subjected to greater forces, all the different types of viscous flow behaviour shown in Figure 5.2 can occur. Tests can be carried out in steady and unsteady flow conditions.

Typical rheological tests, mainly performed by uniaxial 'shear' rheometers, are listed in Table 5.1.

5.4 Rheological parameters

Besides viscosity, elasticity is one of the most important properties which can be measured by rheometry, and which also relates to the texture of the substance. This can be visualized by imagining a sine (deformation or stress) wave being propagated through the test material. The shear stress wave leaving can be different in both phase and amplitude from the deformation wave which is generated initially (Figure 5.4). The change in phase relates to the degree of viscous or elastic behaviour of the material. The smaller the change in phase between the deformation and the related stress wave, the greater its elasticity (phase angle $\delta = 0$ is totally elastic). A factor known as the storage modulus (G') can be measured, which relates to the elastic behaviour.

Any energy loss to the system, due to overcoming its internal viscous friction, will be shown as a reduction in amplitude. The measured viscous quantity is called the 'loss modulus' (G'') and the ratio of G' to G'' is the tangent of the loss angle, δ.

The measurement of these flow parameters can be carried out using the instruments shown in Figure 5.3. Unlike those for the viscosity of thin liquids, where traditionally the cylinder, cone or plate is rotated, the measurement of G' and G'' is normally carried out by oscillating the moving surface. By changing the amplitude or the frequency over a range appropriate to the material, detailed information can be obtained concerning the texture and flow–deformation properties. Most foods are viscoelastic and have both viscous and elastic properties. To make the situation more complicated, they also have a 'memory', which makes their behaviour additionally dependent upon their 'deformation history'.

Table 5.2 lists some of the most frequently measured parameters, whether the measurement requires a large deformation (e.g. rotating cylinders) or a small deformation (e.g. small oscillatory shear) and also whether it relates to the viscous or elastic properties. The table also

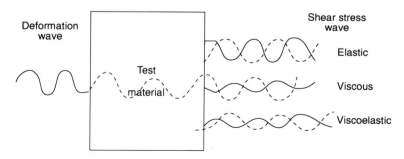

Figure 5.4 Simple representation of elastic and viscous systems.

Table 5.2 Common rheological parameters

Parameter	Large def.	Small def.	Viscous	Elastic	Time dependency	Shear/elong. rate depend.
Flow function, $\tau(\dot{\gamma})$	×		×	—	(×) (thixotropy, rheopexy)	× (pseudoplast., dilatant)
Shear viscosity, $\eta(\dot{\gamma})$	×		×	—	(×) (thixotropy, rheopexy)	× (pseudoplast., dilatant)
Elongational viscosity, $\eta_e(\dot{\varepsilon})$	×		×	—	(×)	×
Loss modulus, $G''(\omega)$		×	×	—	(×)	×
Storage modulus, $G'(\omega)$		×	—	×	(×)	×
Dynamic viscosity coefficient, $\eta'(\omega)$		×	×	—	(×)	×
Loss compliance, J''	×	×	×	—	(×)	(×)
Storage compliance, J'	×	×	—	×	(×)	(×)
First normal stress – diff. $\sigma_1(\dot{\gamma})$	×		—	×	(×)	×
Second normal stress – diff. $\sigma_2(\dot{\gamma})$	×		—	×	(×)	×
'Special' rheological parameters						
Relaxation time/spectra $G(s)$	×		×	×	—	×
Yield value, τ_0		×	—	—	—	—
Loss angle, δ (phase angle)		×	(×)	(×)	(×)	×

shows if the parameter is time-dependent or affected by the rate at which the viscometer is being operated. (×) means that this is a minor property of the measured parameter.

5.5 Food structure and rheological measurements

The flow properties of fluids are closely related to their inner structure. Macromolecules and dispersed matter (which includes solid particles, liquid droplets and gas bubbles) are the major structural components in food systems. For each material there are interacting forces between the various components (chemical bonding, physical interactions such as van der Waals forces, electrostatic forces and fluid–solid interactions, etc.) which dominate the structural behaviour under both static and weak deformation conditions. As deformation increases and movement takes place, the shear or elongational forces may overcome the structure forces. In rheological shear experiments it is assumed that at any particular shear rate the fluid will eventually take an 'equilibrium structure'. At this point the measured viscosity is an equilibrium or steady-state viscosity. It is, in fact, this (equilibrium) viscosity function $\eta(\dot{\gamma})$ which is normally measured [2].

5.5.1 Equilibrium viscosity function, $\eta(\dot{\gamma})$

As the shear rate increases, the viscosity of most foods changes rapidly and they behave in a non-Newtonian manner. Figure 5.5 illustrates this relationship between viscosity and shear rate and shows what is happening in the inner structure related to the viscosity. As can be seen, there are three very different domains, bounded by two critical shear rates $\dot{\gamma}_1^*$ and $\dot{\gamma}_2^*$

Domain A ($\dot{\gamma} < \dot{\gamma}_1^*$). In this low shear-rate region, the hydrodynamic forces can be neglected. Only the Brownian motion is acting against the

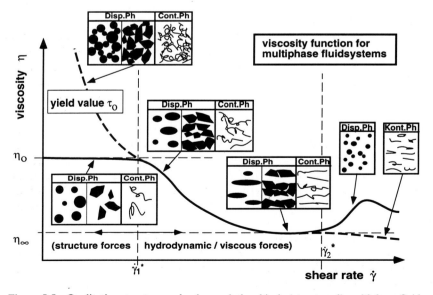

Figure 5.5 Qualitative structure – rheology relationship in 'structured' multiphase fluids

structural forces. If the concentration of the disperse phase, or of the macromolecules in the continuous phase is high, their strong interaction generates a yield value (τ_0). This means a critical force must be applied before any movement takes place. This force can be measured using a viscometer, preferably one where the stress is controlled rather than the shear rate. The apparent viscosity, which is equal to the shear stress/shear rate ratio, is, in theory, infinite in the region until movement takes place.

For low concentrations of structuring components, the Brownian motion dominates and keeps an isotropic structure with negligible interactions between the particles or macromolecules. In this case, the apparent viscosity, η_0, is independent of the shear rate (lower Newtonian regime).

Domain B ($\dot{\gamma}_1^* \leq \dot{\gamma} < \dot{\gamma}_2^*$). In this region, the hydrodynamic forces are of the same order of magnitude as the structural forces. This induces new structures according to the type, direction and time of application of the shearing forces. If this time is long enough, an 'equilibrium structure' is attained. In most multiphase foods, this equilibrium structure has a reduced 'inner' flow resistance than in domain A and hence the viscosity reduces with increasing shear rate (or shear stress). This is called 'shear thinning' or 'pseudoplastic behaviour'. The induced structure (illustrated in Figure 5.5) can completely or partially revert to the state it had in domain A, if the shear rate is reduced again to a low value, or if it returns to a state of rest. This change in structure of the dispersed particles can be viewed experimentally (e.g. by microscopy). Deformable particles, like droplets and bubbles, become deformed and take up an ellipsoidal shape. Coarse particles and their agglomerates are oriented according to their shape. When maximum 'orientation' occurs, there follows a region of minimum viscosity (η_∞) over a range of shear rates (upper Newtonian plateau). If this structure/viscosity is required for the product or for further processing, this state of orientation must be fixed by physical and/or chemical means [3].

Domain C ($\dot{\gamma} > \dot{\gamma}_2^*$). If the second critical shear rate limit $\dot{\gamma}_2^*$ is exceeded, the maximum orientation state may be disturbed. This disturbance may be the result of flow instabilities, particle–particle interactions (particularly for coarse particles) or changes in the particle structure (droplet or bubble break-up, leading to increased dispersion). With these changes, flow resistance and viscosity may increase again. This viscous behaviour is called 'shear thickening' or 'dilatancy'.

In this region, however, the macromolecular solutions may behave completely differently from the particle systems. Having been fully aligned, 'weak' molecular networks may 'break up' and cause a further decrease in viscosity. Molecular degradations were not found in pure shear-flow systems.

The existence of the three shear rate domains A, B and C, presented above, has been proved experimentally. Concentrated emulsions (e.g. mayonnaise) and suspensions (e.g. chocolates), as well as aerated foam systems ('mousse' products) may show similar viscous behaviour. A measured viscosity versus shear rate curve for a 65% oil containing water-in-oil emulsion (mayonnaise) is shown in Figure 5.6. The oil-droplet size and shape in the different equilibrium states can be measured by a microscopic device, combined with a transparent cone and plate gap rheometer (dynamic 'state' measurements) and by a laser diffraction device (to determine size in the final steady state) [4]. For a protein–sugar foam system, the measured bubble-size distributions, in the equilibrium states at four different shear rates, are shown in Figure 5.16 (see section 5.6.4).

As was shown in Figure 5.5, the rheological behaviour of multiphase systems is generally determined by the behaviour of the continuous phase, as well as by the interparticle forces. For a large variety of such fluid food systems, shear-induced structures have a reduced flow resistance, and therefore the rheological behaviour becomes shear thinning and thixotropic.

A numerical description of the rheology–structure relationship for multiphase fluid systems, suitable for predicting viscous flow behaviour, should not only take into account the influence of the shear rate, but must

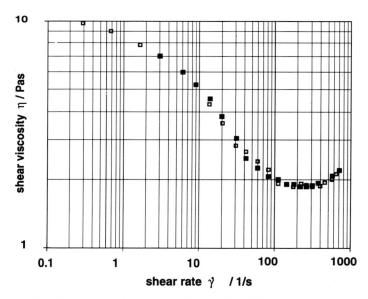

Figure 5.6 Equilibrium viscosity function of a starch-stabilized mayonnaise with 65% oil content.

also include provision for its dependency upon volume fraction of the disperse phase(s). This has led to the use of the so-called relative viscosity, η_{rel}, which is defined according to equation (5.2).

$$\eta_{rel} = \frac{\eta(\text{multiphase system})}{\eta(\text{continuous phase})} \qquad (5.2)$$

Most of the mathematical models and formulae found in the literature [5–7], which describe the relative viscosity of concentrated multiphase fluid systems, do not include the low shear-rate region. This is because the 'structure forces' (e.g. interparticle forces) cannot be taken into account. These models are only quantitatively correct if the shear-flow behaviour is dominated by a 'hydrodynamic mechanism'. This can only be found at higher shear rates (upper B range), which is often inappropriate for the actual conditions found in food processing that they are trying to model.

It is possible, however, to develop a *quantitative* description of the relative viscosity that includes these 'structure forces'. By including the shear rate and the concentration (or volume fraction) dependency of viscosity, this leads to the *shear rate-dependent relative viscosity* $\eta_{rel}(\dot{\gamma})$.

A first step in developing this concept is to look at mathematical descriptions of flow and see how they can be interpreted in terms of the reactions between the individual particles within that flow.

5.6 Modelling of viscous behaviour in disperse systems

5.6.1 $\tau_{0.1}$ 'structure' model

There are many existing equations which try to describe the viscous flow functions of fluid systems (e.g. Casson, Bingham, Herschel Bulkely [8]). These equations include one or more empirical parameters, which are determined by curve fit or by extrapolation. Often good agreement between the model and experimental data can only be obtained over a limited shear rate range (mostly at 'higher' shear rates).

If the rheological behaviour is viscoelastic, constitutive equations, to take into account viscous and elastic aspects, are commonly used to approximate the real flow behaviour. These constitutive equations take the form of either an integral or a differential equation, and many have been proposed in the literature [9].

In this section, a model of multiphase viscous systems is described in some detail, because such systems exist in most complex-structured food systems. This rheological description is based on measurable physical parameters, which will affect the flow/textural characteristics in 'structured' multiphase systems.

As shown in Figure 5.7, many flow functions of concentrated multiphase fluid systems can generally be described by a superposition of a

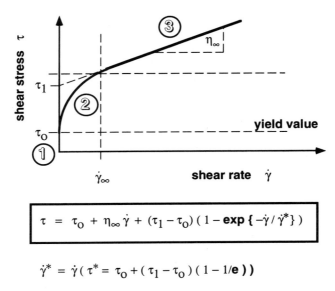

$$\tau = \tau_0 + \eta_\infty \dot\gamma + (\tau_1 - \tau_0)(1 - \exp\{-\dot\gamma/\dot\gamma^*\})$$

$$\dot\gamma^* = \dot\gamma(\tau^* = \tau_0 + (\tau_1 - \tau_0)(1 - 1/e))$$

Figure 5.7 'Superpositioned' flow function.

(measured) yield value (labelled 1), to a flow function in the 'high shear region' (labelled 3) (where the flow function of the suspension has a similar 'shape' to the pure suspension fluid), and an 'intermediate flow function region' $\eta(\dot\gamma)$ (labelled 2), which describes the flow curve when the inner structure changes from the isotropic *state of rest structure* (characterized by the yield value τ_0) to a *shear induced structure*. The upper shear stress limit of this domain is characterized by the so-called 'shear structuring limit' τ_1. This value is determined by linearly extrapolating the flow function from the high shear domain to the τ-axis.

Between the shear stresses τ_0 (yield value) and τ_1 (shear structuring limit) a shear-induced suspension structure is generated. Each shear rate will form its own equilibrium structure and hence a different equilibrium viscosity (see Figure 5.6)

In physical terms, this shear-induced structuring is caused by a combination of mechanisms (e.g. particle orientation, forming of 'microlayers', etc.). Figures 5.7 and 5.8 (and equations (5.3)–(5.6)) show that the shear stress (or shear rate) change during this shear-induced structuring (region 2) can be described with an exponential function between the shear stresses τ_1 and τ_0. The constant $\dot\gamma^*$ can be calculated from τ_1 and τ_0. In order to take more complex shear structuring mechanisms into account, equation (5.3) can be generalized as is demonstrated in Figure 5.8 (and equation (5.5)).

For a large variety of different multiphase systems (e.g. chocolates,

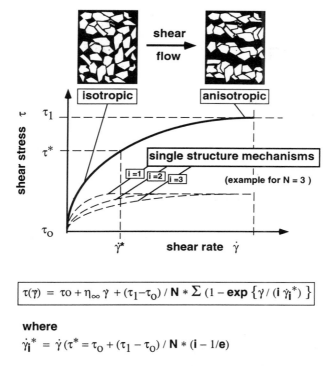

Figure 5.8 Complex approximative description of the 'shear-induced structuring domain' (superposition of structure mechanisms)

starch suspensions, silicone oil–$CaCO_3$ suspensions) is was found that the superposition of between 1 and 3 exponential terms ($N = 1\text{--}3$ in equation 5.5) gave satisfactory agreement between measured and calculated values (<5% relative difference). This means that using the '$\tau_{0.1}$ model' (equations (5.3)–(5.6)), only the values of τ_1 and τ_0 need to be measured. The yield value τ_0 relates to the state of rest structure, whereas the shear structuring limit τ_1 represents the shear stress where the shear-induced structure is fully developed. Between these two values, interparticular forces must be taken into account. At higher shear stresses, however, the viscous behaviour can be described only by the 'hydrodynamic interactions'.

5.6.2 Relative viscosity

Relative viscosity has already been defined in equation (5.2), and is in essence the viscosity of the whole material divided by that of the continuous liquid phase. Various models for relative viscosity can be found in the literature, most of which describe the relative viscosity of suspensions in the 'high shear region' (usually neglecting structural interactions). The Krieger–Dougherty equation, which has often been shown to be the most satisfactory, is given in Figure 5.9 (equation (5.7)). It is also possible to use the $\tau_{0,1}$ model (section 5.6.1) to calculate the relative viscosity η_R, as is shown in equation (5.8).

Both equations have been compared with actual measured relative viscosities using a model suspension consisting of silicone oil (AK 2000, made by Fa.Wacker) and limestone particles. Figure 5.10 shows the comparison for the Krieger–Dougherty equation when the relative viscosity is calculated at different volume fractions for a series of shear rates. The equation is satisfactory at higher shear rates ($\sim \geq 50\,\mathrm{s}^{-1}$) with an intrinsic 'shape factor' k of about 4. It is possible to use the equation at shear rates as low as $1\,\mathrm{s}^{-1}$ but, even when k is 'manipulated', poor agreement is obtained.

When the $\tau_{0,1}$ model (equation (5.8)) is used for the same model suspension (Figure 5.11), an improved agreement is reached at low shear rates, even with only one exponential function, while agreement at higher shear rates is still acceptable. When three or more exponential terms were used to describe the shear-induced structural range (region 2), the calculations were ±5% of the measured values over the whole shear rate range.

5.6.3 Structure parameters for the $\tau_{0,1}$ model

In principle, it is possible to calculate the relative viscosity–shear rate–volume fraction relationship using τ_1, τ_0 and the flow function of the

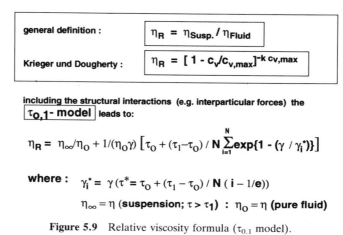

Figure 5.9 Relative viscosity formula ($\tau_{0,1}$ model).

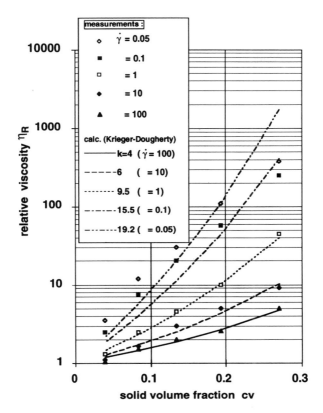

Figure 5.10 $\eta_R(\dot{\gamma})$ 'modified' (variable k) from Krieger–Dougherty equation for silicone oil/CaCO$_3$ system (lines calculated, points measured).

'pure' suspension fluid (without the disperse phase). No further approximation parameter is necessary. τ_1 and τ_0 depend on the solid volume fraction, as shown in Figure 5.12 for the oil–CaCO$_3$ model suspension (they also depend upon particle-size distribution and particle–particle interactions). Their value can best be determined experimentally using controlled stress rheometers [9]

The physical significance of this relationship, between the two shear stresses (τ_0 and τ_1) and the volume fraction, can be calculated using a model based on the following parameters:

- The average distance between the particles λ.
- The number of neighbouring particles **k** (with a distance $\lambda \leq x$ (= average particle diameter)).
- The touching probability w_B.

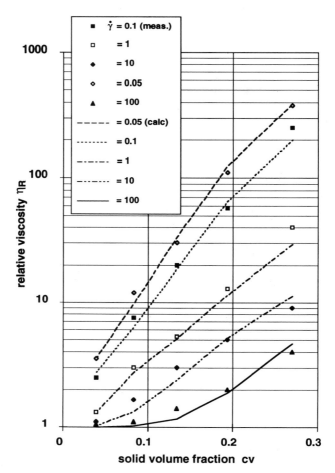

Figure 5.11 $\eta_R(\dot{\gamma})$ calculated from the $\tau_{0.1}$ model (lines) compared with measured data points.

These are illustrated in Figure 5.13, which also shows that the critical shear stresses are related to a dimensionless particle interaction number I^* by a 'simple' power law, which only requires knowledge of the constants C and n. These constants are given in Table 5.3 for the oil–$CaCO_3$ model and for a dark chocolate, whilst Figure 5.14 shows the good fit between the calculated and experimental values (regression coefficient ≥ 0.99).

5.6.4 Structure–viscosity functions

In order to get an improved description of the structure–viscosity relationship, so-called 'structure–viscosity functions' can be determined. This is

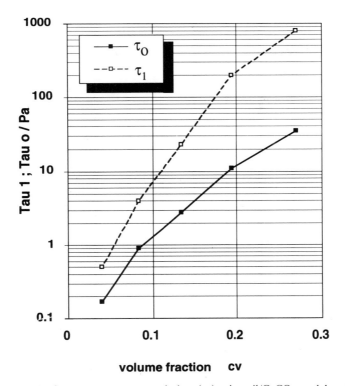

Figure 5.12 Structure parameters $\tau_1(cv)$; $\tau_0(cv)$ using oil/$CaCO_3$ model system.

defined as the viscosity function for a constant structure state. These are measured following the setting up of the required structure, by pre-shearing at a constant shear rate until a steady-state equilibrium viscosity is reached. Then the shear rate is either raised or lowered. The viscosity measurement is taken immediately after this shear rate step, before a significant change to the structure can take place and the viscosity approaches a new equilibrium value. The principle of structure flow function determination is shown schematically in Figure 5.15, and actual results for a protein–sugar foam and a sugar–milk powder–oil mixture in Figures 5.16 and 5.17, respectively. The latter also shows the equilibrium viscosity function, which shows that, over the measured shear rate range, the behaviour of this suspension system was shear thinning.

5.6.5 Intrinsic viscosity

The intrinsic viscosity $[\eta]$ is a structure parameter that characterizes the structure of 'single macromolecules' with respect to the diluted fluid

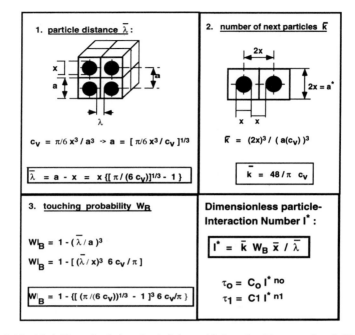

Figure 5.13 Modelling of $\tau_1(c_v)$ and $\tau_0(c_v)$ (a = side length of 'surrounding fluid cube')

Table 5.3 Constants C and n for an oil–CaCO$_3$ model system and for dark chocolate

Suspension	C_0	C_1	n_0	n_1
AX5000/CaCO$_3$	3.6	34.5	1.04	1.50
Dark chocolate (Ma 1.3)	4.0	27.2	1.1	1.40

system. Before determining the intrinsic viscosity it is necessary to define the following parameters:

$$\text{relative viscosity:} \quad \eta_R = \eta_0/\eta_s \tag{5.9}$$

$$\text{specific viscosity:} \quad \eta_{sp} = \eta_R - 1 \tag{5.10}$$

$$\text{reduced viscosity:} \quad \eta_{red} = \eta_{sp}/c \tag{5.11}$$

$$\text{intrinsic viscosity:} \quad [\eta] = (\lim \eta_{red}) \; c \to 0; \; \dot{\gamma} \to 0 \tag{5.12}$$

where η_s = viscosity of the solvent fluid and c = polymer concentration.

Huggins and Schulz-Blaschke/Kramer determined the following relationships [10]:

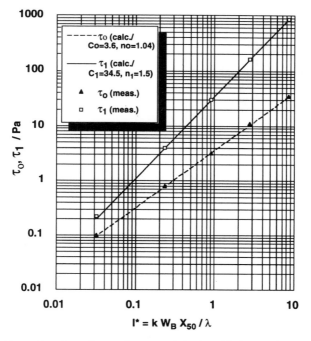

Figure 5.14 Power-law dependency of τ_0, τ_1 and the 'particle interaction number' I^* for a model silicone oil/$CaCO_3$ system (measured and calculated values).

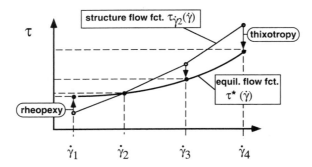

Figure 5.15 Schematic diagram of the principle of structure–flow function determination.

$$\eta_{sp}/c = [\eta] + K_H[\eta^2]c \quad (5.13)$$

$$\eta_{sp}/c = [\eta] + K_{SB}[\eta]\eta_{sp} \quad (5.14)$$

where K_H and K_{SB} are constants.

The intrinsic viscosity can be determined by extrapolating a plot of the reduced viscosity against the concentration (see Figure 5.18).

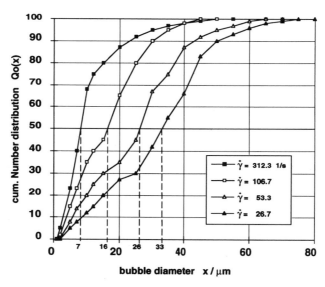

Figure 5.16 Bubble-size distributions for a protein–sugar foam at different equilibrium states of the foam structure after defined shear (in shear flow at variable shear rates).

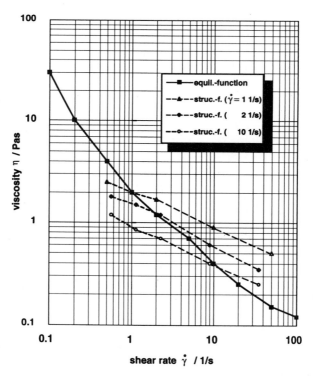

Figure 5.17 Measured structure–viscosity functions for oil–sugar 45 vol%/2.5 vol% milk powder suspension.

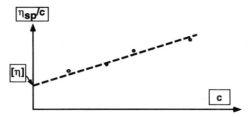

Figure 5.18 Schematic illustration of the determination of the intrinsic viscosity [η] according to the Huggins equation.

For dilute solutions, the η_{sp} versus concentration relationship can be approximated to by a power-law function with an exponent between 1.1 and 1.3. At these concentrations, it can be assumed that there is no significant molecular interaction taking place. As the concentration increases, to the so-called 'semi-diluted regime', the typical power-law exponent is in the range 3.0–3.4. For some polymers, an intermediate regime can also be found. In the semi-diluted regime, the macromolecules interact with each other by molecular entanglements and binding. Figure 5.19 illustrates the three regimes in a plot obtained for the polymer xanthan gum. It can be shown that for high mechanical power input to xanthan solutions (e.g. in rheometers or high pressure homogenizers), changes in the interacting behaviour of the molecules will be detected as changes in $\eta_{sp}(c)$ in the semi-diluted regime. If more sensitive macromolecules are degraded (e.g. guar gum in high homogenizing flow), changes in $\eta_{sp}(c)$ can additionally be detected in the dilute regime [10].

The intrinsic viscosity, [η], is closely related to the molecular configuration and the molecular weight. Such a relationship has been shown on an experimental basis to be [11]:

$$[\eta] = KM^a \tag{5.15}$$

where K and a are constants and M = average molecular weight. This relationship has been confirmed for a large variety of synthetic and biopolymers. The constants, K and a, are empirical. Many are reported in the literature [12].

5.7 Time-dependent viscous behaviour

The structural changes described in section 5.5 are time-dependent according to the structure and fluid system present. If this time exceeds the minimum 'reaction time' of the measuring device, the viscous time effect is measurable and can be interpreted as thixotropic or rheopectic viscous structure behaviour in shear flow. This is illustrated in Figure

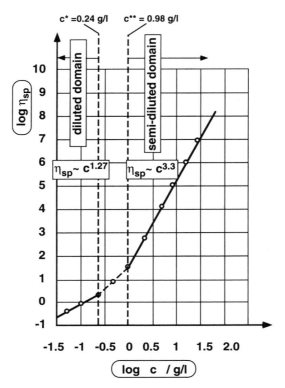

Figure 5.19 $\eta_{sp}(c)$ for aqueous xanthane solutions stabilized with 0.1 M NaCl [10].

5.15, where the arrows indicate the time-dependent behaviour. Figure 5.20 shows the measured viscosity–time functions for the same model suspension used in Figure 5.17. The shape of the equilibrium viscosity function, in Figure 5.17, shows that the mixture is shear thinning, and on that basis a shear rate increase will weaken (thixotropy) the 'structure forces' (and shear flow resistance). The reverse will occur for a shear rate decrease (rheoplexy).

Time-dependent behaviour will depend strongly upon the initial conditions, which must be clearly defined. Two types of initial structure can normally by distinguished, i.e. 'state of rest structure' and 'pre-sheared equilibrium structure'.

Thixotropic and rheopectic behaviour is generally related to the viscous properties of the fluid within the multiphase system (see Figure 5.21). Time-dependent structure changes, which influence the viscous behaviour in shear flow, can be reversible or irreversible (see Figure 5.22).

RHEOLOGY IN FOOD PROCESSING   101

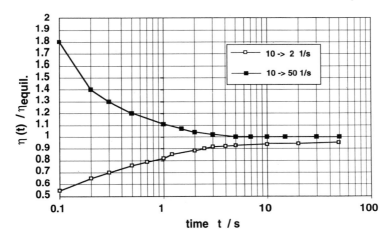

Figure 5.20 Viscosity–time functions for an oil–sugar (43%)–milk powder (2.5%) suspension (shear rate steps to 2 and 50 l/s starting from equilibrium at 10 l/s)

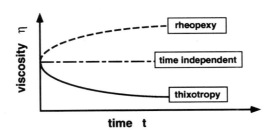

Figure 5.21 General description of thixotropy and rheopexy.

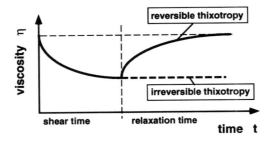

Figure 5.22 Reversibility/irreversibility of thixotropy.

In addition, time-dependent effects are generated by the elastic properties within viscoelastic fluids (e.g. memory effects). Viscous time-dependent behaviour can be interpreted in a similar way as a kind of 'structure memory effect'. The main difference between the two is that elastic deformation is reversible, as is viscous structuring, but viscous deformation is irreversible.

5.8 The influence of temperature on rheological properties

It is well known experimentally that the viscosity of fluids decreases with increasing temperature. The temperature dependency is included in this chapter, because heat treatment is a major aspect of much of food processing.

The temperature dependency of the zero shear viscosity η_0 for a large variety of materials can be described by an Arrhenius function (equation (5.16)), if the temperature change leads to a modification in the Brownian motion (kinetic energy) of the molecules (or macromolecules) without a phase transition or other temperature-induced structural change such as denaturation.

$$\eta_0(T) = \eta_0(T_0) \exp\left\{E_0/R\left(\frac{1}{T} - \frac{1}{T_0}\right)\right\} \tag{5.16}$$

where E_0 = activation energy, R = universal gas constant, T_0 = reference temperature.

The ratio of the zero shear viscosities at the experimental temperature to that at the reference temperature is known as the 'temperature shift factor', a_T:

$$a_T = \eta_0(T)/\eta_0(T_0) \tag{5.17}$$

This temperature shift factor is important when determining the temperature behaviour of a variety of other rheological parameters. For so-called 'thermo-rheologically simple' fluids, the temperature shift factor, derived from the zero shear viscosities (equation (5.17)), is the same as for the non-linear flow regime of viscoelastic fluids:

$$a_T = \eta_0(T)/\eta_0(T_0) = \eta(\tau,T)/\eta_0(\tau,T_0) \tag{5.18}$$

When plotting the shear viscosity against the shear rate for a non-Newtonian fluid, the temperature change will result in a vertical shift by the factor a_T (see Figure 5.23). Figures 5.24 and 5.25 show an example of a temperature viscosity function and its related Arrhenius plot.

RHEOLOGY IN FOOD PROCESSING   103

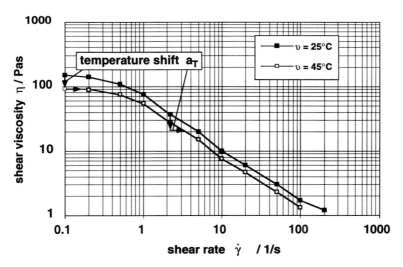

Figure 5.23 Temperature shift for the viscosity function of a non-Newtonian fluid (mayonnaise: starch base, 30% oil).

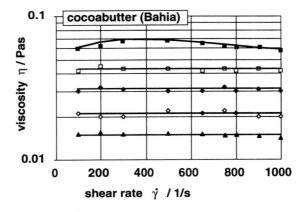

Figure 5.24 Viscosity functions of cocoa butter at various temperatures [14].

5.9 Rheology in food processes

5.9.1 Relationship of process parameters and rheological behaviour

The main process parameters in most food processing unit operations are:

- Mechanical power input
- Thermal power input
- Residence time

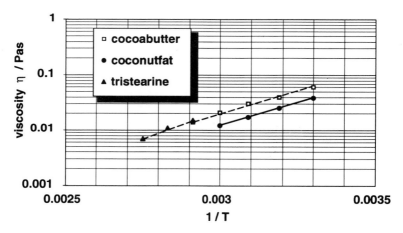

Figure 5.25 Arrhenius plot for cocoa butter (as Figure 5.24) and other fats [14].

Mechanical power input. This leads to deformation and flow of the food system being treated in the process concerned. The material (fluid) behaviour, although often complex in terms of shape and direction, normally correlates well with measurements made on simplified uniaxial rheometric flows (e.g. between the plates or cylinders of a viscometer). This is especially true for highly viscous liquids or plastic food systems (liquids with yield values) under the laminar flow conditions present in pipes, annular gaps, extruders, dies/nozzles, dosing systems, stirring/dispersing apparatus, where the flow can be described on the basis of the rheological parameters described in the previous sections. Structural changes, brought about by mechanical power input, which are important for the resulting food quality (e.g. texture, consistency, mouthfeel, etc.) also correlate with rheological measurements. This means that rheological parameters can often be used to optimize mechanical treatment during food manufacture.

Thermal power input. This is of major interest from the food hygiene aspect (e.g. pasteurization and sterilization) and for heat-induced modifications to functional properties. These changes often influence the rheology of the materials concerned, in which case rheological control and/or optimization is possible.

Residence time. The residence time, or the residence time distribution, in processes where mechanical and/or thermal treatment is carried out, determines the total mechanical and thermal energy input. As shown in section 5.6, structure and its associated rheological properties often change naturally with time, especially for multiphase food systems containing macromolecular components.

If the flow behaviour is the main techno-functional quality parameter (e.g. for coating, filling, dosing and spraying), rheological measurements often play a vital role in controlling the process. In addition, they provide indirect information concerning product structure characteristics such as texture, consistency and mouthfeel. Consequently rheological parameters can be used for process and product development, process and product optimization, and for process control.

Examples of processes, which use rheology for these purposes are described in the following three sections.

5.9.2 Continuous crystallization of fat systems [15]

Figures 5.26 to 5.28 illustrate an example of process control by on-line rheological measurements. The process shown is a precrystallization process for fats or fat-containing food systems such as chocolate. During this so-called 'tempering operation' (chapter 17), a certain percentage of crystals is generated (e.g. 0.5–3% in chocolate). This leads to an increase

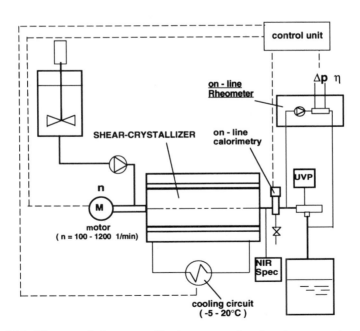

Figure 5.26 Diagram of shear crystallization process for chocolate tempering (NIR = near infrared spectrometer-structure, e.g. fat crystal content; UVP = ultrasonic Doppler velocimetry-flow profile, viscosity function).

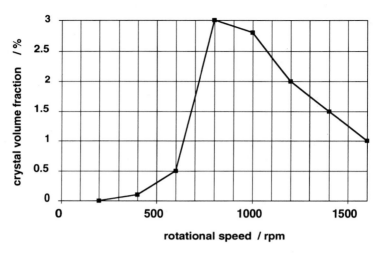

Figure 5.27 Dependency of generated fat crystal content (cocoa butter) on rotational speed of the shear crystallizer.

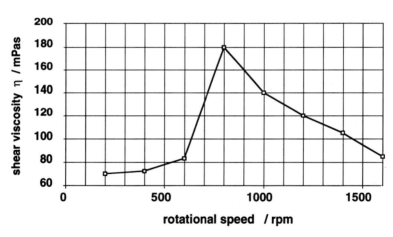

Figure 5.28 On-line measured viscosity (bypass rheometer in shear crystallization process at $\dot{\gamma} = 50\,1/s$).

in solid volume fraction (i.e. structural change) and therefore to an increase in the shear viscosity, $\eta(\dot{\gamma})$, which can be measured either off-line or on-line (bypass capillary rheometer, see Figure 5.26).

The crystallizer operates by shearing the fat-containing material as it is being cooled. Initially, the faster the rotational speed (greater mechanical power input), the more the seed nuclei are able to spread throughout the material and the quicker it sets. Eventually, however, the speed becomes so high that the energy input begins to melt some of the crystals. Further increases in rotational speed will therefore result in a smaller number of crystals. This is shown in Figure 5.27, which shows the crystal volume fraction, as measured off-line by calorimetric means (differential scanning calorimetry) as a function of the speed of rotation of the crystallizer. The crystal content can also be measured on-line by near infrared spectroscopy.

Alternatively, because of the close structure–rheology relationship, on-line viscosity measurements (at a shear rate of $50\,s^{-1}$) can be used. Figure 5.28 shows the shear viscosity versus crystallizer speed relationship, which can be seen to be very similar to the crystal volume relationship.

The on-line by-pass capillary viscometer which was used is shown schematically in Figure 5.29. As the chocolate flows through the U-tube, there is a pressure difference between the inlet and the outlet. From this measurement and the mass flow rate, it is possible to calculate the viscosity.

When a fat is polymorphic (see chapter 7), some crystalline forms are preferable from the point of view of processing or product quality. The crystallizer was able to show that the crystal type is related to the power input per unit volume of material. This in turn has led to the development of an improved tempering process [15].

5.9.3 *Emulsifying process for mayonnaise* (see also chapter 19)

During the dispersion of the oil into the continuous aqueous phase – mayonnaise is an oil-in-water emulsion – the disperse (oil) structure of

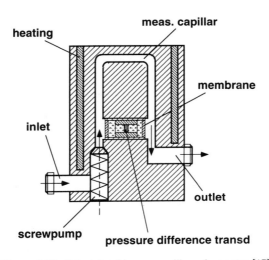

Figure 5.29 Principle of bypass capillary rheometer [17].

the system is significantly changed. The process is usually carried out in a continuous rotor-stator dispersing plant, as is shown schematically in Figure 5.30. The degree of these structural changes strongly depends upon the total power input per unit volume, the residence time in the rotor–stator gap, the gap geometry and the temperature conditions.

As the oil droplets ge smaller, the droplet–droplet interactions increase as do the viscous ($\eta(\dot{\gamma}), G''(\omega)$) and elastic ($G'(\omega)$) properties. Figure 5.31 demonstrates the effect of droplet diameter on these rheological parameters

An increase in power input into the mayonnaise leads to a droplet-size reduction, but at the same time, the 'structure stabilizing' macromolecular network in the continuous aqueous phase (e.g. proteins and polysaccharides) can be destroyed if a *critical power input* is exceeded. Consequently, a mayonnaise may be unstable (phase separation) if the disperse droplets are not small enough (insufficient dispersion – underprocessed) or if the macromolecular network in the continuous phase is not developed enough to prevent the droplets from separation by sedimentation and coalescence. The latter condition is caused by an insufficient amount of

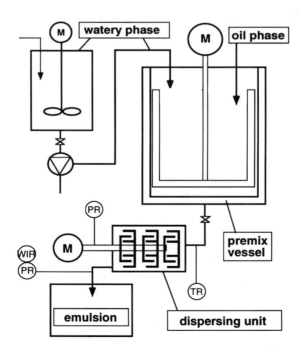

Figure 5.30 Continuous dispersing unit for mayonnaise production (Megatron MT 3–57; Kinematica Ltd, Lucerne, Switzerland).

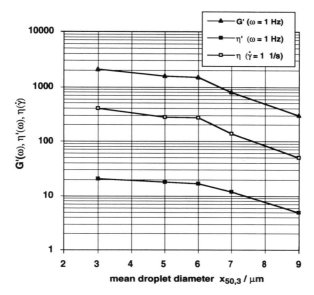

Figure 5.31 Oil-droplet diameter correlation with rheological parameters for mayonnaise processed under different power input conditions (starch-based mayonnaise with 30% oil content).

stabilizers/emulsifiers or by mechanical destruction in the dispersing gap ('supercritical' power input).

To optimize such processes, the rheology and structure of the pure continuous phase and of the whole emulsion system have to be considered separately as they exist in the post-mechanical treatment state. In general, the rotational speed of the dispersing unit, the residence time (which depends upon the mass flow rate), or the dispersing geometry can be changed to optimize the product parameters described.

Figure 5.32 shows the dependency of the mean droplet size on the volume specific energy input (power input × residence time) for different geometries of the dispersing 'shear gap' (G, K, HG). As can be seen, the gap geometry strongly influences the relationship between energy input and droplet size. The main quality parameter is emulsion stability, and both droplet size and rheological measurements are able to be used to define domains where this stability exists (see Figures 5.32–5.34).

The droplet mean diameter ($x_{50,3}$) and a representative rheological parameter (the elastic storage modulus G'') are plotted in Figures 5.33 and 5.34 as a function of the volume specific energy input for three different mayonnaise recipes (A–C). In addition, the energy domains of under- and overprocessing are marked. Emulsions produced in these domains are not stable and the phases will separate.

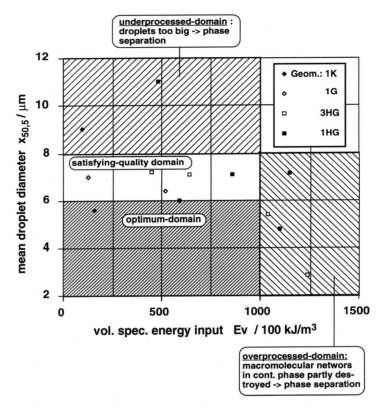

Figure 5.32 Influence of volume specific energy input on mean droplet size (different gap geometries G, HG, K) – mayonnaise A (starch-based with 30% oil).

When obtaining measuremets on this type of emulsion, it is preferable to use oscillatory shear experiments at small deformation amplitudes (G', G''). This avoids additional structural change due to the rheological measurement itself, as would occur for techniques which use high shear rates.

5.9.4 *Sensorial texture optimization of continuously processed ketchups*

Ketchup recipes with varying amounts of starch (4–6.5%) – the stabilizing macromolecular component – were processed in the continuous rotor/stator dispersing device, as was described for the production of mayonnaise (section 5.9.3). The rheological properties ($\eta'(\omega), G'(\omega)$) of the resulting products were measured using small amplitude oscillatory shear tests (Figures 5.35 and 5.36).

RHEOLOGY IN FOOD PROCESSING 111

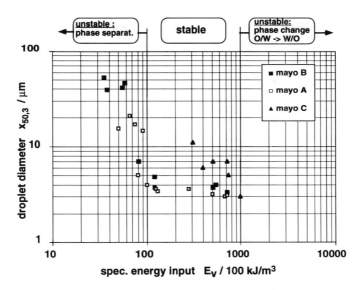

Figure 5.33 $x_{50,3}/E_v$ dependency (different mayonnaise recipes)/stability domains.

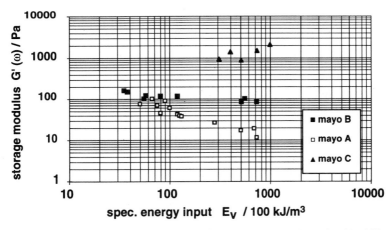

Figure 5.34 $G'(\omega = 1\,\text{Hz})/E_v$ dependency (different mayonnaise recipes)/stability.

In order to optimize the sensorial texture behaviour (particularly mouthfeel) of such complex food systems, both the viscous and elastic properties of the food have to be considered, particularly with respect to how they react to the deformation cycles in the human mouth. Table 5.4 gives an overview of rheological and sensorial (textural) parameters and related popular sensorial terms.

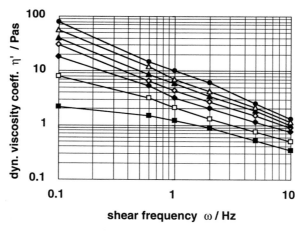

Figure 5.35 Viscous properties ($\eta'(\omega)$) of ketchups (variable starch content; 25°C (77°F)) obtained by oscillatory shear cone and plate gap measurements (Rheometrics DSR).

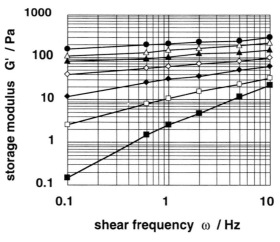

Figure 5.36 Elastic properties ($G'(\omega)$) of ketchups (variable starch content; 25°C (77°F)).

To combine viscous and elastic rheological behaviour the dynamic Weissenberg number W_i' has been formulated [18]. This is defined according to equation (5.19) and is identical to the reciprocal value of the phase angle δ where:

$$W_i' = 1/\tan\delta = G'(\omega)/(\eta'(\omega)\,\omega) = G'(\omega)/G''(\omega) \tag{5.19}$$

Figure 5.37 shows that W_i' correlates not only with the starch content, but also with the sensorial 'thickness' on a 10-point scale, as determined by a specially trained test panel.

Table 5.4 Relationship between rheological measurements and textured parameters

Characteristics			
Primary parameter	Secondary parameter	Measuring techniques	Popular sensorial terms
Mechanical characteristics (whole tomato system)			
Viscosity		Shear rheometry	Thin → viscous
Elasticity		Shear-/elong. rheometry	Plastic → elastic
Hardness		Press. test, penetration	Soft → firm → hard
Cohesiveness	Brittleness	Press. test, penetration	Crumbly → crunchy → brittle
	Chewiness	⎰ tension test, shear-	Tender → chewy → tough
	Gumminess	⎱ and elong. rheometry	Short → pasty → gummy
Adhesiveness		Shear-/ slip rheometry pressure test	Sticky → tacky → gooey
Geometrical characteristics (disperse phase)			
Particle size		Laser diffraction/ scattering, microscopy	Gritty, grainy, coarse
Particle shape and orientation		Microscopy → image anal.	Fibrous, cellular, crystal

Sensory analysis is most frequently performed in order to optimize sensorial quality. The test panel were therefore asked for their preference as well as their impression of 'thickness'. As is shown in the 'texture preference distribution' in Figure 5.38, the 5% starch sample was regarded as being much better than the other two.

In commercial ketchup recipes, there are several components which influence the sensorial texture profile. Furthermore, a variety of different stabilizing agents are used in practice. In addition to the viscosity and elastic-based 'thickness', other important texture parameters for ketchups include adhesiveness and cohesiveness, together with their related sensorial terms (see Table 5.4). Structural properties, like the disperse particle-size distribution, have additional influence on the sensorial texture properties.

5.10 Conclusion

Rheological measurements, made in idealized shear or elongational flows, enable parameters to be calculated which are closely correlated with

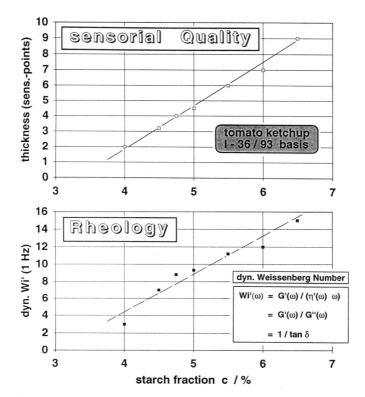

Figure 5.37 Recipe (starch content)/rheology/sensorial texture relationships for tomato ketchups.

the structure of multiphase food systems containing disperse and/or macromolecular components. Furthermore, many typical 'product quality parameters' correspond closely to these measurements.

Another close relationship exists between rheological properties and process parameters which influence fluid structure, for example as has been illustrated by mechanical and/or heat treatment.

Ideally, from the food process engineering point of view, on-line measurements should be made of those rheological parameters which correlate with changes in the structural properties of macromolecular and/or disperse components. This information could be readily used for process control, which in most cases involves adjusting mechanical power, energy input, throughput and temperature.

Rheology is a powerful tool for the development of products with optimized texture quality, as well as for design optimization and control of those processing units, where food structure, and hence rheology, plays a significant role.

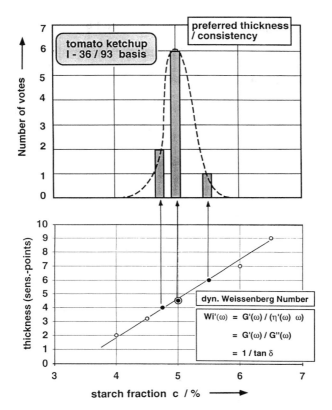

Figure 5.38 Corelation of preferred 'thickness' and rheology for ketchups with different starch contents.

References

1. Nelson, R.B. In *Industrial Chocolate Manufacture and Use*, 2nd edn., Ed. Beckett, S.T. (1994), 167–173.
2. Windhab, E.J. and Wolf, B. *Proc. Food Ingredients Europe 93 Paris* (1993), 267–271.
3. Wolf, B. Dissertation ETH Zürich (1995).
4. Windhab, E.J. *Proc. IVth Int. Congr. on Rheology, Seville, Spain* (1994), 1.
5. Krieger, I.M. and Dougherty, T.J. *Trans. Soc. Rheol.* **3** (1959), 137–152.
6. Maron, S.H. and Pierce, P.E. *J. Colloid Sci.* **11** (1956), 80–95.
7. Eilers, H. *Kolloid Z.* **97** (1941), 313–321.
8. Pahl, M. VDI Verlag GmbH, Düsseldorf (1983).
9. Barnes, H., Hutton, J.F. and Walters, K. *An Introduction to Rheology* (*Rheology series*) Elsevier, Amsterdam (1989).
10. Windhab, E. and zu Höne, T. *Proc. Makromolekulares Symposium TU.* Hamburg 9/91 (1991).
11. Kuhn, R., Krömer, H. and Roszmanith, G. *Angew. Makromol. Chem.* **40/41** (1974), 361–389.
12. Kulicke, W.M. Hütig und Wepf Verlag, Basle (1986).

13. Windhab, E. Progress and trends in rheology. II. *Rheol. Acta Supplement*, Steinkopff Verlag (1988).
14. Weipert, D., Tscheuschner, H.D. and Windhab, E.J. *Rheologie d'Lebensmittel*, Behrs Verlag, Hamburg (1994).
15. Windhab, E., Rolfes, L. and Rohenkohl, H. *Chem.-Ing.Technik* **63** (1991), 4385.
16. Windhab, E., Bollinger, S. and Wagner, T. *Proc. IVth Europ. Congress on Rheol. Seville, Spain* (1994).
17. Windhab, E.J., Grosse Kohorst, W. *Lebensmitteltechnik* **1/2** (1991), 51–53.
18. Windhab, E.J. *Lebensmitteltechnik*. **7/8** (1990), 404–414.
19. Sherman, P. *Industrial Rheology*. Academic Press London (1970).
20. Peleg, M. and Bagley, E.B. *Physical Properties of Foods*. Avi, Westport, Connecticut (1983).
21. Moldenaers, P. and Keunings, R. *Theoretical and Applied Rheology* Elsevier, Amsterdam (1992).
22. Moskowitz, H.R. *Food Texture*. Marcel Dekker, New York (1987).

6 Thickeners, gels and gelling
S.M. CLEGG

6.1 Introduction

Many of the desired properties of food products are undoubtedly related to product texture and, considering the enormous variations in textures between different product types (ranging, say, from drinks through semi-solid desserts, dressings and spreads, to more solid-like meat and confectionery products), it is clear that the food processor has an almost unlimited textural range with which to experiment. One major food ingredient area that significantly controls the textural properties of foods is that of thickeners, stabilisers and gelling agents, and it is probably correct to say that most processed food products utilise these ingredient types in one form or another. Collectively, these food-structuring ingredients can be categorised as food hydrocolloids. Although the manner in which they impart their specific texturising role to any particular food product may vary according to the ingredient in question, there is, nevertheless, a common structural feature amongst food hydrocolloids. This is the fact that they are all high-molecular-weight polymeric compounds, and it is the molecular associations and entanglements of these polymer molecules that give rise to the thickening and gelling properties desired.

This chapter is concerned with the thickening and gelling properties of hydrocolloids specifically designed as ingredients for the food industry, and, as such, will not deal with the functional properties of the hydrocolloids that many foodstuffs contain as part of their intrinsic make-up.

Factors that influence the performance of a food hydrocolloid in any particular application include temperature, shear, pH, ionic strength, etc. and the effect that these factors can have on the functionality of a food hydrocolloid are specific to the hydrocolloid in question. Consequently, care must be taken by food processors when utilising such ingredients to ensure that the maximum functionality of thickening and gelling systems is achieved, while bearing in mind the likely effect that the process itself may have on the system.

Having stated that factors affecting the properties of food hydrocolloids are specific to the particular hydrocolloid, there are, however, many properties that these systems have in common. In particular, the underlying molecular processes, that result in the unique macroscopic solution properties of hydrocolloids, have been widely studied and reviewed [1–4].

These can be broadly used to interpret the observed collective rheological properties of food hydrocolloids. In this chapter, therefore, the aim has been to give the reader a basic understanding of the mechanisms involved in determining the solution and gelling properties of food hydrocolloids, while bearing in mind the likely effects that food-processing conditions might cause in these systems.

6.2 Thickeners

Most commercial thickeners are polysaccharides, and it is the expanded nature of these high-molecular-weight molecules in solution that gives rise to their thickening properties, even when used at relatively low concentrations. The long-chain polysaccharide molecules in solution generally exist as conformationally disordered 'random coils', whose shape fluctuates continually under Brownian motion. However, the precise dimensions of the coil (i.e. how compact or how expanded it is) depend on the monosaccharide composition of the polymer chain and, more specifically, on the types of linkages between residues in the chain. For example, in linear-chain segments, certain types of linkage promote more of a 'sense of direction' from one residue to the next in the polymer chain [4] and, hence, lead to more expanded coils, while, for a given molecular weight, highly branched polymers have more compact structures. Superimposed on these 'type of linkage' effects, the dihedral angles ϕ and ψ (i.e. the angles through which successive residues in the polymer chain can rotate relative to one another) and restriction in their allowed values caused by steric hindrance also play a role in determining the relative orientations of successive residues along the chain, and hence the flexibility of the polymer chain in solution. Clearly, therefore, different hydrocolloids of the same molecular weight, but varying in primary structure, can result in solutions in which the molecules have significantly different coil dimensions which, as will be seen later, are a major factor in determining the viscosity of hydrocolloid solutions.

6.2.1 Intrinsic viscosity and coil dimensions

Polysaccharides, on the whole, have relatively inflexible chain linkages, and this results in their having highly extended structures, with large hydrodynamic volumes and, consequently, good thickening properties. However, before considering the thickening properties of hydrocolloids at typical levels used by the food industry, it is useful to look first at the dilute solution viscosity behaviour of hydrocolloids and how this is related to the molecular size, or more precisely, the effective specific molar volume of the hydrocolloids in solution. A convenient parameter for characterisation of the coil volume of hydrocolloid polymers in solution is the intrinsic viscosity, [η]. This is the fractional increase in viscosity per unit concentration of isolated polymer chains [4]. Experimentally, [η] can be obtained [5] by measurement of the 'specific viscosity' ($\eta_{sp} = (\eta - \eta_s)/\eta_s$, where η is the solution viscosity and η_s is the viscosity of the solvent) over a range of low polymer concentrations (c) and then extrapolation of η_{sp}/c to zero concentration. That is:

$$[\eta] = \lim_{c \to 0} (\eta - \eta_s)/\eta_s c. \qquad (6.1)$$

The extrapolation to infinite dilution is necessary because even in dilute polymer solutions, 'isolated' polymer chains can interact with one another indirectly through their effect on the flow of the solvent. This causes values of η_{sp}/c to be greater than expected if the chains were truly isolated. Quantitatively:

$$\eta_{sp}/c = [\eta] + k'[\eta]^2 c \qquad (6.2)$$

for values of η_{sp} in the range 0.2–1.0, and a plot of η_{sp}/c against c (Huggins plot) [6] thus gives [η] as the intercept at zero concentration and a measure of the polymer interactions through the Huggins' constant, k'. For values of $\eta_{sp} > 1$, higher-order concentration effects cease to be negligible and curvature in the plot of η_{sp}/c against c is observed at higher values of η_{sp} (i.e. higher polymer concentrations).

An alternative extrapolation for determination of the intrinsic viscosity (Kraemer plot) [7] is given below:

$$\ln \eta_{rel}/c = [\eta] + k''[\eta]^2 c \qquad (6.3)$$

where η_{rel} is the relative viscosity = η/η_s and k'' is known as the Kraemer constant. The Huggins and Kraemer constants for a particular polymer and solvent are related by the expression:

$$k' = k'' + \tfrac{1}{2}. \qquad (6.4)$$

The intrinsic viscosity is therefore a measurable parameter that is related to the coil dimensions of polymer molecules in solution and, indeed, is directly related to the radius of gyration (R_g) of the polymer coil through the Flory–Fox equation [8]

$$[\eta] = 6^{3/2} \Phi \, R g^3 / M \qquad (6.5)$$

where Φ is a constant (~2.6×10^{26} kg^{-1}) and M is the molecular weight of the polymer.

Another practical use of intrinsic viscosity is in the determination of the molecular weight of polymers. For any polymer–solvent system, intrinsic viscosity increases with molecular weight according to the Mark–Houwink relationship.

$$[\eta] = KM^\alpha \tag{6.6}$$

in which K and α are constants for the polymer type, solvent and temperature of measurement. Consequently, once K and α have been determined for the polymer–solvent system by calibration of the intrinsic viscosity against molecular weight (determined via some absolute technique), routine molecular-weight determinations can be easily carried out by measurement of the intrinsic viscosity. The parameters K and α in the Mark–Houwink relationship, although empirically determined, give a measure of how compact the polymer coil will be in solution, with large values of K and α obviously indicating more expanded coils. For a hypothetical 'freely jointed' chain, without steric clashes between chain segments, $\alpha = 0.5$ [4]. However, for most hydrocolloid solutions in which the chains are neither freely jointed nor without steric clashes (i.e. ϕ and ψ have restricted values), values of α are more usually in the region of 1 and so intrinsic viscosity is approximately directly proportional to molecular weight. The K parameter is primarily dependent on the geometry of the linkages between residues in the polymer chain [4,5], with higher K values being found for those polymers whose linkages are opposite and parallel to one another (giving the polymer chain more of a 'sense of direction') than for those in which the linkages are not parallel to one another, or for those that are linked through three covalent bonds, such as dextran.

A further factor that alters the intrinsic viscosity (coil dimensions) of polyelectrolytes (such as alginates and carrageenans) is the ionic strength (I) of the solution, and this should be borne in mind if a polyelectrolyte is to be used as a thickener in some food application. For a polyelectrolyte, electrostatic repulsion between the charged groups on the polymer cause the polymer to have an expanded coil volume. However, addition of electrolyte to increase the ionic strength of the solution results in screening of the charged repulsions between the charged groups on the polymer chain, and a consequent decrease in the size of the polymer coil. Quantitatively, intrinsic viscosity decreases linearly with $I^{-1/2}$, extrapolating at infinite ionic strength ($I^{-1/2} = 0$) to the intrinsic viscosity of a neutral polymer of the same primary structure and chain length [5].

6.2.2 Concentration dependency of viscosity and coil overlap

The intrinsic viscosity is a useful parameter for characterisation of the coil volume of polymer molecules in dilute solution, but the dilute concentrations used for intrinsic viscosity measurements are significantly lower than those typically utilised in the food industry. Indeed, the concentration dependencies of viscosity at higher polymer concentrations are of far greater relevance to the food technologist when choosing the correct thickener for his particular application. Nevertheless, the intrinsic viscosity (coil volume) of a polymer is of major importance in helping us to understand the thickening properties of hydrocolloids at higher concentrations.

Most hydrocolloid thickeners, with the notable exception of the ordered rod-like xanthan molecules, exist in solution in a disordered random-coil conformation although, as already discussed, the coil dimensions are specific to the polymer and solvent in question. However, the concentration dependency of viscosities of random-coil polymer solutions usually shows the same general trends.

With increasing polymer concentration, the zero-shear viscosity of the polymer solution, η_0 (i.e. the viscosity of the solution not under the influence of shear), increases with $\sim c^{1.4}$ up to a critical concentration, c^*, at which point there is a marked change in the concentration dependency of η_0 and, above c^*, the zero-shear viscosity increases with $\sim c^{3.3}$. This high concentration dependency of viscosity above c^* is what gives hydrocolloid thickeners their functional role in food systems, with extremely high viscosities being obtainable at relatively low polymer concentrations. The abrupt change in the concentration dependency of η_0 occurs as a result of the polymer solution changing from a situation at low concentrations where the polymer coils are well separated from one another and free to move independently (Figure 6.1a), to a situation at higher concentrations where the coils interpenetrate one another (Figure 6.1c) and movement can only take place by the polymer molecules wriggling through the entangled polymer network [5,9]. This is obviously a much more difficult process and therefore explains the higher concentration dependency of η_0 above c^*. Clearly, between the two extremes, there will be a concentration at which the polymer coils are just touching (Figure 6.1b) and this is the critical concentration, c^*, also known as the concentration for the onset of coil overlap.

From the situation described above, it is clear that, for thickeners displaying random-coil-like behaviour, the onset of coil overlap (c^*) for any given polymer is dependent on the volume of the coil, which can be characterised by the intrinsic viscosity ($[\eta]$), and the number of coils in solution, which is proportional to the polymer concentration (c). So, despite the fact that the size of polymer coils in solution can vary vastly depending on their molecular weight, ionic strength of the solution, polymer primary structure, temperature, etc. the dependency of zero-shear viscosity on concentration can be rationalised in terms of the degree of space occupancy (i.e. $[\eta]c$) of the polymer in solution [5,10,11]. This is illustrated in Figure 6.2, in which the logarithm of the zero-shear specific viscosity (η_{sp}) is plotted against the logarithm of $c[\eta]$ for a number of

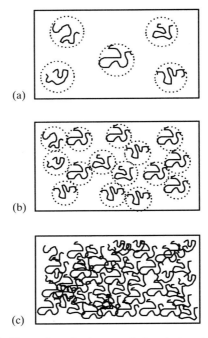

Figure 6.1 Schematic illustration of polymer coils in solution. (a) The independent nature of polymer coils below the onset of coil overlap ($c < c^*$). (b) The onset of coil overlap ($c = c^*$). (c) An entangled polymer network ($c > c^*$).

different random coil hydrocolloids. The points clearly fall onto one master plot, with a sharp increase in the concentration dependency of viscosity at c^*, where the viscosity is about ten times that of water and $c[\eta] \approx 4$.

This unifying concept of polymer space-occupancy determining the zero-shear viscosity of hydrocolloid solutions therefore indicates that any random-coil polymer could be utilised by the food technologist wishing to develop a product with a defined viscosity, providing the correct concentration of the polymer is chosen. This is of course a gross oversimplification as food systems are considerably more complex than simple model hydrocolloid solutions, with different physical structures, ionic strengths, water activities, pH values, etc. all of which can influence the coil dimensions and hence the behaviour of a thickener. Furthermore, in food systems containing more than one hydrocolloid, interactions between the hydrocolloids (both antagonistic and synergistic) are frequently observed and can significantly alter the rheological properties of the system. Nevertheless, the concept of coil overlap and its effect on the viscosity of polymer solutions is an extremely useful starting point for designing food products with desirable rheological properties.

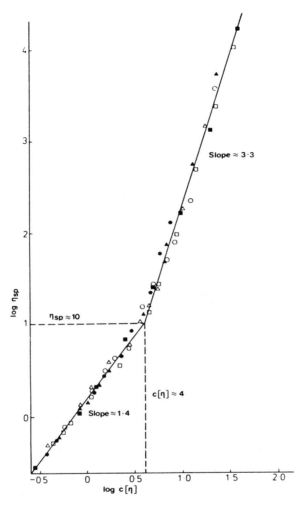

Figure 6.2 Plot of zero shear specific viscosity (η_{sp}) against the degree of space occupancy ($c\,[\eta]$) for different random-coil polysaccharide solutions (represented by different symbols), illustrating the generalised viscosity–concentration behaviour of random coil polymers. (Reprinted from Morris et al. [11], with permission from Elsevier Science Ltd., Kidlington. UK.)

6.2.3 Shear-rate dependency of viscosity

The generalised viscosity–concentration behaviour of polymers in solution described above was concerned with the zero-shear viscosity (i.e. the viscosity of the polymer solution 'at rest'). For solutions of small molecules such as sugars, the viscosity is independent of the rate at which the solution flows (shear rate, $\dot{\gamma}$). That is, the same value would be obtained for the viscosity whether it was measured at high or low shear rates. Such solutions are said to display ideal (or Newtonian) viscosity behaviour (see also chapter 5). In contrast, solutions of hydrocolloid thickeners show non-ideal solution behaviour and their viscosities vary with the shear rate applied to the solution. This shear-rate dependency of viscosity is clearly of major importance considering the different shear rates to which a food product may be subjected, ranging from zero, during in-pack storage, to extremely high values in some food processes, such as valve homogenisation. Of course, the viscosity of a food product at the shear rates experienced in the mouth during mastication and during spreading or pouring processes significantly influences consumers and is therefore also highly relevant.

In general, hydrocolloid thickeners show shear thinning (or pseudoplastic) behaviour. That is, the viscosity of the solution decreases with increasing shear rate of measurement. At low shear rates, there is usually a Newtonian plateau in the viscosity versus shear rate curve where the solution behaves ideally and the viscosity corresponds to the zero-shear viscosity, but, with increasing shear rate, there is a marked reduction in viscosity. The extent of this shear-thinning behaviour is also dependent on the concentration of the polymer in solution and, in particular, on whether it is above the coil overlap concentration or not.

At concentrations below the onset of entanglement (i.e. below c^*), polymer chains are free to move independently through the solvent and, as the shear rate is increased, the viscosity decreases relatively little – typically <30% over several decades of $\dot{\gamma}$ [4]. This decrease in viscosity with increasing shear rate is attributed to the individual polymer coils being stretched out and elongated in the direction of flow, hence offering less resistance to flow (i.e. lower viscosity) [4,9].

Above c^*, the shear-rate dependency of viscosity is far greater and it is not unusual to see a drop in viscosity of several orders of magnitude over the shear rate range of practical importance [4]. The mechanism of the shear-thinning behaviour of hydrocolloid solutions at high shear rates and at concentrations above c^* is quite different from that at concentrations below c^*. Above c^*, the polymer coils are entangled and the Newtonian plateau in viscosity versus shear rate plots at low shear rates reflects a situation in which the number of molecular entanglements is essentially independent of $\dot{\gamma}$ (i.e. there is sufficient time at low shear rates for new entanglements between different polymer chains to replace those pulled apart by the application of shear and, hence, to maintain the overall cross-link density of the entangled network). As the shear rate is further increased, however, there comes a point at which the rate of entanglement reformation falls behind that of entanglement disruption, with the result that the overall cross-link density of the network falls and, hence, so does the viscosity.

Although different random-coil thickeners possess different absolute values of η_0 and the shear rate at which the onset of shear thinning begins, as with the concentration dependency of viscosity, the shear-thinning behaviour of random-coil hydrocolloids is entirely general and can be reduced with reasonable precision to a single master curve. If measured viscosities are expressed as a fraction of the maximum zero-shear viscosity (i.e. as η/η_0) and shear rates are similarly expressed relative to the shear rate, $\dot{\gamma}_f$, required to reduce viscosity to a fixed fraction (f) of η_0, then shear-thinning curves for many random-coil polymers at concentrations above c^* lie on top of one another, as is illustrated in Figure 6.3 [11,12]. The shear-thinning behaviour can be fitted with reasonable precision by the equation:

$$\eta = \frac{\eta_0}{1 + m(\dot{\gamma}/\dot{\gamma}_f)^p} \tag{6.7}$$

in which m is an adjustable parameter dependent on the value of f, i.e.

$$f = \frac{1}{(1 + m)}$$

since by definition $\eta = f\eta_0$ when $\dot{\gamma} = \dot{\gamma}_f$ and so $f\eta_0 = \eta_0/[1 + m(\dot{\gamma}_f/\dot{\gamma}_f)^p]$.

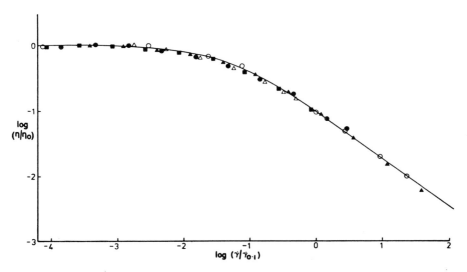

Figure 6.3 Generalised shear-thinning behaviour of random-coil polymers (different symbols represent different polymers), showing the viscosity as a fraction of the zero shear viscosity plotted against the shear rate, expressed relative to the shear rate required to reduce the viscosity to a fraction (in this instance 0.1) of the zero shear value. (Reprinted from Morris et al. [11], with permission from Elsevier Science Ltd., Kidlington, UK.)

If the shear rate required to reduce the viscosity to half its maximum zero-shear viscosity is chosen (i.e. $f = 0.5$), then equation (6.7) further simplifies to:

$$\eta = \frac{\eta_0}{1 + (\dot{\gamma}/\dot{\gamma}_{0.5})^p}. \qquad (6.8)$$

The Newtonian plateau at low shear rates is therefore predicted, since $\dot{\gamma}$ will be much less than $\dot{\gamma}_{0.5}$ (i.e. $\eta \to \eta_0$) while, at high shear rates where $\dot{\gamma}/\dot{\gamma}_{0.5}$ is much more than 1, the equation reduces to:

$$\eta = \frac{\eta_0}{(\dot{\gamma}/\dot{\gamma}_{0.5})^p} \quad \text{i.e. } \log \eta = \log \eta_0 - p \log (\dot{\gamma}/\dot{\gamma}_{0.5}). \qquad (6.9)$$

p is therefore the absolute value of the terminal slope in a double logarithmic plot of η against $\dot{\gamma}$ and has a constant empirical value of 0.76 for samples of wide polydispersity, such as commercial hydrocolloid thickeners [4].

The message from this general shear-thinning behaviour of random-coil hydrocolloids is therefore that the viscosity (η) at any shear rate ($\dot{\gamma}$), can be characterised by two parameters, η_0 and $\dot{\gamma}_{0.5}$. This therefore is of great practical importance to the food technologist interested in knowing the viscosity behaviour of this thickened food system under the different shear rate regimes encountered during food processing and product usage. Determination of both η_0 and $\dot{\gamma}_{0.5}$ can be obtained from a plot of η against $\eta\dot{\gamma}^{0.76}$, which should possess an intercept of η_0 and a gradient of $-(1/\dot{\gamma}_{0.5})^{0.76}$ since equation (6.8) can be rearranged to:

$$\eta = \eta_0 - (1/\dot{\gamma}_{0.5})^p \eta \dot{\gamma}^p \qquad (6.10)$$

6.3 Viscoelastic nature of thickeners and gels

In general, hydrocolloid solutions and gels are viscoelastic in nature (i.e. they possess both liquid-like and solid-like properties) and the rheology that such systems exhibit is dependent on the time scale of the event to which they are subjected.

The measurement of traditional 'steady shear' viscosities, with which previous sections of this chapter were concerned, involves the application of large deformations to the sample under test, and such measurements by their nature are destructive, particularly with regard to the solid-like properties of a sample (i.e. the elastic properties). An alternative, non-destructive rheological technique often utilised for quantifying the viscoelastic properties of hydrocolloid solutions and gels is that of 'mechanical spectroscopy'.

6.3.1 Mechanical spectroscopy

The basic principle in mechanical spectroscopy is to apply a small oscillatory deformation to the sample under test and then to measure the stress generated in resisting the applied deformation. The term 'mechanical spectroscopy' arises as measurements are usually made over a range of oscillatory frequencies (ω) so that information can be obtained on the behaviour of the sample being assessed at different time scales (this relates to the small deformation stress controlled systems described in chapter 5).

The stress generated in resisting the applied deformation can be resolved into components that are in-phase and out-of-phase with the applied deformation. For a 'perfect solid', stress increases with the increasing extent of deformation (strain) and it is thus apparent that the stress generated that is in-phase with an applied deformation represents a solid-like response. This in-phase stress divided by the applied strain therefore gives the modulus, G' (known as the 'storage modulus' or 'elastic modulus'), which is a measure of the solid-like character of the sample. For a 'perfect liquid', by contrast, resistance to flow (stress) increases with increasing rate of deformation, which for an oscillatory system (cosine wave) will be maximum at the mid-point of the oscillation and zero at the extremes (maximum amplitude). The liquid-like response of a sample can therefore be characterised by the component of stress out-of-phase with the applied deformation (see Figure 5.4) and this, divided by the applied strain, gives the modulus G'' (known as the 'loss modulus' or 'viscous modulus'), which is a measure of the liquid-like character of a sample. The ratio of the unresolved 'complex modulus' $G^* = (G'^2 + G''^2)^{1/2}$ (total stress/applied strain) to the oscillation frequency (ω) gives a third parameter, the complex dynamic viscosity ($\eta^* = G^*/\omega$), which is useful for characterisation of the viscoelastic nature of solutions and gels.

6.3.2 Viscoelastic behaviour of random-coil hydrocolloid solutions

Typical mechanical spectra for solutions of random-coil-like hydrocolloid thickeners at concentrations both below c^*, where polymer coils can move almost independently, and above c^*, where the polymer network is entangled, are shown in Figures 6.4(a) and 6.4(b), respectively. For the solution below c^*, the dynamic viscosity (η^*) shows little frequency dependency (i.e. essentially Newtonian behaviour over the accessible frequency range); $G'' > G'$ at all frequencies, indicating the predominantly viscous-like response of the sample, although both increase with increasing frequency, with the value of G' approaching that of G'' at very high frequencies of measurement. The increase in G' relative to G'' at high frequencies has been described as resulting from storage of energy

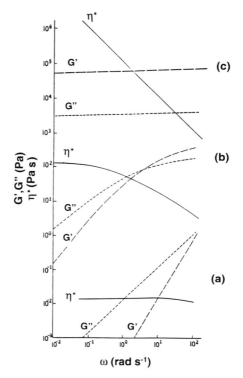

Figure 6.4 Typical mechanical spectra of hydrocolloid solutions and gels, showing the frequency (ω) dependence of G', G'' and η^* for (a) a dilute polymer solution below c^*, (b) an entangled polymer solution above c^* and (c) a cross-linked hydrocolloid gel. (Reprinted from Morris [4] by permission of Kluwer Academic Publishers.)

by contortion of individual molecules into less energetically favoured conformations [4,5].

The mechanical spectra of entangled polymer solutions above the c^* transition deviate significantly from those of dilute solutions below c^* (Figure 6.4b). At low frequencies, where there is sufficient time for entanglements to come apart in the period of an oscillation, the behaviour is similar to that observed for the dilute solution. That is, η^* is effectively independent of frequency and $G'' > G'$, indicating the predominantly liquid-like response of the sample. However, with increasing frequency, G' crosses over G'' and both become flatter in terms of their frequency dependency, representing a changeover from a liquid-like response to a solid- or gel-like response. η^* becomes more frequency-dependent and decreases sharply with increasing frequency at higher frequencies of measurement. This transition from liquid-like behaviour to solid-like behaviour with increasing frequency for entangled polymer solutions arises as a result of there being insufficient time, within the period of an oscillation at high frequencies, for complete disentanglement of the polymer chains. Hence, the system responds as if it is permanently cross-linked at high frequencies, similar to a true gel. The transition from a predominantly liquid-like response to a predominantly solid-like response moves to progressively lower frequencies as the extent of coil overlap increases. Thus, both concentration and the factors that affect the intrinsic viscosity of a polymer solution will influence the transitional frequency through their effect on the coil overlap parameter ($c[\eta]$).

Another point worthy of mention when discussing the mechanical spectroscopy of random-coil polymer solutions is the fact that the frequency (ω) dependency of dynamic viscosity (η^*) is normally closely superimposable on the shear rate ($\dot{\gamma}$) dependency of steady-shear viscosity (η) at equivalent values of $\dot{\gamma}/(s^{-1})$ and $\omega/(rad\,s^{-1})$. This empirical observation was first reported by Cox and Merz [13] and is known as the Cox–Merz rule. It now seems likely that the superposition of the dynamic viscosity and shear viscosity reflects the same dependence of physical properties on the time scale of polymer entanglement and disentanglement under both small deformation oscillatory and large deformation rotational conditions [14].

6.3.3 Mechanical spectra of hydrocolloid gels

Before discussing gelling systems in more detail, it is useful first to look at the type of mechanical spectra expected for a true gel, and this is illustrated in Figure 6.4(c). The value of G' is substantially higher than that of G'' at all frequencies, illustrating the predominantly solid-like response, while both G', G'' and hence G^* are essentially independent of frequency, as would be expected for a perfectly elastic network. The dynamic viscosity $\eta^* = G^*/\omega$ is inversely proportional to ω at all frequencies (as G^* is frequency-independent) and, hence, the slope of the plot of $\log \eta^*$ against $\log \omega$ has the theoretical limiting value of -1.

In practice, hydrocolloid gels are not ideal and do show some variation in G' and G'' with frequency, with the extent of this being a major factor in controlling the texture and properties of gels. Indeed, mechanical spectroscopy raises the question: What is a gel? As already discussed, concentrated random-coil solutions give a gel-like response at high frequencies, where the molecular entanglements become effectively permanent cross-links with respect to the time scale of the measurement. In contrast, traditional gelling systems can show more viscous-like behaviour at very low frequencies of measurement, where the time scale of molecular rearrangements of the 'permanent' cross-links is less than the time scale of the measurement. Consequently, defining a hydrocolloid system as a gel or as a solution can be difficult rheologically, as a result of the viscoelastic nature of such systems and, particularly, the manner in which it changes under different time scales of measurement. Nevertheless, there are a number of hydrocolloid systems that would be obviously classified as a gel by the layperson, which behave in a gel-like manner when evaluated by mechanical spectroscopy over the range of frequencies of practical importance. The formation and properties of these gels will be discussed in the next section.

6.4 Gels and gelling

In contrast to the thickening properties that all hydrocolloids possess, the formation of 'true' gel-like structures is not universal among hydrocolloids. It is in fact shown only by a limited number as a consequence of the manner in which the molecules interact with one another when hydrated under the influence of certain external factors, such as temperature, pH, solvent quality or the presence of specific ions.

Essentially, a gel can be envisaged as an entangled polymer solution in which the entanglements between different polymer chains are more than just transient, and can be classed as permanent cross-links, when considered in relation to the time scales of other molecular events. The polymer molecules are therefore permanently linked together through the cross-links into a three-dimensional network structure, with the solvent molecules being entrapped within the network. In contrast to synthetic polymer gels with their covalently linked networks (i.e. the cross-links between different polymer chains are distinct chemical bonds), the cross-links in hydrocolloid gels are usually composed of ordered 'junction zones', in which chain segments from different polymer chains are packed in an ordered array of non-covalently linked chain segments. The remaining disordered sections of polymer chains then link the ordered junction zones together, as depicted schematically in Figure 6.5. The nature of the ordered junction zones varies between different gelling systems, but they are usually composed of chain segments with conformationally ordered structures, which are the same as those found in the solid state of the hydrocolloid [4,15]. The formation of a hydrated three-dimensional gel network, however, does not merely require that the polymer molecules be capable of existing in a molecularly ordered conformation, as this would simply result in a solid precipitate, rather than a gel. Additionally, there must be some mechanism for terminating the ordered structure, thereby allowing the remaining disordered chain segments of the polymer chains to participate in junction zones with other polymer chain segments, thus building up the three-dimensional network structure.

Formation of ordered structures under hydrated conditions (which occurs when a polymer solution changes from the sol state to the gel

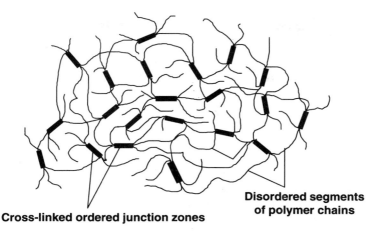

Figure 6.5 Schematic illustration of a cross-linked hydrocolloid gel, with non-covalently bonded ordered junction zones linking the disordered segments of polymer chains into a three-dimensional network structure.

state) involves considerable loss of conformational entropy and this must be compensated for by favourable enthalpic interactions between the residues participating in the ordered junction zones. Normally, end residues in the ordered sequences do not participate fully in the non-covalent bonding pattern. This is due to the lack of residues with which to bond in the neighbouring unordered chain sequences. Consequently, the loss of entropy of the end residues of ordered sequences is not fully compensated by the gain in enthalpy from their non-covalent interactions. The net effect of these 'end effects' is that the junction zones need to be above a minimum critical length for stability, and their formation and dissociation occur as sharp, cooperative processes, brought about by changes in external variables such as temperature, pH, solvent quality, ionic strength and/or the presence of specific ions [4], which can tip the order/disorder equilibrium in either one direction or the other.

6.4.1 Critical gelling concentration and concentration dependency of gel rigidity

The properties of a hydrocolloid gel are dependent on the concentration of the hydrocolloid in the system and, indeed, below a certain minimum concentration (which may vary widely for different gelling systems) no gelation will occur [5,9]. This minimum concentration is known as the 'critical gelling concentration', c_g, and should not be confused with the coil overlap concentration, c^*, although both are clearly going to be dependent on the degree of space occupancy of the polymer molecules.

That is to say, in order for gel formation to occur, junction zones between different polymer molecules must be formed, which naturally requires that the molecules are in close proximity to one another, a condition obviously met at the coil overlap concentration.

Below c_g, the concentration is insufficient to support a three-dimensional gel network and there is a sol phase containing aggregates of polymer molecules [9]. The nature of these aggregates is dependent on the type of ordered structures formed, with, in some instances, microgel-like structures being present in the sol, while in others, precipitation of the aggregates can occur.

Once above the critical gelling concentration, gel-like properties increase rapidly with further increases in concentration, following which, the concentration dependency of gel rigidity (shear modulus) usually falls off to some kind of power-law relationship. For gelatine the shear modulus increases approximately with $(c - c_g)^2$, and a c^2 dependency of gel-like properties has been found for some other gelling systems. However, the precise concentration–gel rigidity behaviour of most gelling systems is specific to the system in question, being dependent on the nature of the gel junction zones, number of junction zones per molecule, etc.

6.4.2 Food gel structures

The types of conformationally ordered structure found in the junction zones of food gels reflect the ordered structure found in the solid state of the hydrocolloid. Broadly speaking, the ordered structures can be split into two classes, i.e. helical and linear. Helical ones may contain one or more chain segments in a helical secondary structure and the helices frequently aggregate together to form junction zones composed of a number of helices. In linear structures, on the other hand, polymer molecules pack together side-by-side to produce the junction zones.

Gels based on helical junction zones. The four main hydrocolloids that form gels with junction zones based on helical structures are agarose, carrageenan, gelatin and starch.

Both agarose and some carrageenans exist at low temperatures in a co-axial double-helical ordered conformation [15] as a result of the alternating repeating sequence, in which every second residue is a 3,6-anhydride. The primary repeating structures of these polysaccharides are given in Figure 6.6. The presence of the anhydride bridge in agarose, ι-carrageenan and κ-carrageenan fixes the sugar ring in the unfavoured chair conformation with C(6) axial, a prerequisite for formation of the helical structure, and hence gel formation. Indeed, it is the absence of the anhydride bridges at certain residues along the chain that limits the size of the helical junction zones and thus promotes network formation rather than

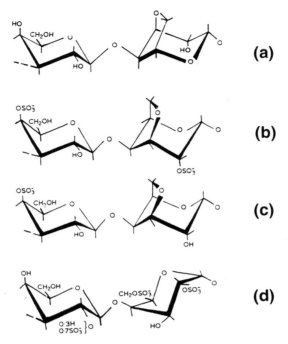

Figure 6.6 Primary repeating structure of (a) agarose, (b) ι-carrageenan, (c) κ-carrageenan and (d) λ-carrageenan.

precipitation on cooling agarose, ι-carrageenan and κ-carrageenan solutions. λ-carrageenan, in which none of the residues possesses the anhydride bridge, does not form gels.

In the gelling carrageenans and agarose, gelation occurs as a result of side-by-side association of the ordered helical chain segments, and various models [16] have been proposed and refined suggesting the precise nature of the gel network. Aggregation of the double-helical structures results in increased thermal stability of the ordered junction zones and, consequently, agarose and carrageenan gels show a marked degree of thermal hysteresis.

Another important factor that affects the gelling properties of agarose and carrageenan is how highly charged the molecules are. At one extreme, agarose is neutral and forms turbid, brittle gels that exhibit the largest degree of thermal hysteresis and that are prone to syneresis. These properties reflect the highly aggregated structure of agarose gels. In carrageenans, on the other hand, and particularly for the most highly charged, ι-carrageenan, the extent of helix aggregation is considerably less as a result of electrostatic repulsion between the charged helical strands. In such systems, gelation can only occur in the presence of cations (typically K^+ or Ca^{2+}) that suppress electrostatic repulsion between the participating chains by packing within the aggregate structure [4,16]. For a more detailed discussion on the gel systems formed by agarose and carrageenans, the reader is referred to the review by Morris [16] on gelation of polysaccharides.

Gelatine is the only proteinaceous biopolymer generally considered as an industrial hydrocolloid, and its wide use in the food industry reflects the novel melt-in-the-mouth properties of gelatine gels. The gels formed by gelatine are composed of helical junction zones, which have the same co-axial triple helical-ordered conformation as the parent protein, collagen, from which gelatine is derived. The triple-helix structure of collagen reflects the rather unusual protein primary structure in collagen, which is quasi-repeating, with every third amino acid residue being glycine and about half the remaining residues being proline or hydroxyproline [17,18]. Both of the last two have ring structures incompatible with the α-helix and β-sheet conformations commonly adopted by protein chains.

Numerous studies have been concerned with the elucidation of the mechanism of gelation in gelatine. All invoke the collagen helical structure as the ordered structural form of gelatine molecules in the junction zones of gelatine gels. However, it is still unclear whether further aggregation of the triple-helical junction zones takes place, and it is possible that aggregation, or lack of aggregation, may be dependent on the concentration of gelatine in the system. Current evidence [19,20] suggests that, on cooling a gelatine solution to below its gelling temperature, junction zones are formed by two strands from one molecule, folded back into a hairpin structure, together with a third strand from another chain. Junction zone termination probably results as a consequence of helix nucleation at various points within the same chains, with any chain therefore being involved in a number of ordered junction zones. For helix nucleation to occur at several points within the same chain, the rate of helix propagation must be relatively slow compared with the rate of helix nucleation. It has been proposed [21] that isomerisation of peptide units from the *cis* form (in which the α-carbons of successive amino acid residues in the gelatine chain are on the same side of the peptide unit) to the *trans* form (in which they are diagonally opposite) limits the rate of helix propagation, as only the *trans* form of the peptide unit can be incorporated into the collagen triple-helix structure.

Starch gels are thermo-irreversible and more complex in nature than the carrageenan, agarose and gelatine gels described above. This complexity arises because of the two structurally distinct polysaccharides, amylose and amylopectin, present in starch. Amylose is an essentially linear polymer of $(1 \rightarrow 4)$ linked α-D-glucose while amylopectin, similarly, contains chains of $(1 \rightarrow 4)$ linked α-D-glucose, but is a very high molecular weight, multiple branched structure, with branches linked at C(6).

Studies on isolated amylose have shown that cooling solutions results in precipitation of the amylose, or gel formation, depending on the concentration of the initial solution, and this behaviour has been termed 'retrogradation' [16]. In the retrograded state, X-ray studies suggest that the amylose is in a partially crystalline state, with side-by-side packing of double helices. For amylopectin, by contrast, only slight retrogradation has been observed and, in the case of amylopectin, this is thermoreversible.

A starch gel formed by cooling a starch dispersion that has been heated to above its gelatinisation temperature is currently envisaged as porous 'amylopectin' granules (from which the amylose has leached out) embedded in an amylose gel matrix. Such a gel structure is extremely complex rheologically, with the resultant rheology of the gel being dependent on the amylose matrix, the rigidity of the amylopectin granule, the volume fraction and shape of the amylopectin granules and the interactions that may be present between the amylose and the amylopectin [22]. All these factors will, of course, be dependent on the processing conditions used during preparation of the starch gel, further illustrating the complexity of the starch gel system.

Gels based on linear-ordered conformations. Some hydrocolloid gels are based on linear-ordered junction zones (i.e. the primary structure does not result in a twist on going from one residue to the next as is the case for helical-ordered structures), and two different types of linear-ordered structures are found in gelling systems. Polysaccharides that are $(1 \rightarrow 4)$ diaxially-linked result in highly buckled structural conformations, while polysaccharides that are $(1 \rightarrow 4)$ diequatorially-linked, result in extended ribbon-like structures (Figure 6.7).

Alginates are charged polysaccharides from brown seaweed and form gels with divalent cations (particularly Ca^{2+}). The basic primary structure of alginate is that of a block copolymer, with blocks of poly-L-guluronate $(1 \rightarrow 4)$ diaxially-linked and blocks of poly-D-mannuronate $(1 \rightarrow 4)$ diequatorially-linked. Also, heteropolymeric regions in which mannuronate and guluronate residues alternate are present in the polymer chain. The poly-L-guluronate blocks, in their highly buckled, 'zig-zag' structure, can pack together, with inclusion of an array of site-bound divalent cations, to form the junction zones of a gel network as illustrated in Figure 6.8 [4]. The heteropolymeric regions and the poly-D-mannuronate blocks of the alginate chain do not have the geometry necessary for the formation of these so-called 'egg-box' junctions and therefore act as junction terminating sequences, linking the ordered junction zones into a three-dimensional gel network.

Calcium ion-binding studies indicate that a minimum critical sequence length of approximately 20 polyguluronate residues, corresponding to a

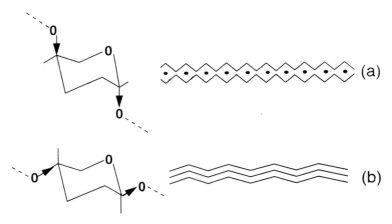

Figure 6.7 Linear-ordered conformations of polysaccharides. (a) $(1 \to 4)$ diaxially-linked residues, resulting in highly buckled structures capable of binding cations. (b) $(1 \to 4)$ diequatorially-linked residues, resulting in extended ribbon-like structures. (Reprinted from Morris [4] by permission of Kluwer Academic Publishers.)

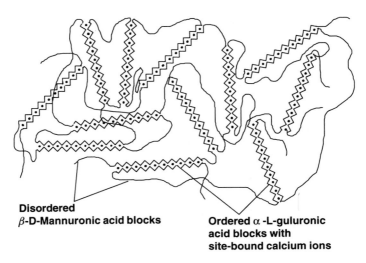

Figure 6.8 Schematic diagram of egg-box junction zones formed by α-L-guluronic acid blocks in calcium-induced gelation of alginate. The junction zones are connected by β-D-mannuronic acid blocks incapable of participating in the junction zones.

minimum array of 10 site-bound calcium ions, is necessary for 'egg-box' junctions to be stable in alginate gels [4].

The relative proportions of poly-L-guluronate and poly-D-mannuronate sequences in alginate molecules also affect the gelling capabilities of alginates and the types of gels formed. Alginates rich in poly-L-guluronate sequences give firmer, more brittle gels, while alginates with lower levels of poly-L-guluronate sequences give less firm, more elastic gels [16]. Gel structures are also dependent on the manner in which the calcium is introduced into the system (i.e. whether it is added rapidly or slowly).

Another polysaccharide that shows calcium-induced gelation is pectin (from the soft tissue of higher plants; see chapter 10), and the gelling mechanism of pectin is essentially the same as that in alginates (i.e. egg-box junction zones). Pectin is a polyelectrolyte and, like the poly-L-guluronate blocks in alginate, has a $(1 \rightarrow 4)$ diaxially-linked backbone, in this case composed of α-D-galacturonate residues (Figure 6.9). Poly-D-galacturonate is the mirror image of poly-L-guluronate, except in the configuration at C(2), and it is therefore not surprising that pectin shows a similar gelling mechanism to alginate.

Unlike alginate, however, pectin contains only one type of uronic acid and junction zone termination occurs as a result of occasional galacturonate residues being present in the form of the methyl ester. These are uncharged and therefore not capable of binding to the calcium. A second primary structure 'defect', which causes termination of junction zones in pectin, is the occurrence of occasional residues of $(1 \rightarrow 2)$-linked L-rhamnose, which are sterically incompatible with the ordered egg-box structure.

The final type of ordered structure that can give rise to gelling properties is the interchain associations of the extended ribbon structures displayed by $(1 \rightarrow 4)$ diequatorially-linked polysaccharides. Unsubstituted polymer chains with this extended ribbon conformation (illustrated in Figure 6.7), such as cellulose and mannan, are insoluble. However, the strong interchain association in such systems can be limited by substituents on the polymer backbone, resulting in partially or fully soluble systems.

The galactomannan family of energy-reserve polysaccharides from plant seeds has a mannan backbone, which is substituted to varying extents, depending on the polysaccharide, by $(1 \rightarrow 6)$ linked α-D-galactose. Galactomannans with low degrees of galactose substitution show a tendency to gel as a result of the galactose residues interfering with the

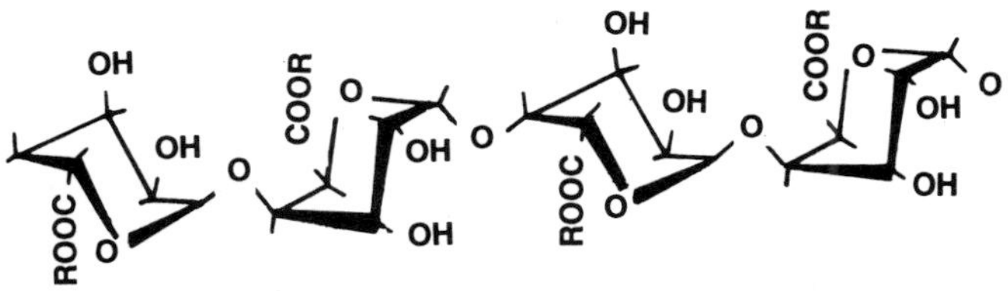

Figure 6.9 Primary repeating structure of pectin, with $(1 \rightarrow 4)$ diaxially-linked α-D-galacturonic acid residues. (R = H or CH_3.)

solid-state packing of the main mannan chain and providing an entropic drive to conformational disorder [4]. In galactomannans with higher ratios of galactose to mannose, the polysaccharide is freely soluble. Similar gel-like properties can be obtained with chemically substituted cellulose derivatives such as carboxymethyl cellulose, depending on the degree of substitution [24].

6.4.3 Weak gels

The final type of rheological behaviour that needs to be discussed, when considering hydrocolloid gels and thickneners, is that of the so-called 'weak gel' systems. These systems essentially possess mechanical spectra similar to those of 'true gels' (i.e. $G' > G''$ at all frequencies and both are essentially frequency-independent) but, in contrast to true gels, weak gels will flow freely under relatively small stresses and are therefore solution-like. This type of behaviour is particularly relevant (and heavily utilised) in products such as pourable dressings and oil-in-water emulsion-based products, where the gel-like properties are strong enough to suspend particulate matter or stabilise oil droplets, while still allowing the system to flow freely. The shear-thinning behaviour of weak gels is different from that found for conventional random-coil thickeners in that there is no evidence of a Newtonian plateau in the viscosity versus shear-rate plot, and the extent of shear thinning is significantly greater, even at high shear rates [4]. Additionally, it appears that weak gels do not obey the Cox–Merz rule, with measured values of steady shear viscosity (η) being appreciably lower than the corresponding values of the dynamic viscosity (η^*) [5,25].

The most notable example of a commercial weak gelling system is the bacterial polysaccharide, xanthan gum, although other not yet commercialised and non-food use bacterial polysaccharides, such as welan and rhamsan, have been found to possess weak-gel properties [14,24].

The weak gel properties of xanthan result from the fact that under most food use conditions, xanthan molecules exist in a rigid, rod-like, conformationally ordered structure, rather than as the disordered random coils, typical of most hydrocolloid thickeners. The primary repeating sequence of xanthan is given in Figure 6.10 and basically consists of a $(1 \rightarrow 4)$ linked β-D-glucose linear backbone, with a charged trisaccharide side-chain attached to every second glucose unit, thus giving a pentasaccharide repeating unit. The trisaccharide side-chain results in the conformationally ordered form of xanthan being a fivefold helix, rather than an extended ribbon-like structure (as might have been expected from the $(1 \rightarrow 4)$ linked β-D-glucose backbone which is basically cellulose). Two helical structures have been proposed that are compatible with the fivefold helix: a co-axial double helical structure [26] and a single-stranded helical

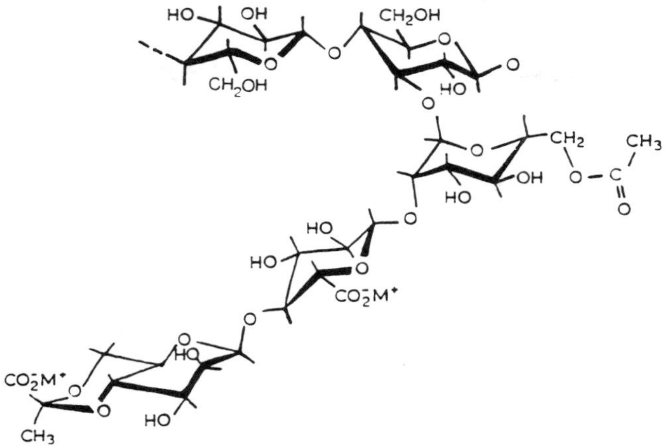

Figure 6.10 Primary repeating structure of xanthan.

structure stabilised by the side-chains packing along the polymer backbone [27].

However, from a rheological point of view, the precise nature of the ordered conformation is not important, and it is the fact that xanthan exists in solution as rigid molecules that gives rise to its rather unique solution and gelling properties. It is generally believed [14] that the weak gel-like properties of xanthan solutions occur as a result of weak side-by-side association of ordered chain sequences from different molecules, to give a tenuous three-dimensional network structure [28]. For 'true' hydrocolloid gels, bonding in the junction zones needs to be relatively strong to counter the considerable loss of conformational entropy as the random coil chain is incorporated into the ordered junction zone, and hence the gels formed are relatively strong. In xanthan, however, no significant loss of conformational entropy occurs in formation of the weak side-by-side associations (as the xanthan molecules are already conformationally ordered), and therefore the weakly associated structure is stable but can easily be broken by application of shear.

The high viscosities at low shear rates and extremely pseudoplastic nature of xanthan solutions are also related to the rigid rod-like structure of xanthan molecules. The rod-like nature of the molecules results in a larger hydrodynamic volume than would be the case for a random-coil polysaccharide of the same molecular weight, and it therefore follows that the c^* transition for xanthan solutions will occur at lower concentrations. Consequently, xanthan solutions, even at low polymer concentrations, can have high viscosities at low shear rates. At higher shear rates, however, the larger shear forces will cause the elongated xanthan molecules to orientate in the applied shear field. Consequently, a greater reduction in viscosity at high shear rates would be expected for xanthan than for a random coil hydrocolloid of the same hydrodynamic volume (i.e. xanthan is more pseudoplastic).

6.5 Conclusion

In chapter 5, the rheological principle of systems having viscous and elastic properties was introduced. In the case of thickeners and gels, the rheological behaviour lies between the two extreme modes of response (perfect solid and perfect liquid), and it is this viscoelasticity displayed by thickeners and gels that is essential in controlling the final texture of many food products. Although each product is different in terms of its components, processing and temperature, etc. various general ideas can be applied governing the behaviour of the molecules which thicken or gel the system. By using rheological measurements, coupled with a knowledge of molecular structure, it is possible for the food technologist to determine which type of structuring ingredient is likely to be most suitable for the particular product.

References

1. Mitchell, J.R. and Ledward, D.A. (Eds) *Functional Properties of Food Macromolecules*, Elsevier Applied Science, London (1986).
2. Harris, P. (Ed.) *Food Gels*, Elsevier Applied Science, London (1990).
3. Glicksman, M. (Ed.) *Gum Technology in the Food Industry*, Academic Press, London (1970).
4. Morris, E.R. Industrial hydrocolloids. In *The Structure, Dynamics and Equilibrium Properties of Colloidal Systems*, Eds Bloor, D.M. and WynJones, E. Kluwer Academic Publishers (1990), 449–470.
5. Morris, E.R. Rheology of hydrocolloids. In *Gums and Stabilisers for the Food Industry 2*, Eds Phillips, G.O. Wedlock, D.J. and Williams, P.A. Pergamon Press, Oxford (1984), 57–78.
6. Huggins, M.L. *J. Am. Chem. Soc.*, **4** (1942), 2716.
7. Kraemer, E.O. *Ind. Eng. Chem.*, **30** (1938), 1200.
8. Flory, P.J. *Principles of Polymer Chemistry*, Cornell University Press, New York (1953).
9. Dickenson, E. *An Introduction to Food Colloids*, Oxford University Press, New York (1992).
10. Launay, B., Doublier, J.L. and Cuvelier, G. Flow properties of aqueous solutions and dispersions of polysaccharides. In *Functional Properties of Food Macromolecules*, Eds Mitchell, J.R. and Ledward, D.A., Elsevier Applied Science, London (1986).
11. Morris, E.R., Cutler, A.N., Ross-Murphy, S.N., Rees, D.A. and Rice, J. *Carbohydrate Polymers*, **1** (1981), 5–21.
12. Morris, E.R. Mixed polymer gels. In *Food Gels*, Ed. Harris, P. Elsevier Applied Science, London (1990).
13. Cox, W.P. and Merz, E.H. Correlation of dynamic and steady flow viscosities. *J. Polymer Sci.*, **28** (1958), 619–622.

14. Morris, E.R. *Pourable Gels*, IFI NR., **1** (1991), 32–37.
15. Rees, D.A., Morris, E.R., Thom, D. and Madden, J.K. Shapes and interactions of carbohydrate chains. In *The Polysaccharides*, Vol. 1, Ed. Aspinall, G.O. Academic Press, New York (1982), 195–290.
16. Morris, V.J. Gelation of polysaccharides. In *Functional Properties of Food Macromolecules*, Eds Mitchell, J.R. and Ledward, D.A. Elsevier Applied Science, London (1986), 121–170.
17. Ward, A.G. and Courts, A. (Eds) *The Science and Technology of Gelatin*, Academic Press, London (1977).
18. Veis, A. *The Macromolecular Chemistry of Gelatin*, Academic Press, London (1964).
19. Busnel, J.P., Morris, E.R. and Ross-Murphy, S.B. Interpretation of the renaturation kinetics of gelatin solutions. *Int. J. Biol. Macromol.*, **11** (1989), 119–125.
20. Busnel, J.P., Clegg, S.M. and Morris, E.R. Melting behaviour of gelatin gels: origin and control. In *Gums and Stabilisers for the Food Industry 4*, Eds Phillips, G.O., Wedlock, D.J. and Williams, P.A. IRL Press, Oxford (1988), 105–115.
21. Bächinger, H.P., Bruckner, P., Timpl, R. and Engel, J. The role of *cis–trans* isomerisation of peptide bonds in the coil–triple helix conversion of collagen. *Eur. J. Biochem.*, **90** (1978), 605–614.
22. Ring, S.G. and Stainsby, G. *Prog. Food Nutr. Sci.*, **6** (1982), 323.
23. Rees, D.A. Polysaccharide gels: a molecular view. *Chem. Ind.* (1972), 630–636.
24. Robinson, G., Manning, C.E., Morris, E.R. and Dea, I.C.M. Sidechain–mainchain interactions in bacterial polysaccharides. In *Gums and Stabilisers for the Food Industry 4*, Eds Phillips, G.O., Wedlock, D.J. and Williams, P.A. IRL Press, Oxford (1988), 173–181.
25. Ross-Murphy, S.B. Rheological methods. In *Biophysical Methods in Food Research*, Ed. Chan, H.W.-S. Blackwell Scientific, Oxford (1984), 138–199.
26. Okuyama, K., Arnott, S., Moorhouse, R., Walkinshaw, M.D., Atkins, E.D.T. and Wolf-Ullish, Ch. Fibre diffraction studies of bacterial polysaccharides. In *Fibre Diffraction Methods*, ACS Symposium Series, **141** (1980), 411.
27. Moorhouse, R., Walkinshaw, M.D. and Arnott, S. Xanthan gum molecular conformation and interactions. In *Extracellular Microbial Polysaccharides*, ACS Symposium Series, **45** (1977), 90.
28. Norton, I.T., Goodall, D.M., Frangou, S.A., Morris, E.R. and Rees, D.A. Mechanism and dynamics of conformational ordering in xanthan polysaccharide. *J. Mol. Biol.*, **175** (1984), 371.

7 Fat eutectics and crystallisation
G. TALBOT

7.1 Glossary of useful terms

cis formation

The C=C double bonds in unsaturated fatty acid chains exist in two conformations – the *cis* formation and the *trans* formation (q.v.). In the *cis* formation, the hydrogen atoms associated with the unsaturated carbon atoms lie on the same side of the double bond, i.e.

$$\underset{R}{H}\diagdown C=C \diagup\underset{R'}{H}$$

cis form

Eutectic

When two dissimilar materials are mixed together they can interact in such a way that the melting point of the blend is lower than the melting points of the individual components. This is a eutectic. One of the most commonly encountered eutectics is when salt is sprinkled onto ice. A eutectic is formed with a melting point below that of pure water and the ice will therefore melt.

Fractionation

Naturally occurring oils and fats are mixtures of triglycerides (q.v.) with different melting points. In many fats, some of these triglycerides are solid at room temperature whilst others are liquid. It is possible to separate the liquid triglycerides from the solid triglycerides by a process known as 'fractionation'. Two types of fractionation are commonly employed – dry fractionation and solvent fractionation. In the dry fractionation process, the fat is held at a temperature at which it is partially liquid – the liquid and solid triglycerides are separated by pressing or by filtration. In the solvent fractionation, the fat is dissolved in a solvent – usually acetone or hexane – and the higher melting triglycerides are allowed to crystallise from solution before being separated by a filtration process. The separation efficiency is better with solvent fractionation than with dry fractionation. The melting profiles of the fractions which are produced are significantly different from those of the starting fat. Palm oil and palm kernel oil are examples of fats which are commonly subjected to fractionation.

Free fatty acid

Triglycerides can undergo hydrolysis (chapter 9) with water, particularly when in the presence of lipase, resulting in the formation of partial glycerides (diglycerides and monoglycerides) together with free fatty acids.

Hydrogenation

Hydrogenation (also known as 'hardening') is a process in which oils and fats are reacted with hydrogen in the presence of a catalyst (usually nickel). During hydrogenation, two competing reactions can take place – saturation in which a hydrogen molecule is added across a double bond of an unsaturated fatty acid group producing a single saturated C—C bond, and isomerisation in which the naturally occurring *cis* double bonds are converted into *trans* double bonds. Both the resulting saturated fatty acids and the *trans* unsaturated fatty acids have a higher melting point than the initial *cis* unsaturated fatty acids – hence the alternative term, 'hardening'.

Isomer

Isomers are alternative distributions of the atoms or functional groups in a molecule. Examples of isomers are *cis* fatty acids and the corresponding *trans* fatty acids. In one case, the hydrogen atoms associated with the C=C double bonds are on the same side of the double bond, in the other they are on opposite sides. However, the molecular weight and the number of carbon, hydrogen and oxygen atoms in the molecule are the same in both cases. Another example of isomerisation is the positional isomerisation which can be found in triglycerides (q.v.). There are three positions for fatty acids on a triglyceride molecule, known simply as the 1-, 2- and 3-position. If a triglyceride, for example, has two stearic acid groups and one oleic acid group, these could be distributed as either 1,2-distearyl-3-oleyl glycerol or 1,3-distearyl-2-oleyl glycerol. These are known as positional isomers.

Polymorphism

Molecules can often pack in a crystal lattice in a number of different ways. This ability to exist in a number of different crystal forms is known as 'polymorphism'. In fats, the most well-known example is shown by cocoa butter which can exist in six different polymorphic forms, i.e. six different forms of crystal packing.

Saturated fat

Hydrocarbon chains which contain no double bonds are said to be 'saturated'. Fats are made up of triglycerides which in turn are esters of glycerol and fatty acids. If the hydrocarbon chain part of the fatty acids contains no double bonds, then the fat is said to be a 'saturated fat'. Saturated fats or saturated triglycerides are commonly found in oils such as palm kernel oil and coconut oil.

Trans *formation*

The C=C double bonds in unsaturated fatty acid chains exist in two conformations – the *cis* formation (q.v.) and the *trans* formation. In the *trans* formation, the hydrogen atoms associated with the unsaturated carbon atoms lie on opposite sides of the double bond, i.e.

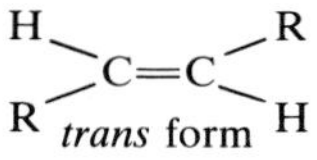

$$\underset{trans \text{ form}}{\overset{H}{\underset{R}{\diagdown}}C=C\overset{R}{\underset{H}{\diagup}}}$$

Triglyceride

Triglycerides are the main chemical species in oils and fats. They are triesters of three fatty acids with the trihydric alcohol, glycerol (see Figure 7.1).

Unsaturated fats

In unsaturated fats, the hydrocarbon chain part of the fatty acids contains one or more double bonds. If there is predominantly one double bond in the fatty acids, then the fat is said to be 'monounsaturated' – olive oil is

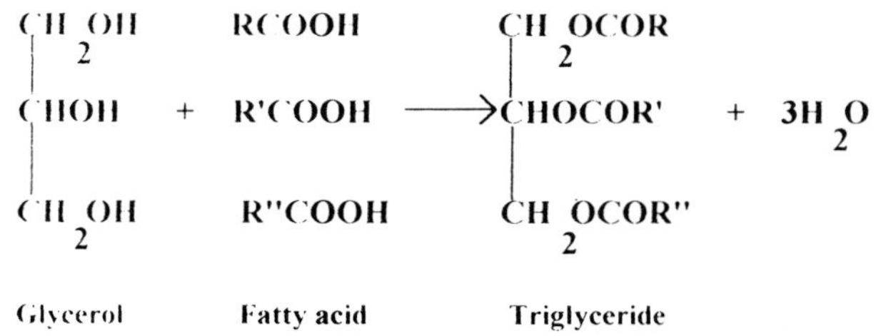

$$\begin{array}{c} CH_2OH \\ | \\ CHOH \\ | \\ CH_2OH \end{array} + \begin{array}{c} RCOOH \\ \\ R'COOH \\ \\ R''COOH \end{array} \longrightarrow \begin{array}{c} CH_2OCOR \\ | \\ CHOCOR' \\ | \\ CH_2OCOR'' \end{array} + 3H_2O$$

Glycerol   Fatty acid   Triglyceride

Figure 7.1 Formation of triglycerides. (Reprinted with permission of Loders Croklaan.)

an oil rich in monounsaturated fatty acids. If there is predominantly more than one double bond in the fatty acids, then the fat is said to be 'polyunsaturated' – sunflower oil and soyabean oil are oils rich in polyunsaturated fatty acids.

7.2 Introduction

Fats form a part of almost all fabricated foods and the physical properties of the fat often play an integral part in the production and, in many cases, the consumption of the food. The crystallisation characteristics of the fat phase are of major importance to manufacturers of chocolate, margarine, spreads, coffee whiteners and ice cream, to give but a few examples.

As foods become increasingly complex and multiphase in their construction, so the number of fats used in a given foodstuff increases. The interactions of these fats (which are often considerably different in their melting and crystallisation behaviour) are also of major importance to both manufacturer and consumer.

This chapter considers these aspects from a practical viewpoint. A large amount of fundamental research has, however, been carried out over the years on the interactions of triglycerides of varying degrees of purity. Some of this research has considerable practical significance in that the combinations of triglycerides can be related to 'real-life' fat systems. Other studies are of much more academic interest. This chapter will concentrate almost exclusively on those fat, or triglyceride systems, which have some further relevance to food manufacture or consumption.

7.3 Triglyceride structure

Fats and oils are mixtures of a number of chemical species, which together can loosely be called 'lipids'. These include phosphoglycerides (such as lecithin), partial glycerides (mono- and diglycerides) and free fatty acids, but, in most natural oils and fats, the species present in by far the greatest amounts are triglycerides.

Triglycerides are esters formed by the combination of glycerol and fatty carboxylic acids (Figure 7.1). In most cases, these fatty acids are straight carbon—carbon chains having an even number of carbon atoms, ranging from as few as four carbon atoms (i.e. butyric acid) in butterfat up to $\geqslant 26$ carbon atoms in some fish oils. The common fatty acids found in oils and fats together with their melting points are shown in Table 7.1. It can be seen from this list of acids and their melting points that:

Table 7.1 Fatty acids commonly found in oils and fats

Chain length: double bonds	Systematic name	Trivial name	Melting point	
			°C	°F
10:0	Decanoic	Capric	31.6	88.9
12:0	Dodecanoic	Lauric	44.8	112.6
14:0	Tetradecanoic	Myristic	54.4	129.9
16:0	Hexadecanoic	Palmitic	62.9	145.2
18:0	Octadecanoic	Stearic	70.1	158.2
18:1	Octadec-*cis*-9-enoic	Oleic	16	60.8
18:1	Octadec-*trans*-9-enoic	Elaidic	44	111.2
18:2	Octadec-*cis*-9,*cis*-12-dienoic	Linoleic	−6.5	20.3
18:3	Octadec-*cis*-9,*cis*-12,*cis*-15-trienoic	Linolenic	−12.8	9.0
20:0	Icosanoic	Arachidic	76.1	169.0

Table 7.2 Melting points of monoacid triglycerides [1]

Triglyceride	Stable (β-form) melting point	
	°C	°F
Trilaurin (LLL)	46	114.8
Trimyristin (MMM)	56	132.8
Tripalmitin (PPP)	66	150.8
Tristearin (StStSt)	73	163.4
Triolein (OOO)	5	41.0
Trielaidin (EEE)	41	105.8
Trilinolein (LiLiLi)	−11	12.2

- as the chain length increases so does the melting point;
- the greater the degree of unsaturation (i.e. the more double bonds there are in the fatty acid), the lower the melting point;
- *trans* unsaturated fatty acids have a higher melting point than the corresponding *cis* unsaturated fatty acids.

In many ways, the characteristics of the fatty acids define the properties of the triglycerides of which they are a part. However, because triglycerides contain three fatty acid groups, the physical properties are more complex than those of the more simple fatty acids.

A triglyceride molecule can have all three of its fatty acids the same (monoacid triglycerides), or two can be the same and the third different (diacid triglycerides), or all three can be different (triacid triglycerides).

Monoacid triglycerides are obviously the simplest in the sense that their melting behaviour basically follows the 'rules' of the fatty acids (see Table 7.2). Diacid triglycerides are more complex and can exist as two distinct isomers: RRR' and RR'R (where R and R' are the two fatty acid chains).

Triacid triglycerides can then obviously exist as an even greater number of isomers:

(a) RR'R" (b) R'RR" (c) R"RR'
(d) RR"R' (e) R'R"R (f) R"R'R

These forms show that chirality or optical isomerism can exist in triglycerides, i.e. forms (a) and (f), (b) and (d), (c) and (e) are pairs of optical isomers. Indeed the asymmetrical diacid triglyceride RRR' can also exist as a pair of optical isomers (RRR' and R'RR).

7.4 Molecular packing of triglycerides

The basic structure of a triglyceride molecule can be likened to a chair and these chair structures can pack together in a crystal lattice in two basic ways, i.e. double-chain length and triple-chain length packing (Figure 7.2).

Triple-chain length structures are found in the symmetrical monounsaturated triglycerides (POP, POSt, StOSt – see section 7.5.1) found in the more stable polymorphic forms of cocoa butter (the fat produced by pressing the cotyledons in cocoa beans). They also exist in saturated triglycerides in which the chain lengths of the fatty acids differ by four carbon atoms or more.

The crystalline structure of fats is complicated further by the phenomenon of polymorphism, i.e. the ability of a fat to crystallise in a number of forms with different types of molecular packing and different thermodynamic stabilities. Four basic polymorphic forms exist in fats – known as sub-α, α, β' and β – although some fats, notably cocoa butter, exhibit more than four polymorphs. The extra structures are, however, generally variants of the four basic forms. The sub-α form is the most transient and least stable of these forms and, in practical applications, the remaining three forms are the most important. The various forms can be distinguished by X-ray diffraction and this method is also used to determine whether the fat crystals are packed in double- or triple-chain lengths

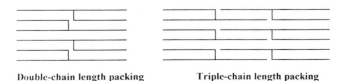

Figure 7.2 Double- and triple-chain length packing configuration. (Reprinted with permission of Loders Croklaan.)

(Figure 7.3). The polymorphic form of a fat is found from its X-ray short spacings (S) whilst the type of chain packing is determined by its X-ray long spacings (L). This is approximately 50% longer for triple-chain packed systems compared with double-chain packed systems. The X-ray short spacings which define the various polymorphic forms are listed in Table 7.3.

The basic structures of the three main polymorphs are shown in Figure 7.4. Triglycerides in the α-form also have their fatty acid chains parallel with each other and perpendicular to the end plane, but, relative to each other, the chains are less ordered than the sub-α form (see below) and when viewed end on, have a more hexagonal configuration, similar to a clump of pencils.

The β' form exists in the form of an $O_\perp$ sub-cell, but, unlike the sub-α form (see below), the fatty acid chains are no longer perpendicular to the end plane, but are inclined.

In the β form, the fatty acid chains are inclined at an angle to the end plane, but differ from the β' form in that all the fatty acid chains are parallel to each other forming a triclinic ($T_\parallel$) sub-cell.

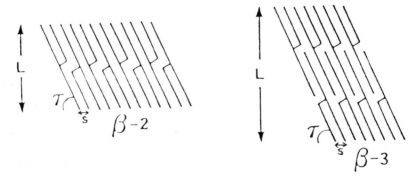

Figure 7.3 X-ray diffraction – long and short spacing. (Reprinted with permission of Loders Croklaan.)

Table 7.3 Assignment of polymorphs [7]

Polymorph	X-ray short spacing characteristics
α	A single, strong, short spacing at about 0.415 nm
β'	Usually two, strong, short spacings at about 0.38 nm and 0.42 nm or three, strong, short spacings at about 0.427 nm, 0.397 nm and 0.371 nm
β	A form which does not satisfy the criteria for α and β', but also usually shows a very strong short spacing at about 0.46 nm
sub-α	A β' form usually melting below an α form

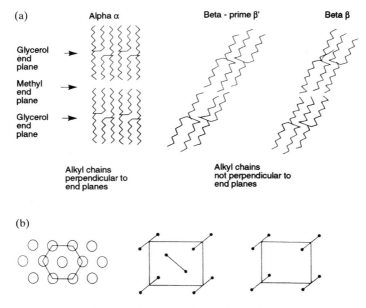

Figure 7.4 Structure of main polymorphic forms. (a) Projection showing arrangement of alkyl chains for α, β and β' polymorphs. (b) Projection parallel to direction of alkyl chain (i.e. arrangement looking onto ends of chains). (Reprinted with permission of Loders Croklaan.)

In the sub-α form (not shown in Figure 7.4), the fatty acid chains are perpendicular to the planes defining layers of glycerides whilst, looking end on to the glyceride chains they form an orthorhombic sub-cell in which the fatty acid chains of adjacent triglycerides are mutually perpendicular (the $O_\perp$ sub-cell). In this respect, this is similar to the β' configuration. Indeed, some workers have considered the sub-α form to be a variant of the β' form.

It has already been mentioned that cocoa butter can exist in six different polymorphic forms. These include two β' and two β forms. The most stable form of each pair is given the suffix 1 and the least stable the suffix 2, i.e. β'_2 has a lower stability than β'_1.

7.5 Composition and structure of natural fats

The remainder of this chapter will concentrate of natural fats commonly used in foodstuffs and will firstly relate their triglyceride composition to their polymorphism and crystallisation characteristics and then discuss the interactions which occur when fats are mixed together.

7.5.1 Cocoa butter

The extensive polymorphism of cocoa butter has already been referred to. Wille and Lutton [2] defined the six polymorphic forms as Forms I to VI and this nomenclature is often used by chocolate and confectionery technologists. The relationship between these six forms and the nomenclature, used earlier in this chapter for polymorphic forms, is shown in Table 7.4 where Wille and Lutton's Forms I to VI are compared with those defined by Larsson [3].

The thermal characteristics of Forms I to V have been determined by DSC (differential scanning calorimetry) heating curves [4].

Form I. This was produced by heating cocoa butter at 100°C (212°F) for 2 minutes before quenching to 0°C (32°F). The fat was immediately heated at 5°C/min (9°F/min) to 40°C (104°F).

Form II. This was produced in much the same way as Form I except that the fat was held at 0°C (32°F) for 15 minutes before measuring the DSC heating curve (5°C/min to 40°C; 9°F/min to 104°F) to allow transformation from Form I.

Form III. This was produced by quenching from 100°C to 5°C (212°F to 41°F) and then holding at 5°C (41°F) for 16 hours. After holding at 0°C (32°F) for a further 5 minutes, the DSC heating curve was measured.

Form IV. This was produced by quenching from 100°C (212°F) to 16°C (60.8°F), holding there for 1 hour and then to 10°C (50°F) for 15 minutes before measuring the DSC curve. This is akin to taking a liquid cocoa butter and putting it through a normal chocolate cooling regime without prior tempering (see chapter 17).

Form V. This was produced by incubating at 20°C (68°F) for 11 days. After holding at 18°C (64.4°F) for 5 minutes, the DSC curve was measured. Properly tempered chocolate produces Form V crystals in the cocoa butter.

Form VI. This needs several months at 20°C (68°F) before it forms.

Table 7.4 Polymorphic forms of cocoa butter

Wille and Lutton [2]	Larsson [3]*	Melting point		Chain packing
		°C	°F	
Form I	β'_2	16–18	61–67	Double
Form II	α	21–22	70–72	Double
Form III	Mixed	25.5	78	Double
Form IV	β'_1	27–29	81–84	Double
Form V	β_2	34–35	93–95	Triple
Form VI	β_1	36	97	Triple

*Suffixes 1 and 2 are used to indicate highest and lowest melting form with similar crystal packing.

FAT EUTECTICS AND CRYSTALLISATION   151

The DSC heating curves of Forms I to V are shown in Figure 7.5. As the stability of the cocoa butter increases (from Form I to Form V) so the melting point of the fat increases. The maximum melting points as measured by DSC heating curves increase from 19.75°C (67.55°F) (Form I) to 33.37°C (92.06°F) (Form V). Other observations have shown that a polymorphic change from Form V to Form VI increases the melting point

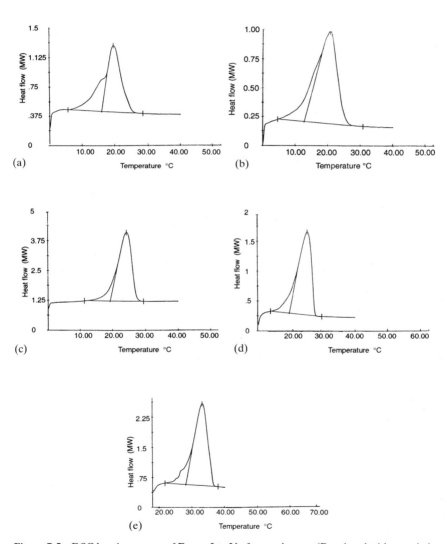

Figure 7.5 DSC heating curves of Forms I to V of cocoa butter. (Reprinted with permission of Loders Croklaan.)

by approximately 2°C (4°F). This polymorphic change and further increase in melting point is often the cause of what is considered to be 'staling' in chocolate which has been stored for a few months. This polymorphic change from Form V to Form VI also results in a change in the crystal structure of cocoa butter which can result in crystals of fat forming on the surface of the chocolate. These have the appearance of mould and are known as 'chocolate bloom'.

Cocoa butter, unlike many other naturally occurring fats, has a relatively simple triglyceride composition. In many cocoa butters, three triglycerides comprise well over 80% of the total. These triglycerides are POP, POSt and StOSt (P = palmitic, St = stearic, O = oleic). These are physically and chemically very similar and might be expected to interact in a predictable, linear fashion. Andersson [5], however, calculated the ternary phase diagram of POP–POSt–StOSt mixtures (Figure 7.6) and showed that there are non-linear interactions between the three triglycerides. The shaded areas show regions where the melting points are reduced below the melting points of the individual triglycerides. At a composition of about 10% StOSt, 40% POSt and 50% POP, a eutectic (*see glossary*) is formed with a melting point of 33.8°C (92.8°F). The ratio of POP:POSt:StOSt found in cocoa butter is shown as point K in Figure 7.6. There are other soft triglycerides in cocoa butter which bring the melting point down below this figure.

Thus, even very similar triglycerides interact in such a way that eutectics

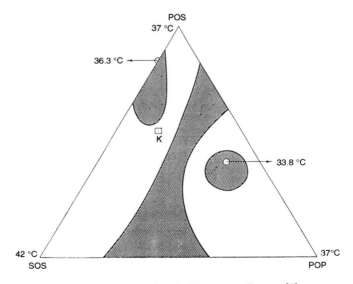

Figure 7.6 POP–POSt–StOSt ternary diagram [5].

are formed. Other triglycerides found in cocoa butter, albeit in smaller quantities, can also have a drastic effect, particularly on crystallisation. When chocolate is processed it is tempered to ensure that the cocoa butter crystallises in a stable polymorphic form (Form V). The major triglycerides in cocoa butter (POP, POSt, StOSt), collectively known as 'SOS' (S = saturated), are usually the highest melting triglycerides and hence crystallise sufficiently during the tempering process to 'seed' the remaining cocoa butter in the same stable polymorphic form. If, however, the cocoa butter also contains some trisaturated triglycerides, then these, being higher melting, will crystallise before the SOS triglycerides. Because these 'extra' crystals effectively reduce the amount of cocoa butter remaining in the liquid phase, the viscosity of the tempered chocolate is increased [6].

The generation of 'seeds' during crystallisation is an important step, whatever the fat, since the rate determining step in a crystallisation process is often the nucleation stage. Classical tempering of chocolate relies on temperature changes firstly to generate both stable β (Form V) and unstable β' (Form IV) polymorphs by cooling below the β' melting point. This is followed by heating to a temperature between the melting points of the two polymorphic forms in order to leave only stable β (Form V) crystals to 'seed' the bulk of the chocolate. Cocoa butter (and hence chocolate) can, however, also be tempered by the external addition of stable β SOS seeds [21]. Because the amount of seed and the polymorphic integrity of the seed can be better controlled than by the cooling–reheating process, this would then be a more robust method of tempering. Higher melting SOS triglycerides, e.g. BOB (B = behenic), have been used in chocolate as a way of reseeding tempered chocolate products which, because of climatic extremes, have melted on storage [22]. This is claimed to improve the heat stability of the product.

Where an unsaturated fatty acid (e.g. oleic acid) is present in a triglyceride chain, there is an angular change in the carbon—carbon chain at the unsaturated carbon—carbon bond. However, it has been shown by de Jong et al. [8] that oleic acid can exist in a number of conformations, all of which show this angular change at the double bond. One conformation, in particular, has a further rotation around an adjacent carbon—carbon single bond giving an overall fatty acid chain showing much greater linearity than was previously considered possible in oleic acid (Figure 7.7). It has been hypothesised from these data [9] that the oleic acid chain in cocoa butter in Form V has the 'straight-chain' oleic acid and that this transforms slowly into the 'bent-chain' configuration producing Form VI cocoa butter. To prevent this transformation, a triglyceride which crystallises in a straight-chain, triple-packed, β-configuration can be introduced into the cocoa butter crystal lattice. This is the basis of a patented product ('Prestine' from Loders Croklaan) which prevents the

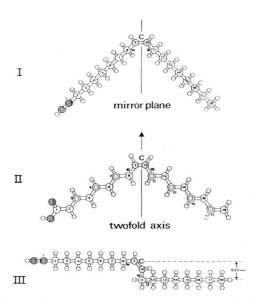

Figure 7.7 Alternative conformations of oleic acid [8].

Form V to Form VI change and hence staling and bloom formation in chocolate [10].

7.5.2 Milk fat

The crystallisation characteristics of milk fat are clearly of importance to the dairy industry in general and butter-making in particular. The phase behaviour has been studied in depth by Timms [11,12] who showed that triglycerides in milk fat can be divided into three broad groups based on their melting points. He called these HMF (high melting), MMF (middle melting) and LMF (low melting) fractions. The DSC heating curve of milk fat shows three distinct peaks which correspond broadly to these three fractions. The low-melting fraction is normally liquid at ambient temperatures and acts as a solvent for the other two fractions. The two solid fractions differ, however, in their polymorphism with HMF being β-stable and MMF being β'-stable. Both fractions exhibit double chain length packing.

7.5.3 Palm oil

Because of its commercial importance in a wide range of foods, palm oil is probably one of the most characterised oils in terms of its triglyceride composition (Table 7.5). If palm oil is cooled rapidly from liquid to

Table 7.5 Triglyceride composition of Sumatran palm oil [11]

Saturated		Mono-unsaturated		Di-unsaturated		Tri(+)unsaturated	
PPP	6.1	MOP	1.3	POO	18.9	OOO	3.2
PPSt	0.9	POP	25.9	StOO	2.6	POLi	2.6
Others	1.5	POSt	3.1	OPO	1.2	PLiO	4.3
		PPO	6.0	PLiP	6.8	Others	8.5
		Others	1.6	PLiSt	1.9		
				Others	3.6		

−15°C (5°F) a sub-α polymorph is produced [12]. Reheating to −10°C to −5°C (14°F to 23°F) results in a transformation to an α form. Further heating to between 5°C and 7°C (41°F and 45°F) gives a β' form. In practice, palm oil is not subjected to the drastic cooling necessary to generate the sub-α form, but the more stable polymorphs are important in, for example, margarine production.

Palm oil can also be fractionated using either dry fractionation or solvent fractionation (*see glossary*) [13]. Usually three fractions are collected (although dry fractionation processes are often used to produce only palm oleine and palm stearine). These fractions result in a concentration of the trisaturated triglycerides (PPP etc.) in the 'top fraction', disaturated triglycerides (POP, POSt, PLiP; PPO Li = linoleic) in the 'mid-fraction' and the more unsaturated triglycerides (POO, OOO etc.) in the 'oleine fraction'. Monoacid saturated triglycerides such as PPP and, indeed, those trisaturated triglycerides where the fatty acid chain lengths differ by two carbon atoms (i.e. PPSt, StPP) have the β_{-2} form as their stable polymorph. Since, in many applications, the oleine fraction is liquid in use, its polymorphism is perhaps less critical, although it has been shown to proceed from sub-α through α to a β' form. The mid-fraction is perhaps the most interesting fraction in terms of crystallisation and polymorphism, partly because of the combinations of triglycerides in that fraction and partly because of its use as a constituent of cocoa butter equivalents (section 7.7.2 [14]). It has been shown that palm mid-fraction produced by acetone fractionation of palm oil normally crystallises in the β'_{-2} form.

The major triglycerides in palm mid-fraction are POP and POSt with smaller but significant levels of PPO, PPP and PLiP. Rossell, in his review of phase diagrams of triglyceride systems [15], considers the phase diagrams of binary mixtures of triglycerides such as these. Of importance in the context of palm mid-fraction are POP–PPP, POP–POSt and POP–PPO. As is often the case where triglycerides of widely different melting points are studied, the POP–PPP system is essentially monotectic. The POP–POSt binary system shows a eutectic composition at about 30% POP and Rossell [15] also suggests the probability that a two-phase region exists between 15% and 45% POP. This region is removed from the POP–POSt composition found in palm mid-fraction. Finally the POP–PPO system has been extensively studied by Moran [16], Rossell [17] and Timms and Bessell [18]. Moran [16] showed a eutectic composition at 15–22% POP. Timms and Bessell [18] showed that compound formation occurred with POP–PPO blends. POP in the β'_{-2} form and PPO in the β'_{-3}, form produced a polymorphically stable compound in the β_{-2} form. From this and other similar binary studies, Timms formulated conditions under which compound formation occurred [7]:

- There should be a mixture of symmetrical and asymmetrical mono-unsaturated triglycerides.
- The symmetrical triglyceride should contain only one sort of saturated acid.
- The asymmetrical triglyceride may contain two different saturated acids provided that the acid common to both glycerides forming the compound is at the 2-position.

When these conditions are fulfilled, the fatty acid chains pack to give three identical acids side by side.

7.5.4 Palm kernel and coconut oils

Both palm kernel oil and coconut oil are rich in lauric acid (approximately 50% in both cases) and the predominant triglyceride is trilaurin. Trilaurin is one of the few triglycerides on which single-crystal X-ray studies have been carried out [19]. These studies showed that pure trilaurin crystallises in the β form. X-ray studies on palm kernel oil and coconut oil products, however, have shown these to be β'_{-2}-stable. It is also known that there is a relationship between crystallisation rate and fatty acid chain length – the shorter the chain length, the faster the crystallisation rate [13]. Thus lauric fats, such as these, will crystallise more rapidly than their non-lauric counterparts. Since fast crystallisation also gives smaller crystals, the end result is a fat containing small β' crystals. This makes them ideal for use in analogues of whipped dairy products, such as toppings and non-dairy creams, where these types of crystals provide the greatest surface area to stabilise the whipped system.

7.5.5 Animal fats

The main animal fat used in food products is probably butterfat (milk fat); this has already been considered in section 7.5.2. Carcass fats such as tallow and lard are less widespread in their use, especially with consumer trends towards vegetable fats on the grounds of healthier nutrition. Their polymorphism is quite complex with beef tallow being mainly β'_{-2} with a small amount of β (possibly from POP–PPO compound formation). At ambient temperatures, lard is composed of both β and β' crystals, the latter probably from the StPO triglyceride. When heated to 31°C (87.8°F), lard transforms into a pure β phase [20].

7.5.6 Hydrogenated fats

Fats can either be hydrogenated to completion, in which all the unsaturated bonds are converted to saturated bonds, and hence only trisaturated triglycerides are present, or they can be partially hydrogenated, in which case some of the *cis* unsaturated bonds will also be converted to *trans* unsaturated bonds. Apart from a few specialised applications (stabilisers, hardstocks for spreads and margarines) the only fats which are extensively used in a fully hydrogenated form are palm kernel oil and coconut oil. This is simply because longer chain oils (palm oil, soyabean oil, etc.), when fully hydrogenated, have a melting point which is greatly in excess of mouth temperature. The crystallisation characteristics of fully hydrogenated lauric oils are essentially the same as those already referred to in section 7.5.4, i.e. their stable polymorphic form is the $β'_2$ form.

The presence of *trans*-fatty acids, such as elaidic acid in partially hydrogenated fats, does tend to complicate their polymophism, but again they have a tendency to crystallise in the $β'_{-2}$ form.

7.6 Mixtures of fats and eutectic effects

Mixtures of fats in food products are usually there deliberately, because they impart some functionality to the product, but they can also occur as a result of fat migrating from one part of the product to another. In the increasingly complex multiphase structure of foodstuffs of today, this 'migration mixing' is becoming more of a problem.

This mixing of fats can often result in the formation of a eutectic composition. The simplest definition of a eutectic composition is one which has a melting point below that of the components making up the mixture. Reference has already been made to eutectics occurring in binary mixtures of triglycerides and there are many more examples where blending of fats can result in melting points below those of the constituent fats. It has also been shown already that different fats have differing polymorphic stabilities and blending of these, whether deliberate or because of migration, can then result in polymorphic instabilities, phase separation, etc.

Although all fats exhibit polymorphism to some extent, particularly when shock-cooled to low temperatures and then allowed to warm up, it is fats like cocoa butter and the constituent fats of cocoa butter equivalents (palm oil, shea butter, illipe) which have been considered to be 'polymorphic'. These all have the property of being β-stable. Other fats (lauric fats, hydrogenated fats, etc.) are much easier to crystallise into a stable polymorphic form, and in fact do so to such an extent that they have been labelled 'non-polymorphic' but are more accurately called β'-stable. Of this latter group, fats based on lauric oils (palm kernel and coconut) and fats resulting from hydrogenation of liquid oils (palm, soyabean, rapeseed, etc.) also behave quite differently and, despite their polymorphic similarities, can also give crystal structure 'problems' when mixed together.

Thus, in very simple terms, we can divide fats into three main groups:

1. β-stable fats which need tempering (cocoa butter, cocoa butter equivalents).
2. β'-stable, non-lauric fats (hydrogenated palm, soyabean, rapeseed oils).
3. β'-stable lauric fats (palm kernel, coconut oils).

In the following section we will consider the interactions between fats from these three groups when they are blended with each other.

7.7 Blends with β-stable fats

Since the most common 'polymorphic' fat is cocoa butter and, indeed, most work has been done with cocoa butter, this section will consider the effects of blending other fat systems with cocoa butter.

7.7.1 Cocoa butter–milk fat

This is the basis of milk chocolate and, as a consequence, has been extensively studied. The melting profiles of blends of cocoa butter and milk fat are shown in the form of an iso-solids diagram in Figure 7.8. This type of diagram joins points on a composition–temperature diagram with equal solid fat contents. If the lines are parallel and approximately horizontal, then the two components exhibit good compatibility. It is quite obvious from Figure 7.8 that this is not the case with cocoa butter–milk fat blends. Even small amounts of milk fat soften cocoa butter quite dramatically. For example, the solid fat content at 20°C (68°F) of cocoa butter is about 75%. When 20% milk fat is added to the cocoa butter, the solid fat content drops to <60%. Many milk chocolates contain this amount of milk fat (calculated on the fat phase). The reasons for this sharp reduction in solid fat content are two-fold:

- milk fat is in itself a softer fat than cocoa butter and therefore addition (as with any soft fat) will reduce the amount of solid fat present, and

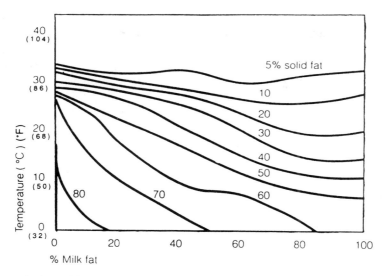

Figure 7.8 Iso-solids phase diagram of cocoa butter–milk fat. (Reprinted with permission of Loders Croklaan.)

- the triglycerides in milk fat exhibit a different crystal structure to that found with cocoa butter.

7.7.2 Cocoa butter–cocoa butter equivalent

Cocoa butter equivalents (CBEs) are commercially available vegetable fats which contain the same basic triglycerides – POP, POSt and StOSt – as are present in cocoa butter. The iso-solids phase diagram for blends of cocoa butter with a typical CBE, Coberine, is shown in Figure 7.9 (from [23]). In this case, the lines are generally parallel and horizontal showing that the two fats are fully compatible with each other. Because CBEs contain the same triglycerides as cocoa butter, they also exhibit the same polymorphism and crystallisation behaviour. Thus, all combinations of cocoa butter and Coberine crystallise in a β_{-3} (Form V) crystal form and no eutectic is observed.

7.7.3 Cocoa butter–lauric fats

Substitute chocolate can be produced from lauric fats, particularly hydrogenated and/or fractionated palm kernel oil. In order to make this look and taste like chocolate, cocoa powder (ground cocoa beans with the majority of the cocoa butter removed) is used in the composition. This also contributes some cocoa butter to the total coating and so the inter-

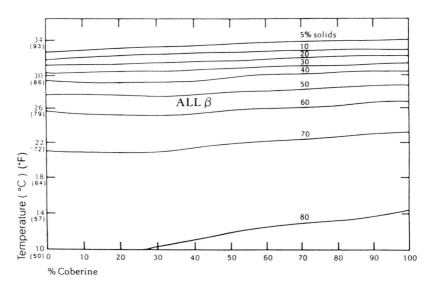

Figure 7.9 Iso-solids phase diagram of cocoa butter–Coberine (CBE). (Reprinted with permission of Loders Croklaan.)

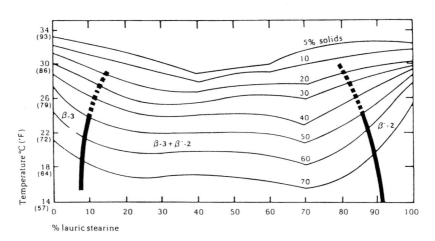

Figure 7.10 Iso-solids phase diagram of cocoa butter–lauric fat. (Reprinted with permission of Loders Croklaan.)

action between cocoa butter and lauric fats becomes important. The iso-solids phase diagram of this system is shown in Figure 7.10 (from [23]). It has already been noted that cocoa butter crystallises in a stable β_{-3} structure and lauric fats in a stable β'_{-2} structure. When either of these fats predominates in the composition, the crystal structure will be dictated by the predominant fat. However, when ca. <95% of the predominant fat is present, a mixed-crystal system is observed, which not only shows crystal instability but also considerable softening.

7.7.4 Cocoa butter–hydrogenated fats

Substitute chocolate coatings can also be produced from hydrogenated non-lauric oils (e.g. palm, soyabean, cottonseed). The iso-solids phase diagram of this system is shown in Figure 7.11 (from [23]). There are some similarities with the cocoa butter/lauric fat system in that:

- Both the lauric fat and the hydrogenated fat have a stable β'_{-2} crystal structure.
- There is an unstable mixed-crystal structure when neither fat is predominant.
- Softening is observed when one fat is added to the other, although this is not as great as with the lauric fat system.

The main difference between the two systems is that, with the cocoa butter–hydrogenated fat system, the hydrogenated fat will allow a much greater inclusion of cocoa butter before it moves away from stable β'_{-2} structure into a mixed β_{-3}–β'_{-2} system. The effect of this, in practice, is to allow a greater inclusion of cocoa butter into these types of substitute coating.

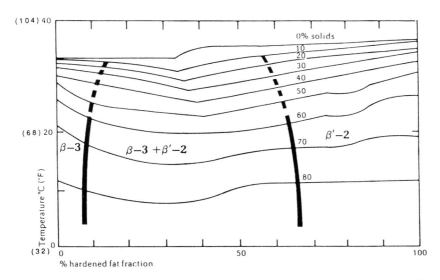

Figure 7.11 Iso-solids phase diagram of cocoa butter–hydrogenated non-lauric fat. (Reprinted with permission of Loders Croklaan.)

7.8 Migration of fats in composite products (see also chapter 21)

Many products consist of one phase containing a hard fat adjacent to a second phase containing a softer fat. One of the problems associated with products like these is that the softer, more mobile, fat can migrate into the harder fat. Thus, the interactions between the two fats can be of importance in determining, at a practical level, the shelf-life and consumer acceptance of the product, and at a more fundamental level, the crystal structure and melting profile of the product.

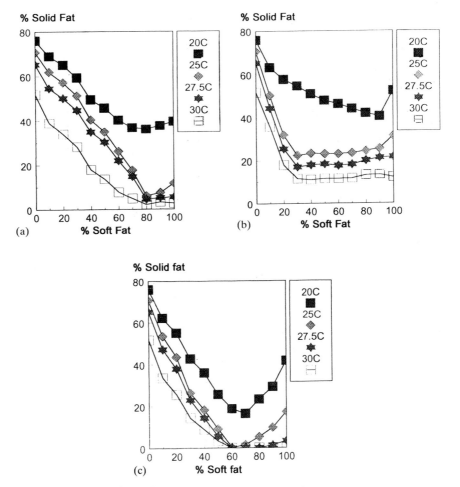

Figure 7.12 Interactions of cocoa butter with soft fats. (Reprinted with permission of Loders Croklaan.) (a) Soft β-stable fat. (b) Soft β'-stable non-lauric fat. (c) Soft β'-stable lauric fat.

FAT EUTECTICS AND CRYSTALLISATION 163

Distinction has already been made between β-stable, cocoa butter-like fats, β′-stable lauric fats and β′-stable non-lauric fats. Using that distinction, the practical interactions between all possible pairs of these three groups will be examined [24].

The three 'hard fats' are (a) cocoa butter, (b) a fractionated lauric fat and (c) a hydrogenated and fractionated non-lauric fat. The three 'soft fats' are (a) a β-stable, i.e. cocoa butter-like soft fat, (b) palm kernel oil and (c) a soft hydrogenated, fractionated non-lauric fat. The interactions

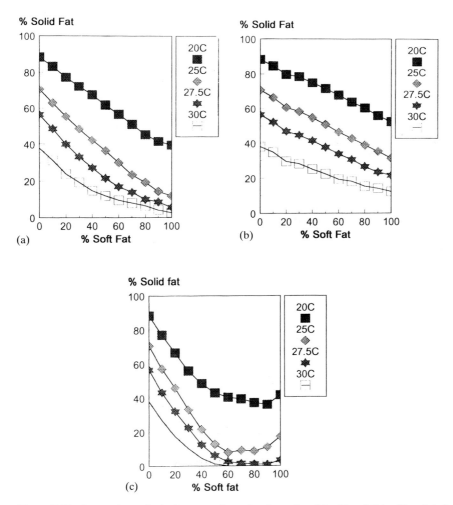

Figure 7.13 Interactions of a hydrogenated non-lauric coating fat with soft fats. (Reprinted with permission of Loders Croklaan.) (a) Soft β-stable fat. (b) Soft β′-stable non-lauric fat. (c) Soft β′-stable lauric fat.

between each of the three soft fats and cocoa butter are shown in Figure 7.12. When cocoa butter is mixed with a softer, but still, β-stable fat, a slight eutectic is observed at about 20% cocoa butter. Between 100% and 20% cocoa butter, the decrease in solid fat content is fairly linear; between 20% and 0% cocoa butter, a slight rise in solid fat contents at 20°C and 25°C (68°F and 77°F) is observed.

The interactions between cocoa butter and a soft hydrogenated non-lauric fat show distinct phase differences. Between 100% and 70% cocoa

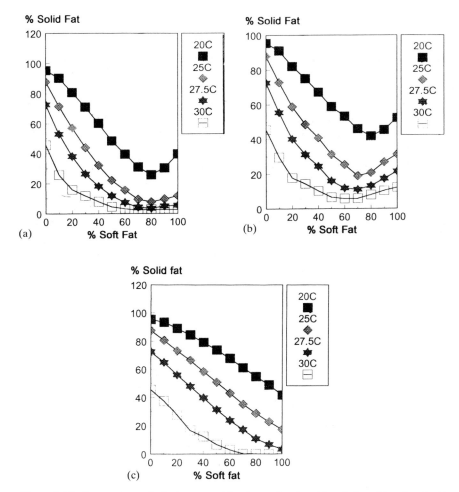

Figure 7.14 Interactions of a hydrogenated lauric coating fat with soft fats. (Reprinted with permission of Loders Croklaan.) (a) Soft β-stable fat. (b) Soft β'-stable non-lauric fat. (c) Soft β'-stable lauric fat.

butter, considerable softening, particularly at 25°C (77°F) and above, is observed as the level of soft fat increases; between 90% and 100% soft fat, an increase in solid fat at 20°C and 25°C (68°F and 77°F) is observed. It is likely, therefore, that phase boundaries exist at about 70% and 10% cocoa butter. Between these phase boundaries the solid fat content, particularly at ⩾25°C (77°F) is relatively unaffected by differences in composition.

The third case, cocoa butter mixed with a soft lauric fat, shows a deep eutectic at 60–70% soft fat and demonstrates the incompatibility between these two fats.

The interactions between a high-solids, hydrogenated, non-lauric coating fat and three softer fats are shown in Figure 7.13. Here we see that when the soft fat is 'polymorphic' and, particularly, when it is also a hydrogenated, non-lauric fat, the interactions between the hard and soft fats are fairly linear, indicating a good degree of compatibility between the fats. The mixtures with the soft lauric fat, however, again show a eutectic composition between 60% and 90% soft fat (although this is not as pronounced a eutectic as that observed with cocoa butter).

When the hard fat is a hydrogenated lauric coating fat, the interactions are as shown in Figure 7.14. These show eutectics with the β-stable fat and with the hydrogenated non-lauric fat, both at about 70–80% soft fat, but relatively linear mixing with the soft lauric fat.

The compatibilities of fats from these three groups can be summarised as shown in Figure 7.15. When fats from the same basic group are mixed together, the compatibility between them is usually very good. Polymorphic (β-stable) fats and hydrogenated non-lauric fats show a limited (denoted as 'fair') compatibility. Lauric fats have a poor compatibility with any fats, apart from other lauric fats.

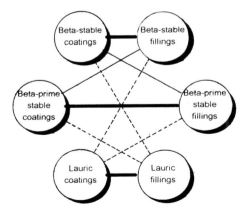

Figure 7.15 Compatibility of fats. (Reprinted with permission of Loders Croklaan.)

7.9 Conclusion

The way in which fats crystallise, particularly the polymorphic changes which can take place in a fat, at what may be critical processing temperatures, is of paramount importance in the processing of foods. Equally, as composite foods become more complex, the interactions between fats in the different parts of the food is also critical. This is especially so if the textural differences, which can be so important for consumer acceptability, are to be maintained.

References

1. Hagemann, J.W. Thermal behaviour and polymorphism of acylglycerides. In *Crystallisation and Polymorphism of Fats and Fatty Acids*, Eds Garti, N. and Sato, K. (1988), Marcel Dekker, New York.
2. Wille, R.L. and Lutton, E.S. *J. Am. Oil Chem. Soc.* **43** (1966), 942.
3. Larsson, K. *Acta Chem. Scand.* **20** (1966), 2255–2260.
4. Peilow, A. Unpublished Unilever results.
5. Andersson, W. *Rev. Intern. Chocolat.* **18** (1963), 49.
6. Cebula, D.J., Dilley, K.M. and Smith, K.W. *Manuf. Conf.* (1991) (May), 131–136.
7. Timms, R.E. *Prog. Lipid Res.* **33** (1984), 1–38.
8. de Jong, S., van Soest, T.C. and van Schaick, M.A. *J. Am. Oil Chem. Soc.* **68** (1991), 371–378.
9. Talbot, G. *International Food Ingredients* (1995) (January–February), 40.
10. Cain, F.W., Hughes, A.D. and Talbot, G. European Patent Application 530864.
11. Jurriens, G. In *Analysis and Characterisation of Oils and Fats and Fat Products*, Ed. Boekenoogen, H.A. (1968), Interscience, London.
12. Persmark, U., Melin, K.A. and Stahl, P.O. *Riv. Ital. Sost. Grasse* **53** (1976), 301–306.
13. Loders Croklaan, *Specialty Fat Technology* (1990).
14. Talbot, G. Vegetable fats. In *Industrial Chocolate Manufacture and Use*, 2nd edn., Ed. Beckett, S.T. (1994), 242–257.
15. Rossell, J.B. *Adv. Lipid Res.* **5** (1967), 353–408.
16. Moran, D.P.J. *J. Appl. Chem.* **13** (1963), 91–100.
17. Rossell, J.B. *Chem Ind. (London)*, **17** (1973), 832–835.
18. Timms, R.E. and Bessell, S. Unpublished Unilever results.
19. Larsson, K. *Ark. Kemi* **23** (1965), 1.
20. Riiner, U. *Lebensm. Wiss. Technol.* **3** (1970), 101–106.
21. Cebula, D.J., Cain, F.W. and Hargreaves, N.G. European Patent Application 521205.
22. Koyano, T., Hachiya, I. and Sato, K. *Food Structure* **9** (1990), 231–240.
23. Gordon, M.H., Padley, F.B. and Timms, R.E. *Fette Seifen Anstrichmittel* **81** (1979), 116–121.
24. Talbot, G. and Bennett, J. Unpublished results.

8 Surface effects including interfacial tension, wettability, contact angles and dispersibility
C.A. MOULES

8.1 Introduction

The subject of chapter 3 was emulsions, i.e. the liquid–liquid interface, particularly from a chemical-surfactant point of view, whereas chapter 6 was about gels, and chiefly involved molecular structures. This chapter is mainly concerned with the understanding of physical rather than chemical processes, e.g. wetting and dispersing. In the food industry, this relates to such problems as dissolving milk powder in water, or the dispersion of cocoa solids in milk. This chapter therefore includes details of the physics involved where more than two phases meet, one of which is a solid. When two phases come together they form an interface and four types are readily defined: liquid–gas, liquid–liquid, liquid–solid and solid–gas. The first three of these are examined in detail, as being crucial to food processing. The important physical parameters, such as surface tension, surfactant efficiency and contact angle are considered, both as to how they can be measured, the instrumental methods themselves and how the data can be used. Pitfalls, due to making the wrong measurement, are also noted. The theory of wetting is very complicated, but a résumé is included as an aid to understanding the processes which are taking place. The chapter concludes by examining some of the physical parameters which are important when dispersing powders in liquids.

The use of the words 'surface' and 'interface' can often be misleading to the newcomer to surface chemistry, because in essence the words have the same meaning. Normally the boundary between solids and/or liquids is called an 'interface' and when one of these phases is a gas, this interface is commonly referred to as the 'surface'. For the purpose of this chapter the 'interface' will be used to describe all boundaries in general and 'surface' will be used specifically when the boundary has a gas phase.

With the formation of any type of interface, there is an associated free-energy change, i.e. the molecules at the boundary have an excess of free energy due to being in the interface. The units (per unit area) of specific excess interfacial free energy are the same as those for interfacial tension (surface tension being one special case), but are only numerically equal for pure liquids in equilibrium with their vapour, or when two pure immiscible liquids make contact at a plane interface, and are:

168   PHYSICO-CHEMICAL ASPECTS OF FOOD PROCESSING

or
$$\text{work/area,} \quad \text{mJ m}^{-2} \quad (\text{dynes} \cdot \text{cm}^{-1})$$
$$\text{force/distance,} \quad \text{mN m}^{-1}$$

(Tabulated values in dynes/centimetre must be multiplied by 6.8523×10^{-5} to convert to pounds-force per foot.)

Despite having equivalent units, the two parameters are in fact quite different. Interfacial tension is a tensor acting perpendicularly to the interface, whereas interfacial free energy is a scalar quantity without directional values and is a property of an area of the surface. It is the quantity involved in the thermodynamics of the interface. Some consider it sufficient to define both interfacial tension and interfacial free energy as the work required to increase the area of surface isothermally and reversibly by a unit amount.

8.2 Surface/interfacial activity

Surfactants (surface active agents/emulsifier) are used to modify the behaviour of liquids, because by definition they exhibit a high degree of surface activity (adsorption). These materials consist of molecules containing polar and non-polar parts, which adsorb at the interface as an orientated monolayer. This is more favourable in energy terms than the complete solution of the surfactant in one or other of the phases. The effect is to expand the interface, but this is balanced by the tendency for the interface to contract under normal interfacial (surface) tension forces. This means that if a surface-active agent is used to make fat form droplets in a water system, or vice versa, work is required to form an interface leading to small drops with a spherical shape. The result is a lowering of interfacial tension, which can be expressed by the equation:

$$\sigma = \sigma_0 - \pi \tag{8.1}$$

where σ_0 = initial interfacial tension before addition of surfactant, σ = final interfacial tension and π = interfacial pressure (the expanding pressure of the adsorbed layer of surfactant). *Note*: The ideal conditions for spontaneous emulsification and miscibility of liquids are reached when $\pi > \sigma_0$.

Surfactant systems reach equilibrium (static interfacial tension) when all the surfactant molecules are aligned and orientated at the interface. Once the monolayer is saturated, excess molecules aggregate to form micelles. At this critical micelle concentration (CMC), the interfacial tension remains constant and cannot be further reduced by the addition of more of the surfactant (see Figure 8.1). The CMC gives the minimum interfacial tension that can be reached for a particular surfactant or surfactant mix. The CMC value determines a surfactant's potential use as

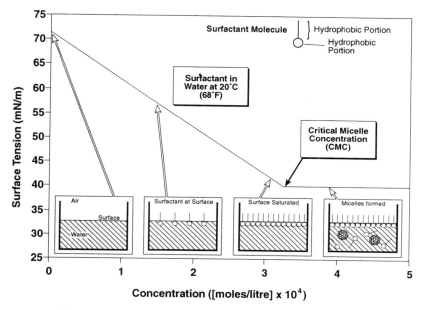

Figure 8.1 Determination of critical micelle concentration.

a wetting agent for a particular application and acts as a measure to prevent unnecessary additional amounts of these, sometimes expensive, ingredients from being used in a formulation.

An interfacial measurement that detects change over time must, by definition, be dynamic. At the constant area of interface between two so-called immiscible pure liquids, some of each phase will dissolve into the other until saturation is reached, thus lowering interfacial tension, e.g. water–hexane falls from about 49 to 47 mN m^{-1} in 1 hour.

The rate at which new interfaces are formed can be controlled and changed (by speeding up/slowing down bubble formation; see sections 8.3.3 and 8.5.2) to study the efficiency with which individual surfactants and mixtures of surfactants can reach the surface to lower interfacial tension. For pure liquids, the measurement of interfacial tension is independent of the rate at which the interface is created, as it remains constant.

8.3 Liquid–gas interface

8.3.1 Static/equilibrium measurements

Static or equilibrium surface tension of water is sensitive to as little as 100 ppb of surfactant. The highly reproducible measurements from the

Du Nouy Pt/Ir Ring and lesser known Wilhelmy Pt Plate techniques are routinely used to determine the equilibrium surface tension and CMC for different surfactants. Industrial companies and research groups conduct experiments for applications such as the following:

- Development of new surfactants.
- Testing surfactant blends to determine synergistic effects.
- Checking storage stability of surfactant solutions.
- Evaluating the effects of pH and salt on surface tension or CMC.

Instruments, which measure both the surface and interfacial tension, come in a range of sophistication, from simple manual tensiometers with large dials for teaching purposes, to PC-controlled processor tensiometers which can automatically perform the following:

- Measurement of surface and interfacial tension down to $1\,\text{mN}\,\text{m}^{-1}$.
- Control of dosing devices to calculate the CMC.
- Measurement of dynamic contact angles and surface free energies on a range of solids, e.g. metal, glass, plastics, loose powders, films, foils and fibres.
- Study of adsorption behaviour (in some industries known as 'wicking behaviour').
- Measurement of the density of liquid or mix under test (by recording its displacement by a known weight).
- Calculation of an approximate sedimentation rate.
- Comparison of texture by a simple penetration method.

Figure 8.2 shows a simple manual tensiometer and Figure 8.3 the top-of-the range processor tensiomenter. Both can be operated using a method developed by Lecomte du Nouy in 1919 [1]. His ring tensiometer technique has now become the most popular method for measuring surface and interfacial tension. A clean platinum/iridium (Pt/Ir) ring is wetted by totally immersing it in the liquid under test and then slowly pulling it through the surface. The force exerted by the liquid on the ring as it is raised is greatest at a point just prior to the liquid lamella breaking, i.e. when only the vertical force vector exists, $F_{\max}$ in Figure 8.4. This maximum force is directly proportional to the surface tension of the liquid

$$\sigma = F/L \cos \theta \tag{8.2}$$

where F = force, L = wetted length (twice the circumference, i.e. inner and outer edges of ring) and θ = contact angle ($\theta = 0$ for the clean ring).

The measurement requires a correction due to the weight of liquid under the ring which is above the liquid surface and is dependent on the ring circumference, diameter and density of the liquid(s). This method can also be used for liquid–liquid interfaces. Harkins and Jordan [2]

SURFACE EFFECTS 171

Figure 8.2 Simple tensiometer using Du Noüy ring method.

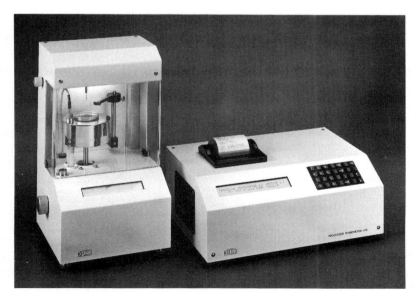

Figure 8.3 Processor tensiometer model K12 manufactured by Krüss (Krüss GmbH, Borsteler Chaussee 85–99a, D22453, Hamburg, Germany).

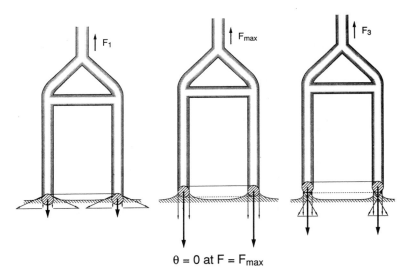

Figure 8.4 Force vectors during a surface tension measurement.

published tables for the empirically corrected values in 1930 and Zuidema and Waters [3] derived a universal equation for a correction factor.

8.3.2 Time-dependent measurements

If a series of readings is required over time, it is necessary to partially return the ring into the surface between readings to prevent the lamella breaking. The true time-dependence measurements are, however, distorted by the disturbance of the surface due to the test procedure, i.e. the surface expands as the ring is pulled up and decreases as the ring is re-immersed. Molecules diffuse towards the newly generated surface at a rate depending on the amount of surfactant present; for example, above CMC, because of the abundance of excess molecules, the time will be fractions of seconds and below CMC it may be several minutes. It is therefore more usual to use the Wilhelmy plate technique to measure medium- to long-term effects at the interface.

With the Wilhelmy plate technique, the ring is replaced by a plate, which can be made of roughened platinum or glass. In the former case it must be flamed, like the Pt/Ir ring as part of the cleaning process, whereas the glass must have a special surface treatment, to ensure a zero contact angle between the plate and liquid. The plate is dipped a short distance into the surface to ensure it is wetted before returning it to sit at the surface. The wetted (contact) line is equal to the plate circumference and, as with the ring, the contact angle of the liquid on the plate is

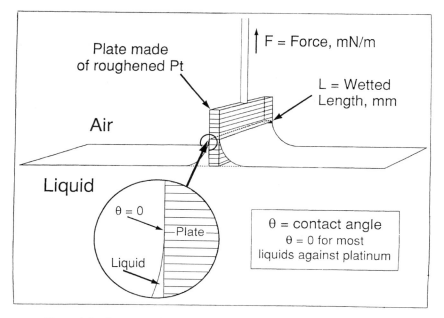

Figure 8.5 Contact angle at a completely wetted Wilhelmy platinum plate.

assumed to be zero (Figure 8.5) and so equation (8.2) can be applied. In this case, however, there is no volume of liquid above the surface and hence no correction is required. As the surface is not disturbed, once the plate is in position, this geometry is ideal for studying time effects.

8.3.3 *Dynamic measurements*

In many processes the time which a liquid has to perform its function is much faster than the time required to carry out the types of static surface tension measurements described above (e.g. a spraying process takes fractions of seconds). Other food processes that involve rapidly changing surface properties include coating, wetting, equipment cleaning and foaming. Therefore, the surfactant molecules must reach the surface during the time required to carry out the process, in order for the quality of the process to be achieved.

The speed at which surfactant molecules move to the surface is dependent on the kinetics of the surfactant and the concentration, i.e. different surfactants or surfactant blends exhibit different time dependencies and surface-tension values compared with their equilibrium values. Good surfactants, or surfactant blends, are those which show little, if any, time dependency hence they can perform their role during the life of the process. Poor surfactants will show a marked time dependency with the surface tension of the liquid, rising to the surface tension of the solvent as the surface age decreases, which in the case of an aqueous base is about $72\,\text{mN}\,\text{m}^{-1}$ at 25°C (77°F).

A simple method of monitoring dynamic surface tension was described by Miller and Meyer [4]. This works by generating gas bubbles in a liquid via a capillary at varying rates of bubble formation, thereby enabling surfaces of different ages to be produced, i.e. as the bubble rate increases so the surface age decreases (Figure 8.6).

Dynamic surface tension is related to the maximum pressure obtained during the formation of a bubble, which is at the point when the bubble radius is at a minimum:

$$\sigma = (P_{\max} - P_0)r/2 \tag{8.3}$$

where σ = dynamic surface tension, $P_{\max}$ = maximum bubble pressure, P_0 = atmospheric and hydrostatic pressure and r = radius of bubble.

There is now general recognition that it is essential to use dynamic surface tension as the criterion for the determination of a surfactant blend which performs best under dynamic conditions imposed by the application and where new surfaces are rapidly being created. There are several instruments on the market, all of which calculate surface age from bubble

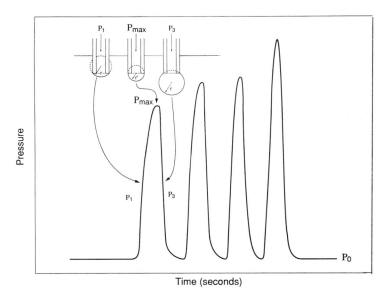

Figure 8.6 The principle of the maximum bubble pressure technique.

frequency. A bubble pressure tensiometer can sweep through a range of bubble frequencies from 0.5 to 10 Hz (~0.2–0.01 s surface age) and can be used to demonstrate the large difference between static and dynamic values and to show how static values can lead to mistaken correlations (see section 8.4.2). This can be very important where legislation may limit the amount of additive that can be used, or indeed where a surfactant is very expensive.

8.4 Applications of static and dynamic surface tension measurements

8.4.1 Equilibrium tests

The amount of foam is an important sensory quality for some drinks, such as stout, and if surfactants are left on the glass after the cleaning process, customers will complain about the lack of a good head. Equilibrium surface tension measurements can check the rinse liquid for excess surfactants which cause the foam to collapse rapidly.

8.4.2 Dynamic tests

In a spraying process (e.g. a herbicide), the optimum surface tension of the droplet must exist at impact with the leaf surface. If there are no surfactants present, then that surface tension is the same as the static value, but with the use of 'environmentally friendly' media (aqueous-based), surface-active agents are required. The surface tension at impact will therefore be somewhere between water and the CMC value, dependent on the rate of migration of the surfactant molecules to the interface, during the time of formation of the droplet and impact with the leaf. If it is too high, the droplets will bounce off the leaves and the expensive ingredients for weed control will fall to the ground.

Figure 8.7 shows that a non-ionic surfactant with a low CMC value (33 mg l^{-1}) and a static surface tension of 26 mN m^{-1} requires 8000 mg l^{-1} to produce a curve which is almost time-independent but still has a surface tension of 39 mN m^{-1}, well above the static value. Thus, if a static value had been used, the predicted surface tension and wetting ability would have been very wrong.

Obviously, in the above case, a different type of surfactant or surfactant blend is needed to produce a solution which is less sensitive to the creation of new interfaces and lower in dynamic surface tension. The same environmental restrictions, in the coatings and ink industries, have led to a large upsurge in waterborne formulations, which of necessity require surfactants to obtain the same performance levels as solvent-

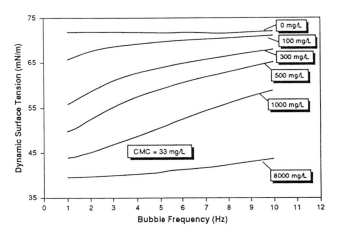

Figure 8.7 Dynamic surface tension of nonylphenol ethoxylate solutions.

based systems. In all those cases, reference to static surface tension would result in unsuccessful formulations.

A further example is taken from food packaging. To ensure there is no contamination of food and drink, the inside of the packaging is treated to ensure it is not wetted by the product. If a minute untreated fibre from the packaging protrudes into the container, it is sufficient to allow the food to absorb into the fibre and seep into and through the package. This creates the conditions where harmful bacteria can thrive. However, the outside of the package requires a surface which can be wetted very rapidly during a fast printing process with the product details, etc. Dynamic surface tests are required to check the suitability of inks for this application.

8.5 Liquid–liquid interface

8.5.1 Static/equilibrium measurements

At the interface between two liquids, there is an imbalance of forces which can be readily measured. Applications requiring knowledge and control of interfacial tension include determining the miscibility of two liquids, the testing of emulsifiers for food and the characterisation of emulsions and microemulsions.

A high interfacial tension means the two liquids have little tendency to mix spontaneously. As the interface ages, the interfacial tension will change due to the mutual diffusion of the liquids into each other. Surfactants or impurities (such as oxidation products of organic oils) will also diffuse to the interface and change its properties with time. Dependent on the range of interfacial tension, the Du Nouy Ring ($1-500\,\text{mN}\,\text{m}^{-1}$) or Spinning Drop ($10^{-6}-5\,\text{mN}\,\text{m}^{-1}$) method can be used to determine equilibrium interfacial tension.

The theory of the spinning drop tensiometer was derived by Vonnegut [5] and the interfacial tension is given by:

$$\sigma_i = (\omega^2 r^3 \Delta\rho/4)(1 + 2r/3L) \tag{8.4}$$

where σ_i = interfacial tension, ω = angular velocity ($\text{rad}\,\text{s}^{-1}$), r = radius of drop, L = length of drop and $\Delta\rho$ = density difference between the two liquids.

Figure 8.8 shows a schematic diagram of the spinning drop method. The limitation to the minimum interfacial tension, which can be measured, depends on there being a measurable density difference between the two liquids. Normally the capillary holds the liquid containing surfactant and a drop of the light phase is injected through a septum via a hypodermic syringes. Oil is used as a thermostatting medium and, because the drop can be observed over long periods, both temperature and time-dependent effects can be observed.

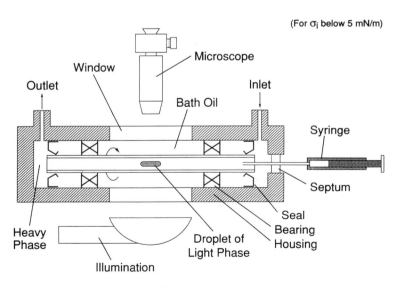

Figure 8.8 Schematic diagram of a spinning drop tensiometer.

8.5.2 Dynamic measurements

Dynamic interfacial tension characterises diffusion and orientation rates of surfactants and other soluble materials on very short time scales. Some processes, such as emulsification and cleaning, require materials to lower interfacial tension in seconds or even fractions of seconds.

The drop volume technique of dynamic interfacial tension measure change in the growing liquid–liquid interface as new area is being formed. One liquid, the light phase, is pumped continuously through an orifice into the other liquid (usually water-containing surfactant), while monitoring the volume of each drop produced. Interfacial tension is given by:

$$\sigma_i = V_{drop}(\rho_h - \rho_l)g/\pi d \qquad \text{for } \sigma_i > 0.2\,\text{mN}\,\text{m}^{-1} \qquad (8.5)$$

where σ_i = interfacial tension, V_{drop} = volume of the drop, ρ_h = density of heavy phase, ρ_l = density of light phase, g = gravitational constant and d = diameter of capillary.

As is shown in Figure 8.9, when a droplet forms, a balance of forces exists at the orifice until the drop increases sufficiently to detach. At that instant, the volume of the drop is directly proportional to the interfacial tension between the two phases.

Hool [6] described an instrument for measuring dynamic interfacial tension, in which the flow rate of the pump can be varied to change the rate at which new interface is formed. Neglecting mutual diffusion of the liquids, which is minimised as fresh liquid is constantly being used, then any variation in interfacial tension with flow rate is due to the effects of diffusion and orientation of surfactants present. It should be noted that interfacial tension between two pure liquids is independent of the rate of interface formation.

8.5.3 Applications

One of the most important food industry processes, which requires the study of both static and dynamic interfacial tension, is emulsification. Addition of emulsifiers (surfactants, water-soluble polymers) lowers interfacial tension and facilitates the development of emulsions and enhances their stability. Static tests can put a value to the interfacial tension necessary to aid the process and dynamic tests will show the best emulsifiers for the application. Figure 8.10 gives results which compared the suitability of different water-soluble polymers, which adsorb at the interface as emulsifiers for edible oil. This figure highlights the large differences in lowering interfacial tension between the various polymers. It should be noted that the process of demulsification (e.g. creaming, breaking and inversion of milk to butter), can equally be studied by these techniques.

SURFACE EFFECTS

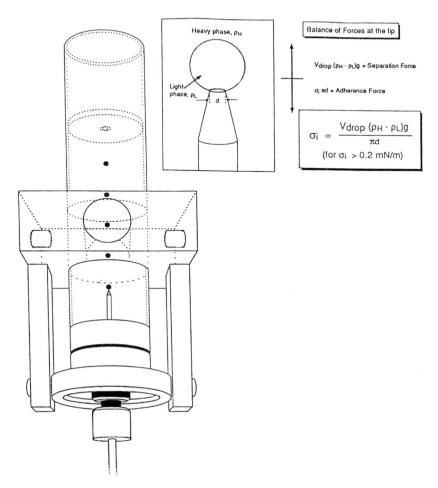

Figure 8.9 Illustration of drop volume technique.

8.6 The liquid–solid interface

8.6.1 Contact angle and wetting

What is meant by contact angle? A contact angle occurs at the boundary between phases. It can be visualised by considering two droplets of rain on the surface of a car. One droplet is seen on a freshly cleaned and polished section of a car and the other on a part waiting to be cleaned. The profile of the droplets will be very different. The polished section has

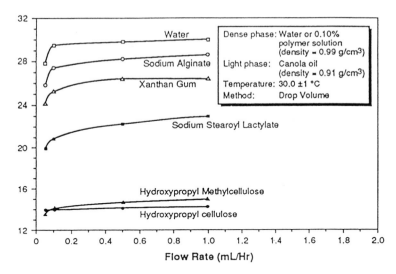

Figure 8.10 Results obtained by the drop volume method for testing the effectiveness of emulsifiers. From: Conklin, J.R. Surface activity and interfacial elasticity of methylcellulose and derivatives by a novel and improved drop-time liquid–liquid interfacial tensiometer. Presented at Institute of Food Technology Annual Meeting, June 4, 1991, Dallas, TX, Paper 478.

its paintwork protected from the rain by a layer of wax which prevents the rain from wetting the surface. The contact angle, θ, of the rain droplet is high, $90° < θ < 180°$. However, on the dirty part, the droplet has a much lower profile and the contact angle is $0° < θ < 90°$.

This example highlights the importance of the contact angle in demonstrating the wetting ability of a fluid where wetting, which involves three phases, is defined as the displacement from the surface of one phase by another, i.e. rain (liquid) displacing air (gas) on the car (solid). Surface-active agents are used to promote or inhibit wetting depending on the application in question.

Wetting can be looked at in another way, i.e. when a drop of liquid is placed on a solid, the liquid either spreads to form a thin, nearly uniform film (wetting) or remains as a discrete drop (non-wetting). Figure 8.11 illustrates water droplets the spectrum of good to poor wetting surfaces.

The reason why a liquid spreads on a solid has puzzled people for hundreds of years and led to a vast amount of literature on the subject. At a basic level, the phenomenon of wetting can be described by spreading, adhesion and immersion. A brief résumé of these three types of wetting is given below.

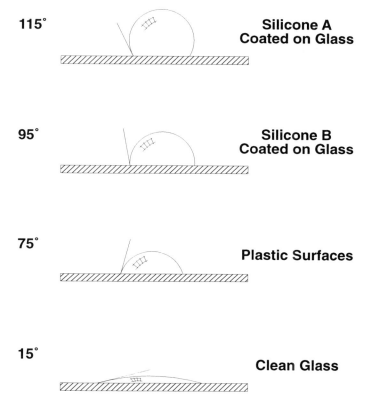

Figure 8.11 Spectrum of wetting of a liquid drop on a solid surface.

8.6.2 Spreading wetting

If a fluid, which is in contact with a solid, spreads further across that surface, it can be seen that the liquid–solid and liquid–gas interfaces increase and the solid–gas interface decreases. The spreading coefficient, S, is defined as;

$$S = \sigma_{sg} - (\sigma_{ls} + \sigma_{lg}) = -\Delta G_s/A \qquad (8.6)$$

where σ_{sg} = surface free energy between solid and gas or surface tension of the solid in equilibrium with the vapour pressure of the wetting fluid; σ_{ls} = interfacial free energy between liquid and solid; σ_{lg} = surface free energy between liquid and gas (surface tension of liquid); ΔG_s = free energy increase due to spreading; and A = interfacial area.

In other words, the fluid remains as a drop with a finite contact angle against the solid surface when S is negative, but spreads spontaneously when S is positive or zero. The equilibrium angle is obtained when the total free energy of the system is at a minimum:

$$\sigma_{sg}A_{sg} + \sigma_{ls}A_{ls} + \sigma_{lg}A_{lg} \quad \text{is a minimum.} \tag{8.7}$$

When a fluid with an equilibrium contact angle of θ spreads further to cover an area dA of the solid, the increase in liquid–gas interfacial area, $dA \cos \theta$ and the increase in free energy is:

$$dG = \sigma_{ls}dA + \sigma_{lg}dA \cos \theta - \sigma_{sg}dA \tag{8.8}$$

and when $dG = 0$

$$\cos \theta = (\sigma_{sg} - \sigma_{ls})/\sigma_{lg} \quad \text{Young's equation} \tag{8.9}$$

where σ_{sg} is the surface tension of the solid in equilibrium with the vapour of the wetting fluid.

If the surface tension of the solid against its own vapour is σ_s, then the spreading pressure, π_{sg}, represents the decrease in surface tension of the solid due to adsorption of the vapour

$$\sigma_s - \sigma_{sg} = \pi_{sg}. \tag{8.10}$$

The interfacial tension theory of Fowkes [7], when combined with Young's equation for non-polar liquids at a solid surface, shows that the cosine of the contact angle is:

$$\cos \theta = -1 + 2(\sigma_s^d/\sigma_{lg})^{1/2} \tag{8.11}$$

where $\sigma_{lg}^d = \sigma_{lg}$. Hence θ decreases as σ_{lg} decreases and becomes zero below a critical value.

8.6.3 Adhesional wetting

In this type of wetting, the area of liquid–gas interface decreases as the liquid, which is not initially in contact with the solid, makes contact and adheres. The work of adhesion is described by the Dupré equation:

$$W_a = \sigma_{sg} + \sigma_{lg} - \sigma_{ls} = -\Delta G_s/A. \tag{8.12}$$

Combining equations (8.9) with (8.12), the Young–Dupré equation is formed with:

$$W_a = \sigma_{lg}(1 + \cos \theta). \tag{8.13}$$

Therefore, when $\cos \theta = 1$, $W_a = 2\sigma_{lg} = W_c$, the critical wetting of adhesion. Zero contact angle occurs when the attractive forces between the liquid and solid are equal to, or greater than, those between liquid and liquid. The solid is completely wetted at $\cos \theta = 1$ and partially wetted at a finite contact angle. Complete non-wetting when $\theta = 180°$

(seen as a single contact point of a sphere of liquid on a solid) does not occur as there is always some liquid–solid attraction.

8.6.4 Immersional wetting

In this case, the solid is immersed in the liquid, hence the liquid–air interface remains constant. The free-energy change for immersion is:

$$-\Delta G_i = \sigma_{sg} - \sigma_{lg} = \sigma_{lg} \cos \theta. \quad (8.14)$$

Immersional wetting is spontaneous when $\theta < 90°$, i.e. $\sigma_{sg} > \sigma_{lg}$, but work needs to be done if $\theta > 90°$, i.e. $\sigma_{sg} < \sigma_{lg}$.

8.7 Theories of wetting

According to published theories, wetting by a liquid of a solid is preferred if certain conditions are met:

- The solid molecules should have a high surface free energy, σ_{sg} (a high self-affinity).
- The liquid molecules should not have a high surface free energy, σ_{lg} (a low self-affinity).
- The degree of repellency, σ_{ls}, between liquid and solid should be low.

Of these, only σ_{lg} can be measured directly.

How strongly molecules attract one another can be correlated roughly to their physical state; gases are expected to have a low self-affinity and metals a high self-affinity. As gases can be cooled and metals melted, a direct measurement of the surface tension and therefore surface free energy can be made (e.g. helium = $0.098 \, \text{mJ m}^{-2}$; glass = $150 \, \text{mJ m}^{-2}$; mercury = $470 \, \text{mJ m}^{-2}$, with most common liquids between 20 and $40 \, \text{mJ m}^{-2}$ but with water high at $72.8 \, \text{mJ m}^{-2}$ (at 25°C, 77°F)). Accordingly, it can be predicted that mercury would not wet glass and in fact the contact angle at room temperature is $\sim 136°$, i.e. the liquid has a greater affinity for itself, which can be seen from the spherical shape it assumes.

Despite the fact that some parameters in Young's equation (equation (8.9)) cannot be measured directly, useful information can be obtained from contact angle data alone. Wetting is essential for adhesion and calculation of the work of adhesion between a liquid and solid requires knowledge only of contact angle, $\cos \theta$, and surface tension of the liquid, σ_{lg}, as does the energy to bring about spreading. Therefore, $\cos \theta$ is the property which characterises the effect of the liquid–solid interaction and σ_{lg} is a fundamental property of the liquid.

Zisman [8] has developed a technique for finding the critical surface tension of wetting for solids, σ_c, i.e. the surface tension of a liquid which on contact with the solid has $\theta = 0$. By measuring the advancing contact angle of an homologous series of liquids on a given solid and plotting $\cos\theta$ vs. the surface tension of each liquid (Figure 8.12), σ_c is obtained by extrapolation to $\cos\theta = 1$. In practice, however, it can be seen that, depending on the liquids used, differences can be found in the value of σ_c and in the slope of the line drawn through the experimental points. Hence the Zisman plot should be taken as an aid to the wetting characteristic of the solid, although it has been found to be particularly useful in evaluating the wettability of plastic substrates.

Additionally, Fowkes [7] recognised that dispersive forces must play a part in the interaction between molecules and proposed that the surface free energy of any substance is the sum of the dispersed and non-dispersed parts (polar or hydrogen bond forces), for example, water molecules are highly hydrogen-bonded and the surface free energy (surface tension) of water can be divided into dispersion forces and those which contribute to hydrogen bonding. Using alkanes, which have only disperse forces,

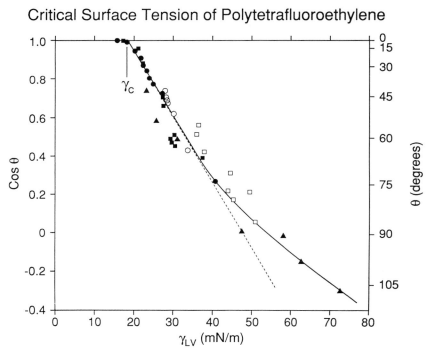

Figure 8.12 A typical Zisman plot.

Fowkes measured the interfacial tension against other liquids to obtain the dispersed and polar components of their surface-tension values. He derived a geometric mean function, which related the dispersive force contribution to interfacial free energy.

Others such as Owens and Wendt [9] have extended this theory to apply the method to solids. Contact angles of liquids, with known polar and disperse forces, are measured on a given solid and, from the intercept and slope on a plot of $(\sigma_{lg}^p/\sigma_{lg}^d)^{1/2}$ versus $(1 + \cos\theta) \cdot \sigma_{lg}/2(\sigma_{lg}^d)^{1/2}$, σ_{sg}^d and σ_{sg}^p can be found (where 'd' denotes dispersive and 'p' polar.) Wu [10] conducted his research in the field of molten polymers and suggested that the harmonic mean equation produced more realistic results. The equation enables σ_{ls}, σ_{sg}^p and σ_{sg}^d to be calculated. There are other theories published, and no doubt many more will follow, but $\cos\theta$ and σ_{lg} are likely to be fundamental to all of them.

8.8 Other factors affecting wetting

Viscosity and roughness of the substrate can affect the process of wetting. Although these factors will not be discussed in detail here, they must always be taken into account with each new wetting study. It should be noted that:

- Energy is required to wet a surface and the rate of spreading of a liquid will be modified by the dissipation of energy to overcome viscous liquid friction.
- Surface roughness does not notably alter the level of displacement of gas by a low-energy liquid, but penetration of the liquid into voids or pits depends on the radii of curvature of the pits and the viscosity of the liquid, i.e. the pits may not be completely filled by the liquid.

The roughness factor can be expressed as $\cos\varphi/\cos\theta$ for the same liquid on the same material, where φ = contact angle on the rough surface and θ = contact angle on the smooth surface.

8.9 Measurement of contact angles

Over the years, various methods of measuring the contact angle have evolved and this section describes the principles adopted by the most common types of commercial instruments on the market today, i.e. the sessile drop technique, the Wilhelmy plate method and the Washburn [11] or modified Washburn methods.

8.9.1 Sessile drop method

The traditional method for measuring the equilibrium contact angle of a droplet of a liquid placed on a solid surface is by the use of a simple manual system which consists of:

- a flat stage on which to place the solid sample,
- a syringe to provide a droplet of the test liquid and
- an optical device to magnify the droplet with an in-built 'protractor' for measuring the angle of the liquid at the interface with the solid.

This set-up is commonly referred to as a 'goniometer'. More sophisticated instruments use a video camera to capture the picture and convert it to a digital format in a PC for image analysis. By leaving the tip of the syringe in the liquid droplet, the volume can be increased to allow the advancing contact angle to be measured and by sucking back the liquid into the syringe, the volume of the drop can be decreased to obtain the receding angle.

With a video camera and associated computer control and analysis package, the data can be evaluated according to the desired model. Figure 8.13 shows a schematic diagram showing such a system, with its numerous accessories. In the one illustrated, up to six syringes can be connected to automatically place drops of different liquids onto a solid for measurement and analysis. Commercial databases containing a large number of researched liquids and analysis methods for the well-known

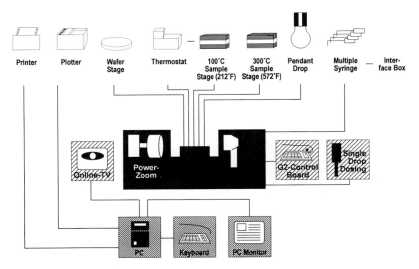

Figure 8.13 Individual components of a Krüss G2 angle meter (Krüss GmbH, Borsteler Chausee 85–99a, D22453 Hamburg, Germany).

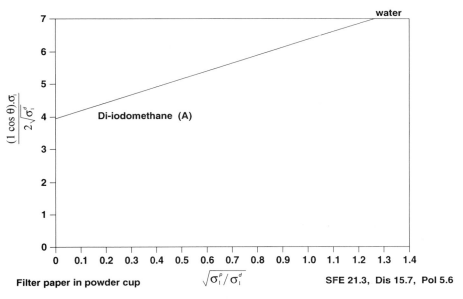

Figure 8.14 An Owens–Wendt–Rabel–Kaelble plot.

theories are available to make this technique a powerful tool for the characterisation of products and processes. Figure 8.14 shows a typical plot according to Owens *et al.*

The sessile drop can be used when the solid has a different surface on each side, for example treated on one side, a leaf, etc. This technique is particularly applicable to working at extremes of temperature.

8.9.2 *Wilhelmy technique*

If a solid sample is dipped into a liquid, the contact angle of the liquid at the interface can be measured during immersion. This is known as the 'dynamic advancing angle'. Likewise, if the solid is pulled out of the liquid, the liquid recedes from the surface of the solid and a dynamic receding contact angle can be measured. This principle can be applied by using an interfacial tensiometer (as described earlier for interfacial tension measurements) which has been adapted so as to hold a solid sensor shaped into a known geometric form (e.g. rectangle, cylinder, etc.). A PC controls the immersion and removal of the solid at known speeds. Solids which are flat, non-porous, easily fashioned into a plate and identical on both sides and edges, can be used for dynamic contact-angle measurements by the Wilhelmy technique. This method is more sensitive than the goniometer type, due to the sensitivity of the balances involved and the sophistication of the encoders, which measure distance. On the other hand, sample preparation can be more time-consuming. Single fibres down to a diameter of 5 μm (0.2 × 10^{-3} in.) and resolution of immersion depths <1 μm (0.04 × 10^{-3} in.) can be measured. Special holders are available for foils, films and single fibres.

Provided the surface tension of the liquid is known (the tensiomenter can, in any case, easily obtain this value), the wetted length is measured (for example, with a micrometer); equation (8.2) can be rearranged to provide the contact angle:

$$\cos \theta = F/L\sigma_{lg}. \tag{8.15}$$

The force generated by a liquid, of known surface tension, acting on a solid, is used to determine contact angle. When the liquid advances across a solid surface, a different value is usually obtained from that when the liquid recedes from this surface. The advancing angle can be used to calculate the surface free energy, the receding angle gives a measure of surface roughness, whilst the hysteresis provides information on adsorption, dissolution behaviour and surface homogeneity.

Force is measured directly from the integral balance in the tensiometer. Depending on the solid this might be a milligram or microgram system, the latter being used when a single fibre is the solid under test. A single dynamic advancing and receding angle can be measured by extrapolation of the force versus immersion depth slope to zero immersion. This counteracts the buoyancy effect. Provided the density of the liquid is known, the contact angle at each measurement point can be calculated. Again density of the liquid and wetted length of the solid can be obtained directly by measurement from commercial tensiometers (milligram and microgram versions) in situations where the values are unknown or difficult to measure. Figure 8.15 shows a plot of the measurement of a single human hair, one part of which is as nature intended and the other bleached.

This type of measurement has been used to determine the wetting of pasta by various sauces (liquids), and the dipping of food into batter or chocolate.

8.9.3 Washburn technique

If the solid to be tested is a bundle of fibres or loose powder then special geometries are available to hold the specific sample under test. Figure 8.16 shows the loose powder holder. The method for measuring the contact angle relies on the test fluids penetrating the porous sample. By relating the fibres and powder to capillaries, the theory of Washburn, who described the flow of liquid in a capillary, can be applied:

$$l^2/t = (\sigma r \cos \theta)/2\eta \tag{8.16}$$

SURFACE EFFECTS

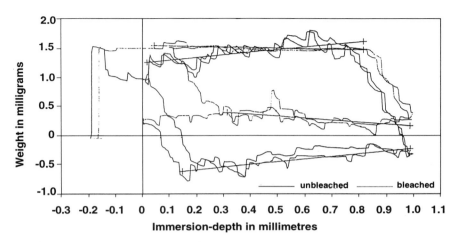

Figure 8.15 Measurements of dynamic contact angle on bleached and unbleached human hair.

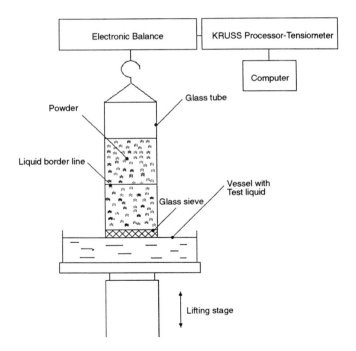

Figure 8.16 Diagram of apparatus used for measurements on loose powder.

where l = rise height, σ = surface tension of liquid, r = radius of capillary, θ = advancing contact angle, η = viscosity of liquid and t = time of flow.

For the fibres and packed powder, this equation can be modified such that they are seen as a bundle of capillaries, with a mean radius of capillary, r. A modified Washburn equation can be used:

$$l^2/t = (cr)\sigma\cos\theta/2\eta. \tag{8.17}$$

c is a constant to estimate the tortuous path of the capillaries and (cr) is a material constant, depending on the corresponding bundle of fibres or powder with a dimension of a length.

For a given bundle of fibres or packed powder, (cr) remains constant and the value can be calculated from equation (8.17) provided the liquid totally wets the solid, $\cos\theta = 1$. It is possible to determine whether this assumption is valid by plotting $2\eta l^2/t$ against σ. If this gives a straight line through the origin with an increasing (cr), then the liquid totally wets the sample. The relationship becomes non-linear at higher σ values, i.e. the rise height decreases.

Once (cr) is determined for a given sample, the advancing angle of liquids with $\cos\theta < 1$ can be calculated by replacing the rise height of the liquid with the weight gain of the sample (remembering that the instrument uses an integral balance as a means of measuring force for surface tension and dynamic contact angles). Hence the equation becomes:

$$W^2/t = K\rho^2\sigma\cos\theta/2\eta \tag{8.18}$$

where W = weight of penetrating liquid, ρ = density of measuring liquid and K = the geometric factor related to the dimensions of the instrument and the porosity of the material. K can be found by conducting a preliminary test on the sample using a totally wetting non-polar liquid such as n-hexane. When working with powders, it is important to ensure that there is laminar flow and a constant packing of the powder. This can be achieved by weighing a fixed amount and tapping it down to a constant volume in the sample tube.

It should be noted that unlike a bundle of fibres, for which an approximate geometry can be ascribed, powders have an irregular structure. For complete wetting of all internal surfaces, air must escape and theoretical work suggests that zero contact angle is a necessary prerequisite for total displacement of air. In this case, the pressure to force the liquid into the powder, or to prevent its entry is given from the Laplace theory:

$$P = -2\sigma_{lg}\cos\theta/r \tag{8.19}$$

which shows that flow is spontaneous when the contact angle $<90°$. Combining equation (8.19) with equation (8.9) and because σ_{sg} is practically constant if the contact angle is zero, then:

$$P = -2\sigma_{lg}/r \tag{8.20}$$

Therefore, in order to achieve maximum penetration, σ_{lg} should be large and σ_{sg} small. At the same time, the rate of air removal is relevant and from the linear rate of laminar flow in a cylinder, as described by Poiseuille and assumed by the Washburn equation, rapid penetration occurs when $\sigma_{lg} \cos \theta$ is high, θ and η is low and K as large as possible, i.e. loosely packed. This contradicts equation (8.9), so obviously there are incompatibilities in these arguments. In practice conditions are optimised by the use of surfactants, and it is the contact angle which is kept as low as possible.

Within the food industry, there are many processes where powders are dispersed in liquids or where a powder is used to coat a product (e.g. for some colours), in addition to agglomeration procedures. All these can benefit from a fundamental knowledge of the surface properties of the materials concerned, which can be obtained using the instruments described above.

8.10 Dispersion

Many food processes involve the dispersion of a solid within a liquid. Sometimes the solid is soluble, as in the case of sugar and water, at other times some or all the particles may be required to remain in suspension, e.g. drinking chocolate or cocoa. In almost every case, there is a need for the process to take place as rapidly as possible. Part of this is the removal of air and its replacement by the liquid. It has already been shown that surface energy and surfactants can play a vital role and that measurements of factors, such as the contact angle and interfacial free energy between the liquid and solid, can aid in the choice of surfactant and processing conditions. There are, however, many other important physical factors, one of which is the particle size and shape and another, for non-soluble materials, is the surface charge.

The particle-size distribution can play different roles depending upon the process concerned. Where a solid is being dissolved in a liquid, the bigger the surface area of the solid, the more likely it is to dissolve rapidly; hence, a very fine powder is normally desirable. This is also true in a relatively dilute suspension of insoluble particles. In drinking chocolate, for example, it is important that the cocoa particles disperse rapidly into the water and remain in suspension for as long as possible. A small particle will sediment much more slowly than a larger one, so once again a finer powder is desirable. The need for a slow sedimentation also makes particle shape important. Fibres or platelets sediment at a much slower rate than a spherical one of the same mass. Very approximately, a fibre will sediment at a rate proportional to its diameter and independently of its length, so extremely long fine fibres should give relatively stable dispersions. There remains, however, the problem of agglomeration, where particles become attached to one another, thereby increasing their overall sedimentation rate. This is where surface charge can play an important role. For certain products it is possible to use treatments such as alkalising to change the surface conditions and improve the dispersibility of the system.

Water is not the only liquid medium found in food processing, and very often it is necessary to disperse a solid in a fat, for example in chocolate or in many bakery products. In many of these cases, the liquid is only present in a limited amount, and is required to coat the surfaces of the solid particles. In this case, it is desirable to have a minimum surface area of solid material to coat and so the particles should be as spherical and as large as is possible, whilst retaining the correct mouthfeel of the product (see also chapter 17).

8.11 Conclusion

Physical phenomena such as wetting and dispersing play a vital role in the food processing industry. Much of the basic data for the ingredients involved, however, are not available in the literature. Modern computerised instruments, based on classical physical principles, are now available to enable the food technologist to obtain much more of this information and thereby optimise processing and ingredient usage.

References

1. Lecomte du Nouy, *J. Gen. Physiol.* **1** (1919), 521.
2. Harkins, W.D. and Jordan, F. *J. Coll. Int. Sci.* **52** (1930), 1751.
3. Zuidema, H. and Waters, G. *Ind. Eng. Chem.* **13** (1941) 312.
4. Miller, T.E. Jr. and Meyer, W.C. Method for the measurement of dynamic surface tension. *American Laboratory* (1984), February.
5. Vonnegut, B. *Rev. Sci. Instr.* **13** (1941), 6, included as Appendix I.
6. Hool, K. and Schuchardt, B. A new instrument for the measurement of liquid–liquid interfacial and dynamics of interfacial tension reduction. *Meas. Sci. Technol.* **3** (1992), 451–457.
7. Fowkes, F.M. Attractive forces at interfaces. *Ind. Eng. Chem.* **56** (12) (1964), 40–52.
8. Zisman, W.A. Relation of the equilibrium contact angle to liquid and solid constitution. *Adv. Chem.*, Series 43 (1964), 1–51.
9. Owens, D.K. and Wendt, R.C. Estimation of surface free energy of polymers. *J. Appl. Polym. Sci.* **13** (1969), 1741–1747.
10. Wu, S. Polar and nonpolar interaction in adhesion. *J. Adhesion* **5** (1973), 39–55.
11. Washburn, E.D. *Phys. Rev.* **17** (1921), 347.

9 Fermentation
R.M. GIBSON

9.1 Abbreviations and glossary

ADP	Adenosine diphosphate
ATP	Adenosine triphosphate
DNA	Deoxyribonucleic acid
Aerobic	Conditions in which oxygen is present
Anaerobic	Conditions in which oxygen is absent
Biosynthesis	The synthesis of biological molecules
Doubling rate	The time in which the number of cells multiplies by a factor of two
Hydrolysis	Cleavage of chemical bonds by water
Metabolism	The sum of all the chemical processes in living organisms
Metabolite	A chemical product of metabolism
Osmosis	Flow of water between two solutions of differing ionic strength through a semi-permeable membrane
Proteolysis	Decomposition/hydrolysis of proteins by enzymes

9.2 Introduction

Microbiology has catalogued many thousands of micro-organisms and all may, in the broadest sense of the term, display fermentative properties. Fermentation may be defined as the process by which micro-organisms propagate themselves utilising their external medium as a source of nutrients. The physical and chemical changes to that medium during fermentation, which we recognise and manipulate in fermented foods, are the result of microbial metabolism. The number and diversity of fermentative processes, that are employed in the generation of food products, may appear to be quite numerous, but these only amount to a small fraction of the microbiological fermentations occurring in the biosphere.

It was not until the mid-19th century that the role of micro-organisms in food fermentations was identified. Louis Pasteur [1], through studies on fermentations, discovered that micro-organisms could survive in the absence of air and also showed that the presence of air actually inhibited alcoholic fermentation by yeasts (the major product of such a process being carbon dioxide). This led Pasteur to the conclusion that 'fermentation is life without air' which became the accepted definition of the term for many years. More recently, fermentation has come to describe a wider range of processes involving microbial growth, largely due to the aerobic growth of micro-organisms in vessels generically known as 'fermentors'. However, when considering food fermentations, the original definition of Pasteur will largely hold true.

The understanding of fermentation was furthered in 1897 when Buchner, using disrupted (broken) yeast cells, demonstrated that cell-free alcoholic fermentation was possible. This showed that the nature of fermentation was chemical and not biological. The modern science of biochemistry can be traced to this understanding and the description of natural phenomena in physico-chemical terms had begun.

9.3 Microbiology of fermentation

All living organisms, whatever their complexity, are based on the cell which is basically of the same composition throughout nature, i.e. it contains the same groups of chemicals and functions in accordance with common biochemical principles. However, organisms can be classified into groups according to their cellular structures and functions. The most fundamental classification in cellular terms is that between bacteria, animal and plant cells. Bacteria and related micro-organisms do not possess a true nucleus, lack other complex internal structures and are referred to as the 'prokaryotes'. The more complex plant and animal cells are known as the 'eukaryotes'. Several organisms of great importance to a consideration of food fermentations belong to the fungi. *Saccharomyces* species (bread and alcohol), *Aspergillus* species (soy sauce) and *Penicillium* species (cheese) are all classified as fungi and fall into the eukaryote classification of cells. Although they display many functional similarities to bacteria, structurally they bear a close resemblance to plant cells.

One of the functional similarities between fungi such as *Saccharomyces* and many bacteria is their ability to grow in both aerobic and anaerobic conditions. Plant and animal cells are strictly aerobic and require oxygen to sustain viability. Prokaryotes, on the other hand, can be subdivided according to their ability to tolerate different conditions: aerobic, anaerobic and those which can survive in either environment (facultative).

Other criteria are also used for the classification of micro-organisms, i.e. their ability to grow on certain substrates, their response to pH, temperature and their sensitivity to certain dyes (known as the Gram method) among others.

9.3.1 Microbial growth

For a successful fermentation process, the medium in which the micro-organisms will grow must provide essential nutrients and the correct environment to promote respiration and growth. Many of the steps involved in preparing a food material for a fermentation process are designed to optimise the nutrient sources available to the fermenting micro-organism and provide the optimum conditions for growth. An example of this is the complex series of procedures leading to the fermentation of beer (section 9.7.2). Considering the complex nature of biochemical molecules, it is not surprising that there are substantial nutritional requirements which must be met to support microbial growth.

9.3.2 Carbon source

While photosynthetic organisms can utilise inorganic carbon in the form of carbon dioxide as a source of energy, the majority of organisms require a source of organic carbon to provide energy and the basic material for cell growth. The ability to utilise a carbon source in the absence of oxygen is a characteristic of many food micro-organisms and the basic mechanism of this process will be discussed in section 9.4. The most common sources of carbon/energy in food fermentations are polysaccharides, particularly starch, a plant storage polysaccharide, which is composed of many covalently linked glucose molecules. Glucose is the precursor required by the major metabolic pathways involved in the generation of energy and the utilisation of carbon in biosynthetic pathways. Many micro-organisms are able to utilise starch and related polysaccharides such as cellulose directly by excreting hydrolytic enzymes (e.g. amylases and cellulases) into the growth medium. These enzymes break down the starch molecules into smaller units which can then be transported into the microbial cell and utilised. Other organisms, notably *Saccharomyces*, do not have this capability and rely on other processes to provide a suitable carbon source. While plant storage polysaccharides are the main source of carbon in food fermentations, almost all organic materials can be utilised by specialised micro-organisms and a wide variety of hydrolytic enzymes exist to utilise varied carbon sources such as lipids and proteins as well as carbohydrates.

9.3.3 Nitrogen/sulphur source

Cell growth is dependent upon the synthesis of proteins, both as structural elements and as enzymes which control the various metabolic pathways. In addition to carbon, proteins are composed of nitrogen, sulphur, oxygen and hydrogen. The latter two elements are usually present in the molecules utilised as a carbon source as well as water in the growth medium. Nitrogen and sulphur are usually available in inorganic forms such as ammonium salts, or derived from amino acids such as lysine and cysteine.

9.3.4 Other nutritional requirements

In addition to the elements listed above, micro-organisms require a range of other constituents in the growth medium to achieve viability. Vitamins such as biotin, riboflavin and thiamin are required as co-enzymes and phosphorus is essential for the biosynthesis of DNA and related compounds as are purines and pyrimidines. Elements such as iron, calcium, magnesium, sodium and potassium are essential and even heavy metals such as copper, cobalt, zinc and manganese may be required. Any material which cannot be synthesised by the micro-organism itself, but is a requirement of the metabolism of that micro-organism, must be available in the growth medium.

Food fermentations based on plant or animal matter generally contain most of the elements necessary to sustain microbial growth. However, low levels of essential substances in the growth medium (e.g. usable nitrogen sources) may significantly restrict growth and prevent fermentation progressing to its full extent. Fermentation will slow and eventually stop when one or more essential nutrients become limiting in the medium. It may, therefore, be necessary to augment a fermentation with materials deficient in the medium to extend the fermentation process. This is demonstrated in modern brewing practice when hydrolysed corn syrups are added to supplement the supply of usable sugars and produce greater amounts of alcohol. The baking industry also uses so called yeast foods which augment the sources of nitrogen and vitamins, which are deficient in flour, and thus promote more vigorous yeast activity during proofing (chapter 12).

9.3.5 Environmental conditions

To achieve the optimum conditions for fermentation, it is not only necessary for the essential nutrients to be present in the medium, but the physical conditions must also be conducive to the growth of the particular micro-organism. The temperature of the fermentation will determine the rate of growth of a micro-organism. Lower temperatures will inhibit the rate of growth and respiration; higher temperatures may cause rapid initial growth but lead to increased cell death and thus effectively prevent fermentation. In general terms, the optimum temperature is that which produces the most rapid growth and maximum yield of cells. This parameter can, however, be manipulated to the advantage of the fermentation process to delay fermentation, as in the case of frozen bread doughs, or to stop fermentation, as in the use of pasteurisation of beers.

Micro-organisms are sensitive to pH and have an optimum pH at which the best rates of growth are achieved. Most bacteria have an optimum pH between 6.0 and 8.5, whereas most yeasts can tolerate acidic pH levels as low as 3.5. Alcoholic fermentation by yeast produces a progressively more acidic environment (due to increasing levels of dissolved carbon dioxide) which most micro-organisms cannot tolerate. This phenomenon was the basis for the widespread use of fermented alcoholic beverages as relatively sterile staple drinks in former times.

The presence of moisture is essential for the growth of micro-organisms, whilst their tolerance to low water activity depends upon the organism itself. Most micro-organisms require a water activity >0.98. However, a number of species, including yeasts, can survive at lower values. The osmotic potential of the medium can affect the viability of cell growth. Hypertonic solutions, in which the ionic strength of the medium is greater than the internal ionic strength of the micro-organism, can lead to loss of water from the cell and inhibit activity. Hypotonic solutions (with their lower ionic strength) can also have a detrimental effect on the cell, as its constituents are lost to the lower ionic-strength medium and water flows into the cell, which in extreme cases may cause the cell to rupture.

The presence of oxygen in an anaerobic fermentation process is detrimental to the production of desirable fermentation products and should be avoided. Oxygen may be introduced into a fermentation by stirring or otherwise disturbing the medium.

9.4 Anaerobic fermentation

During fermentation, micro-organisms use a variety of metabolic processes to extract energy and biosynthetic precursors from the nutrient sources present. Not all fermentative organisms employ the same metabolic pathways and a description of all possible mechanisms is beyond the scope of this book. Therefore, the process by which yeasts assimilate the available substrates will be described as an example. For those interested in fermentations involving other micro-organisms, descriptions of alternative metabolic processes can be found in general microbiological textbooks [2-4].

The yeast most commonly associated with food fermentation is *Saccharomyces cerevisiae*. Different strains of this species provide the yeast used in baking and many alcoholic fermentations. This yeast utilises the Embden-Meyerhof or glycolytic pathway (Figure 9.1) to convert the carbon source (glucose) to pyruvic acid, a common product of the major

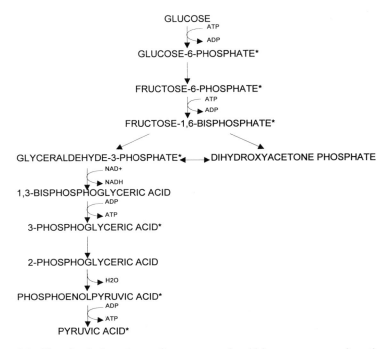

Figure 9.1 The glycolytic pathway. * = compounds which are precursors for other biosynthetic pathways.

pathways of carbohydrate metabolism. If one molecule of glucose is completely converted to pyruvate the glycolytic pathway has a net balance as follows:

$$1 \text{ Glucose} \rightarrow 2 \text{ Pyruvate} + 2 \text{ ATP} + 2 \text{ NADH} \qquad (9.1)$$

Adenosine triphosphate (ATP) is the molecule used to transfer energy between processes within the cell and NADH is used as a hydrogen donor in oxido-reduction reactions. Thus the organism has generated both a source of utilisable energy and reducing power for further metabolic processes. However, during the conversion of glucose to pyruvate, a number of metabolites are formed which provide precursors for other biosynthetic processes (see Figure 9.1) and, therefore, the conversion of glucose to pyruvate will be utilised by the micro-organism to provide both energy and precursor molecules depending upon the requirements of the cell.

Further metabolism of pyruvate is the source of many of the varied products observed in different microbial fermentations. In aerobic conditions pyruvate enters the tricarboxylic acid (TCA) cycle via acetyl CoA a precursor molecule and further ATP and precursor metabolites are formed. This process is not available to micro-organisms in an anaerobic environment and a variety of alternative pathways complete the degradation of pyruvate. Alcoholic fermentations convert the pyruvate to ethanol and carbon dioxide, thus:

$$\text{Pyruvate} \rightarrow \text{Acetaldehyde} + CO_2 \rightarrow \text{Ethanol} \tag{9.2}$$

The overall conversion of glucose to ethanol can be summarised by the following equation:

$$C_6H_{12}O_6 \rightarrow 2CO_2 + 2C_2H_5OH \tag{9.3}$$

Other conversions of pyruvate, of significance to food fermentations, include the production of lactic acid by species of *Lactobacillus* and *Streptococcus* in the production of cheeses and butter.

9.5 Characteristics of microbial growth

If a fermentation is allowed to progress to completion, the micro-organisms involved will display a characteristic pattern of growth. This is generally referred to as the 'growth curve'. Figure 9.2 shows the classical behaviour of micro-organisms in a batch culture in which the initial fermentation medium is not altered by addition of nutrients, agitation, etc.

Four phases of growth can be seen. Initially the *lag phase* represents the period in which the micro-organisms achieve their maximum rate of growth. Several factors are important in determining the length of time

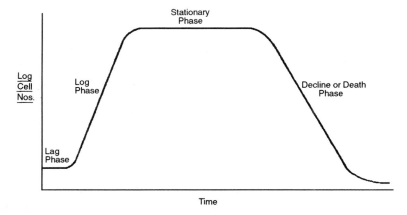

Figure 9.2 The microbial growth curve.

occupied by the lag phase. One of these is the condition of the inoculating – cells. If these are in a stationary phase growth, they will require longer to adapt to the new environment than cells in an exponential phase. This can be overcome by the use of starter cultures in which the inoculum is pre-incubated in a separate process. The condition of the fermentation medium is also important. The presence of the correct nutrients in suitable concentrations at the optimal physical conditions in terms of temperature, pH, etc. will shorten the lag phase, whilst deficiencies in these factors may prolong this stage and have an inhibitory effect on the exponential phase which follows.

Once all the conditions for cell growth are met, the growth curve enters the *exponential* or *log phase* in which the micro-organism devotes its energy to cell division and increasing biomass. The rate of growth of the cells can be followed in terms of its doubling rate, which will depend on the conditions of the fermentation and the particular organism involved. In an alcoholic fermentation, such as beer, this stage will be characterised by vigorous carbon dioxide production and consequent frothing of the fermentation.

A consequence of rapid cell growth and utilisation of nutrient sources is the gradual depletion of those nutrients. As the concentration of one or more essential nutrients falls to a certain level, it will begin to have an inhibitory effect on the micro-organism's ability to continue to grow and the growth curve will enter the *stationary phase*. In this way, the nutrient is said to become limiting and an example of such a situation would be the depletion of sugars during alcoholic fermentation. Other factors which might lead to a limitation on the rate of growth are the gradual increase in toxic products of metabolism or high population counts of cells. Once the cells have entered the stationary phase they will maintain this state as long as sufficient energy can be gained from cell storage materials. The *death phase* is entered when these stores of energy are depleted.

9.6 Secondary metabolite production

Those metabolites produced during microbial fermentations which are not directly related to the primary metabolism of a micro-organism, i.e. those processes essential for the maintenance of cell viability and growth, are referred to as 'secondary metabolites'. A vast number of such chemicals are produced by micro-organisms and they are, to a large extent, responsible for the industrial importance of many micro-organisms.

The role of secondary metabolites in microbial metabolism is still not clearly understood. Their production in many instances would seem to be a burden on the resources of the micro-organism and the benefit to the micro-organism is questionable. Indeed, in some cases a secondary metabolite can be toxic to the producing organism. The view that secondary metabolic pathways are examples of redundant processes in the cell, which have evolved and not been discarded during further evolutionary development, is possibly the best explanation to date for their existence. Whatever the role of secondary metabolism, it is a phenomenon which can be exploited to the great benefit of many industrial processes. For instance, antibiotics, discovered by Fleming in 1928, have had a major impact on medical science. Many other drugs and chemicals are manufactured by the fermentation of micro-organisms (see for example Table 9.1). The importance of secondary metabolites in food fermentations is not nearly as well studied as in antibiotic production. However, many

Table 9.1 Some industrially important micro-organisms

Micro-organism	Product	Reaction involved
Saccharomyces cerevisiae var. ellipsoides	Wine	Glucose/Fructose $\rightarrow$ Ethanol + CO_2
Kloeckera spp.	German wine	
Saccharomyces cerevisiae Saccharomyces carlsbergensis Saccharomyces uvarum	Beer	Glucose/Fructose $\rightarrow$ Ethanol + CO_2
Saccharomyces cerevisiae	Bread	Glucose/Fructose $\rightarrow$ Ethanol + CO_2
Aspergillus oryzae	Sake	Conversion of rice starch to fermentable sugars
Aspergillus oryzae Mucor rouxii	Soy sauce	Production of proteases, amylases, lipases, etc. in starter culture (koji)
Lactobacillus acidophilus Lactobacillus bulgaricus Streptococcus thermophilus	Yoghurt/buttermilk	Formation of lactic acid
Streptococcus thermophilus Streptococcus lactis Streptococcus cremoris Lactobacillus helveticus Lactobacillus bulgaricus Lactobacillus lactis	Cheese	Formation of lactic acid and subsequent curdling of casein
Penicillium camemberti Geotrichum candidum	Soft cheese	Hydrolysis of proteins and lipids
Penicillium roqueforti Penicillium glaucum	Blue cheese	Lipolysis and pigmentation
Propionibacterium spp.	Swiss cheese	Formation of holes due to CO_2 production as a result of the proprionic acid fermentation of lactic acid/flavour
Leuconostoc mesenteroides Lactobacillus plantarum Lactobacillus brevis	Pickling	Preservation due to fermentation of sugars to lactic acid

important flavour compounds are produced by yeasts in the baking and brewing/wine industries, etc. Their usefulness is evident in the practice of maturing beers and wines in the presence of yeast, often for prolonged periods.

Secondary metabolites are generally produced towards the end of the exponential phase and during the stationary phase of growth, when cellular resources are not wholly devoted to cell growth and the toxic effects of some secondary metabolites on the growing cells are still at a low level.

The following section contains a summary of several important food fermentations. In doing so, it is not the intention to give a full description of the process, which will be covered in context in the individual chapters dealing with the particular food types, but to summarise the microbial aspects of the fermentations and indicate how the products of fermentation influence the final product.

9.7 The application of fermentation in the beverage industry

9.7.1 Wine

Grape juice has a high content of fermentable sugars and can support microbial fermentation without supplementation or elaborate preparation of the raw material. Wild yeast varieties are present on the skins of grapes at harvesting and, during the subsequent crushing and pressing, these yeasts are transferred to the juice (or 'must') which will be used for wine fermentation. Modern processes of wine production may also employ inoculations of specific yeast cultures to standardise the qualities of particular wines. The character of wines from different regions will depend on many factors, such as the climate, grape variety, soil conditions and also the microbial flora present on the surface of the grape. *Saccharomyces cerevisiae* var. *ellipsoides* is commonly associated with wine production, however, other yeasts, such as *Kloeckera* spp., associated with German wines, have been identified.

The primary fermentation of wine is usually carried out in open vats in which the grape sugars are converted to ethanol and carbon dioxide as well as other metabolic products, which will contribute to the characteristics of the wine. Due to the high levels of sugar in the initial fermentation medium, alcohol content can rise to ca. 15%. Yeast growth will be perturbed when the supply of fermentable sugars is exhausted and/or the concentration of alcohol inhibits yeast metabolism. The fermented material is matured, once again in open vats in which oxidative reactions take place. This will have a bearing on the final wine quality; for example, malic acid is converted to lactic acid and other reactions contributing to colour, flavour and aroma occur.

Secondary fermentation of wine can occur in open or closed systems, to produce still or sparkling wines, respectively, and this can be initiated by further addition of fermentable sugars and yeast. In closed vessels, carbon dioxide generated during the secondary fermentation cannot escape from the wine and this results in a carbonated product. Secondary fermentation of champagne is actually carried out in the bottle resulting in a highly carbonated wine. This method is known as the 'methode champenoise'.

Final maturation of wines occurs in the bottle, once again specific chemical changes occur which affect the character of the wine and the extent of maturation will depend on the composition of the wine following fermentation. Red wine contains relatively high levels of tannins which give a bitter flavour. During maturation the tannins are converted to other compounds and the wine develops a more mellow flavour. White wines, on the other hand, contain fewer tannins and can be drunk after a shorter period of maturation than is generally the case for red wines.

9.7.2 Beer

As the brewing process is described in detail in chapter 20, this section will only highlight the parts of the process which are important for yeast activity during fermentation.

The generic term 'beer' is used to describe a fermented drink produced from malted barley and hops. There are two major processes employed in the production of beer and the distinction between them depends to a large part on the method of fermentation employed. The two types of beer are generally referred to as 'ale' and 'lager' and, though they bear many similarities, they differ in several significant respects. Ales, such as bitters and stouts, are fermented using 'top yeasts' while lagers, such as pilsner, are produced using 'bottom yeasts'. Microbiologically, these yeasts are very similar; however, to the brewer they behave in a significantly different manner during fermentation. Ale yeasts, *Saccharomyces cerevisiae*, grow rapidly at 15–20°C (60–68°F) and produce large amounts of carbon dioxide, which has the effect of carrying the flocculating yeast to the surface of the fermentation, hence the name 'top yeast'. Lager is traditionally brewed at lower temperatures (10–12°C, 50–54°F) than ale and consequently the metabolic rate of the yeast is lower. In addition, the yeast strains used are different. *Saccharomyces carlsbergensis*, also known as *Saccharomyces uvarum*, used in lager brewing tend to fall to the bottom of the fermentation vessel and are known as 'bottom yeasts'. The choice of yeast will not only be determined by the type of beer to be produced, but will also have a bearing on the flavour characteristics of the beer; thus brewers will maintain particular strains for the beers they produce and this contributes to the varied character of beers of the same type.

The mashing of malted barley produces the mono- and disaccharides, such as glucose and maltose, which can be utilised by yeast as a carbon source. These compounds, along with other essential nutrients, are extracted to produce the wort which is used as the medium for the yeast fermentation. Additional sources of nutrients can be added, such as corn syrup to increase the initial sugar content of the medium and replace a portion of the more expensive malt.

Fermentation quality will also be determined by the state of the yeast when added to the wort. If yeast is used in several consecutive fermentations with little or no exposure to air, it will lead to a deterioration in the viability of the yeast and production of undesirable flavour compounds. In addition, yeast growth is affected by the type and shape of the fermentation vessel and whether or not the fermentation is agitated. This is because aerating the medium increases the rate of yeast growth and can also affect the flavour of the resulting beer.

Following the first fermentation of the wort into beer, a secondary fermentation is carried out. When brewing lager, this process traditionally can take long periods of time (>1 month). The 'lagering process', as it is known, allows further interaction of the yeast with the medium, resulting in additional flavour development. In addition, other chemical changes also occur in the beer, such as the conversion of α-acetolactate (produced by yeast metabolism) to diacetyl and the removal of this and other undesirable volatiles by the evolution of carbon dioxide in the beer. Secondary fermentation in ale brewing often takes place in the vessel (e.g. bottle or cask) in which the final product will be sold. The beer, still containing active yeast, is placed in the vessel often with additional fermentative material and further alcohol is produced, as well as carbon dioxide which pressurises the beer in the vessel.

A comprehensive review of the brewing process is given by Briggs *et al.* [5].

9.7.3 *Vinegar*

Vinegar is included in the section on alcoholic beverages as traditionally its production shares many common features with alcoholic fermentations. There now exist, however, purely chemical means of producing vinegar, which do not include fermentation processes. There are many types of vinegar produced and in many cases these are named for the raw material, which contributes the source of fermentable sugars. The name may also reflect the traditional alcoholic fermentations used for this type of product. Thus we find malt vinegars, wine vinegars and cider vinegars as well as vinegars produced from many types of fruit, honey, etc.

Once again, the major microbial agent used in the conversion of fermentable sugars to alcohol is *Saccharomyces cerevisiae*. Modern methods usually employ a cultured inoculation of yeast to preserve a consistent character in the product although, traditionally, for example, spent brewer's yeast might have been used in the case of malt vinegar. Fermentation is carried out for 2–3 days at around 30°C (86°F) to produce a liquid with the required alcohol content. The fermentation may then be filtered or centrifuged to remove the yeast and the resulting alcoholic solution used for the process of acetification (formation of acetic acid). This process also employs micro-organisms to achieve the conversion of ethanol to acetic acid. The cultures used are less clearly defined and often mixed, i.e. containing several species of micro-organism. However, *Acetobacter* species such as *Acetobacter aceti* have been identified in many vinegar processes. The microbial production of acetic acid differs from that of ethanol in that it is an aerobic process, i.e. it requires oxygen, as shown in the following equation:

$$C_2H_5OH + O_2 \rightarrow CH_3CO_2H + H_2O \qquad (9.4)$$

If an alcoholic fermentation is left to stand, a film will often form on its surface. This is composed of acetic acid-producing bacteria, which will convert ethanol to acetic acid at the interface between the anaerobic fermented liquid and the air, and is known as 'surface culture' when used to produce vinegar. The Orleans process, as studied by Louis Pasteur, replaces this process by a semi-continuous method in which a portion of the acetified material is periodically replaced with fresh vinegar stock to maintain the acetification. More modern techniques of vinegar production employ aerated vessels, which greatly increase the surface area available for the acetic acid bacteria to interact with the alcoholic medium.

Once the acetification process is complete, the vinegar may be allowed to mature to improve its flavour and clarification may be necessary to remove cell debris. The vinegar may also be pasteurised or treated with a chemical sterilising agent.

9.8 The application of fermentation in the food industry

9.8.1 Dough fermentation

Traditionally, Western bread was prepared using yeast (*Saccharomyces cerevisiae*) obtained from the top yeast produced by ale fermentations. With advances in technology, this practice has been superseded by the adoption of strains which are propagated specifically for bread making. It is now known, however, that many specific types of bread are the result of mixed cultures containing not only yeast, but also bacteria, which contribute to the flavour profile and properties of the bread. Thus many 'speciality' breads require complex starter cultures, composed of a number of different micro-organisms, to achieve their individual character. A description of the microbiology of several specific dough fermentations is given by Sugihara [6].

The most striking effect of dough fermentation is the leavening which occurs during proving (see chapter 12). This is due to the production of carbon dioxide by the yeast, which gradually fills and enlarges the microporous structure produced in the dough by the incorporation of air during kneading. This results in a gluten foam structure. The volume of the dough increases with time during the proving and to optimise this process the baker can employ a number of techniques. Wheat flour is a poor source of nutrients for the growth of yeast and it is common practice to provide additional sources of utilisable sugar, along with other essential nutrients, to promote the rate of yeast metabolism and thus shorten the proving time. The preparation of a starter culture, as in sponge dough fermentations, allows the yeast to overcome the lag phase of growth and also provides time for the production of flavour compounds, such as organic acids produced by *Lactobacillus* species, which are an important aspect of sour doughs for example. Other aroma compounds associated with white bread fermentation include dimethyl sulphide, acetaldehyde, dimethyl disulphide and 2-methylpropanol, etc. These should not be confused with the baking aromas associated with bread, largely due to Maillard reactions which occur during formation of the crust.

The other major product of the primary metabolism of yeast, ethanol, is lost from the dough by evaporation during baking.

An exhaustive account of all aspects of the process of bread manufacture can be found in the book edited by Pomeranz [7].

9.8.2 *Fermented milks*

There are a wide variety of fermented milk products produced in many parts of the world, many of which will not be familiar to the Western consumer. Similarly the raw material used in their manufacture can be varied; for example, milk from cows, goats, sheep, buffaloes and camels are all used to produce fermented milk products. Those wishing to study this subject in greater detail are referred to Obermann [8], as a starting point for further investigations.

There are a number of micro-organisms associated with milk fermentations. However, the predominant species found in many cultures are lactic acid bacteria, particularly *Lactobacillus bulgaricus* and streptococcal strains such as *Streptococcus thermophilus*. These two microbial species are responsible for yoghurt fermentations as well as many other traditional fermented milk products. During yoghurt fermentation organic acids such as lactic, formic and acetic are produced, as well as compounds such as diacetyl and acetaldehyde. In addition, proteolysis of the milk proteins occurs leading to heightened levels of amino acids in the finished product. Yoghurt has a final pH of around 4.2 due to the production of organic acids and it has a smooth gel-like texture resulting from the proteolytic action on the milk proteins. The low pH enhances the keeping properties of the product over the original milk and the hydrolysed protein is more easily digestible.

Beneficial dietary claims are made for some fermented milk products; acidophilus milk, for example, is fermented using the lactic acid bacteria *Lactobacillus acidophilus* which survives in the human digestive system and is said to have a therapeutic action. Similarly acidophilus yoghurts, fermented using cultures containing combinations of *Lactobacillus acidophilus*, *Streptococcus thermophilus*, *Streptococcus lactis* and *Bifidobacterium bifidum*, replace intestinal flora and have antibiotic and antiviral activities. The term 'probiotics' has been used to describe such products used in a therapeutic role. Further information on fermented milk products can be found in Kurmann et al. [9].

9.8.3 Cheese fermentations

As with fermented milks, the number and variety of cheeses is very large and traditionally it is a very regional product, with, in many cases, the cheese being named after its place of original manufacture. Similarly, the starting material is varied and the type of milk used can be from a wide range of sources. A comprehensive review of cheese production has been edited by Fox [10].

The formation of curd can be brought about by organic acids resulting from the fermentation of milk due to the naturally occurring bacteria present in raw milk or the addition of a starter culture of sour whey, clotted milk or buttermilk. However, the advent of pasteurisation and rigorous quality control has led to the use of more clearly defined starter cultures. Also, the use of rennet (a proteolytic enzyme), in modern cheese manufacture, has greatly reduced the time required to form the curd and the initial role of the microbial fermentation of cheese (i.e. formation of the curd) has therefore been reduced. Starter cultures are still employed in cheese manufacture, however, as they are essential in producing the flavour and textural characteristics of a cheese.

The composition of the starter culture in cheese manufacture varies depending on the type of cheese being produced. Generally, a starter culture will contain lactic acid-producing bacteria, such as *Streptococcus cremoris*, *Streptococcus lactis*, and lactobacilli such as *L. helveticus*, *L. bulgaricus* and *L. lactis*. In addition, specific micro-organisms may be added to the culture to produce desirable properties in the final cheese.

Flavour and aroma may be enhanced by the inclusion of micro-organisms with lipolytic and proteolytic activities and strains of moulds and yeasts may be added to produce textural changes. Soft cheeses such as Camembert and Brie are surface ripened by the addition of moulds such as *Penicillium camemberti* and *Geotrichum candidum*. These aerobic micro-organisms grow on the surface of the maturing cheese and release enzymes which enter the body of the cheese and bring about textural changes due to their hydrolytic action. Blue cheeses are inoculated with moulds which produce a characteristic blue or green colour. Examples of such moulds are *Penicillium glaucum* and *Penicillium roqueforti*. During ripening of these types of cheese, the body of the cheese is often pierced to allow the mould to grow within the cheese and produce the characteristic veined effect. One other significant micro-organism, used to modify the texture of cheese, should be mentioned; Emmental and related cheeses are inoculated with species of *Propionibacterium* which convert lactic acid to propionic acid, acetic acid and carbon dioxide. It is the evolution of carbon dioxide during maturation which gives rise to the characteristic holes found in such cheeses.

A wide range of flavour and aroma compounds are produced by the action of micro-organisms during cheese manufacture. These include products of proteolysis, a large number of short- and medium-chain length fatty acids (produced by the action of lipases) and many other compounds such as aldehydes, esters and alcohols.

9.8.4 Cocoa

Although originating in Central America, the cocoa bean, *Theobroma cacao*, is now an important cash crop in tropical countries around the world. The manufacture of cocoa products such as chocolate and cocoa drinks has become a major industry and yet, in many cases, the early stages of cocoa processing are still carried out by the traditional methods used by tropical farmers for centuries, as described in chapter 17. It is only recently that concerted attempts have been made to modernise the process and introduce industrial scale methods for the harvesting, fermentation and drying of the beans in some countries.

Cocoa bean fermentation is carried out by one of two methods:

The heap method. Piles of beans are removed from the pod and placed in heaps, covered with leaves and allowed to stand.
The box method. The beans are placed in one of a series of boxes and the beans are regularly transferred between boxes until fermentation is complete. This allows considerably greater quantities of beans to be handled simultaneously.

Fermentation lasts for 6–7 days and brings about changes within the beans which are important for the subsequent generation of cocoa flavour during roasting. Microbial activity takes place in the mucilaginous tissue surrounding the beans, which is left in place when the beans are removed from the pod. The initial stage consists of yeast growth on the sugar-rich mucilage, producing ethanol. Subsequently, the ethanol is used as a carbon source by acetic acid bacteria, which become active as increasing amounts of ethanol are produced by the yeast. Lactic acid bacteria are also associated with cocoa fermentations and convert sugars to lactic acid. Wild strains of yeast, acetic acid and lactic acid bacteria are responsible for this microbial activity and as such, the species encountered will depend on the locality of the fermentation. Not surprisingly, therefore, a wide variety of species have been identified in different fermentations. For instance, many yeast species such as *Saccharomyces* spp., *Hansenula* spp. and *Candida* spp. have been isolated, similarly many species of *Acetobacter* and *Lactobacillus* have also been found.

The exact role of fermentation in cocoa processing is still unclear. If the integrity of the cocoa bean shell is maintained, then it is not possible for the micro-organisms to directly interact with the material within the bean and they cannot, therefore, be responsible for the degradation of this material. During fermentation, the temperature within the mass of fermenting material rises considerably and this was thought to lead to the death of the bean at around 3 days into the fermentation, allowing the degradative enzymes of the bean itself to break down the storage materials present. The results of this breakdown will be the release of sugars and amino acids from the bean's polysaccharide and protein reserves. These will become the precursors of Maillard-type reactions during subsequent bean roasting. However, the process may be more complicated than this and involve the diffusion, into the beans, of the products of microbial activity, e.g. ethanol, acetic acid and lactic acid. The resulting drop in pH has also been implicated in the death of the bean. Differences in cocoa bean flavour quality are also apparent depending on the treatment of the beans during fermentation (for example, turning the beans, as is carried out during the box method of fermentation, will aerate the fermenting mass, thereby stimulating the oxygen-dependent production of acetic acid by the *Acetobacter* and inhibiting the anaerobic production of ethanol by yeasts). It is apparent that a complex balance of environmental conditions will be necessary to achieve the optimum cocoa fermentation, as judged by flavour quality of the roasted beans. The large variations in perceived quality between beans from different geographical locations may indeed be largely due to both the distribution of micro-organisms present in a given region and local fermentation practices.

Further aspects of cocoa/chocolate production are described in the works of Beckett [11] and Minifie [12].

9.8.5 Vegetable fermentations

Preserving many types of vegetable crops, to prolong their usefulness following harvesting, has been carried out for centuries and is still in use today despite the advent of modern techniques such as freezing and canning. This is in large part due to the changes brought about to the product by the process of preservation, in effect creating new, desirable products such as pickles and sauerkraut (see also chapter 19).

Pickling of cucumbers, for example, is carried out by placing the vegetable in brine, the strength of which must be carefully controlled to prevent inhibition of growth of desirable micro-organisms. Initially yeast and lactic acid bacteria metabolism leads to a drop in the pH of the brine solution and ideally this low pH will prevent the growth of undesirable micro-organisms. The concentration of lactic acid produced will rise as fermentation progresses and the 'pickling' of the vegetable will continue to completion.

Manufacture of sauerkraut also involves a brine process and once again lactic acid fermentation occurs, however, the initial stages of fermentation are dominated by the growth, not of yeasts, but a lactic acid bacterium named *Leuconostoc mesenteroides*. As fermentation progresses, other lactic acid bacteria increasingly play a role in the production of lactic acid. These are mainly lactobacilli, such as *Lactobacillus plantarum* and *Lactobacillus brevis*. Other major end products of fermentation include acetic acid, ethanol and carbon dioxide. Once again, the decrease in pH during fermentation prevents the growth of other less-tolerant micro-organisms. A number of other compounds have been identified as important to the flavour and aroma of sauerkraut, some of which are present in the fresh cabbage and some of which are generated during fermentation [13].

9.8.6 Other food fermentations

The preceding sections have outlined specific fermentations involved in the production of important food groups, some of which will be covered in more detail in subsequent chapters. There are, however, many more such fermentations which are not covered in this chapter. Fermented sausages, such as salami, are enjoying an increasing popularity and there are large numbers of fermented foods produced in Africa and Asia which, in many cases, have a very different microbiological basis to those described above [14]. For the reader wishing to investigate this subject more thoroughly, there are many excellent texts available both of a general nature and on specific food groups which give far more detail than has been possible here (see for example *Prescott and Dunn's Industrial Microbiology* [15]).

Table 9.1 summarises many of the micro-organisms involved in important food fermentations.

9.9 Conclusions

When studying food fermentations, it soon becomes apparent that many of the processes used today have their foundation in centuries-old practices of food preparation and preservation. It is indeed fortunate that there exists within nature a wide range of micro-organisms, which human civilisation has learned to utilise, either by accident or intent, to the great benefit of both its nutritional and hedonistic requirements. Fermentation provides a means of preserving fresh foods for much longer periods than would be possible in the absence of modern preservation techniques. In many cases it produces foods which are free from or have much reduced numbers of pathogens and it is responsible for many of the most highly valued textural and flavour attributes of foods such as cheese and wine.

As our knowledge of fermentative processes and microbial biochemistry continues to grow, it is unlikely that food fermentations will remain unchanged. Indeed, they are ideal subjects for the attentions of biotechnologists and genetic engineers. There is every likelihood that one of the greatest challenges facing the food scientist and producer in the coming years will be the extent to which these new technologies should be embraced.

References

1. Pasteur, L. *Études sur la Bière*, Macmillan (1879).
2. Schlegel, H.G. *General Microbiology*, 7th edn. Cambridge University Press (1993).
3. Stanier, R.Y., Ingraham, J.L., Wheelis, M.L. and Painter, P.R. *The Microbial World*, 5th edn. Prentice Hall (1986).
4. O'Leary, W. *Practical Handbook of Microbiology*, CRC Press (1989).
5. Briggs, D.E., Hough, J.S., Stevens, R. and Young, T.W. *Malting and Brewing Science*, 2nd edn., Vols 1 and 2, Chapman Hall, London (1981).
6. Sugihara, T.F. In *Microbiology of Fermented Foods*, Vols 1 and 2, Ed. Wood, B.J.B., Elsevier (1985), 249–261.
7. Pomeranz, Y. (Ed.) *Wheat: Chemistry and Technology*, 3rd edn. American Association of Cereal Chemists (1988).
8. Oberman, H. In *Microbiology of Fermented Foods*, Vols 1 and 2, Ed. Wood, B.J.B., Elsevier (1985), 167–195.
9. Kurmann, J.A., Rasic, J.L. and Kroger, M. *Encyclopaedia of Fermented Fresh Milk Products*, Van Nostrand Reinhold (1992).
10. Fox, P.F. *Cheese: Chemistry, Physics and Microbiology*, 2nd edn., Vol. 1, Chapman and Hall, London (1993).
11. Beckett, S.T. (Ed.) *Industrial Chocolate Manufacture and Use*, 2nd edn. Blackie A&P, Glasgow (1994).
12. Minifie, B.W. (Ed.) *Chocolate, Cocoa and Confectionery*, 3rd edn. Van Nostrand Reinhold (1989).
13. Reddy, N.R., Pierson, M.D. and Salunkhe, D.K. *Legume-based Fermented Foods*, CRC Press (1986).
14. Campbell-Platt, G. *Fermented Foods of the World*, Butterworths (1987).
15. *Prescott and Dunn's Industrial Microbiology*, 4th edn. Macmillan (1983).

10 Change in cell structure
M. EDWARDS

10.1 Introduction

Fruit, vegetables, meat, poultry and fish are all derived from living organisms which are composed of cells. It is this cell structure which is one of the major contributors to the characteristic texture of a food. The breakdown pattern of the food as it is chewed, and the shape, size and composition of the particles produced by chewing, are responsible for the food texture which is perceived in the mouth [1]. These factors are determined by the initial structure of the food material and the effects of processing and cooking in modifying the planes of weakness in the tissues. It is part of the aim of food processing to maintain the characteristics of that structure through the treatments which are applied, including freezing, thermal processes or other methods such as chemical treatments. The texture at which the food processor is aiming will also depend on the expectations of the consumer; for example, the desired texture of a frozen soft fruit may be very close to that of the fresh fruit, whilst in processing meat the aim will be to produce a tender piece of meat after final cooking.

Many of the changes discussed in the earlier chapters of this book have an influence on the behaviour of cells during food processing, but it is primarily their structure which determines how this influence is manifested. This chapter will examine changes in the cell structure caused by food processing of plant and animal tissues, considering both freezing and wet and dry heat, and will then attempt to draw some general conclusions.

10.2 Plant tissues

All living organisms are composed of cells comprising a living protoplast bounded by a fragile semi-permeable membrane, the plasmalemma. The protoplast contains the cytoplasm of the cell, in which are the cell nucleus, mitochondria and other organelles, and the cell vacuole. Each of the various organelles has a particular function in life, and to this end is surrounded by a semi-permeable membrane, which controls the movement of molecules into and out of the organelle. One of the defining characteristics of the plant kingdom is that the plasmalemma is further surrounded by a cell wall. The wall is composed of cellulose and hemicellulose fibres surrounded by water. Neighbouring cell walls are cemented to each other along the junction between them (the middle lamella) by polymers of galacturonic acid or pectins (Figure 10.1). Cell walls cemented together impart a degree of rigidity, but the main means of maintaining shape in plant tissue is the turgor pressure within individual cells. Turgor pressure is the cell pressure produced by osmotically active constituents (i.e. able to drive moisture across the semi-permeable membrane), principally sugars and ions. When turgor pressure is lost, the structure collapses to a greater or lesser degree, as illustrated in a plant which wilts owing to lack of water. Different plants and plant organs have different degrees of susceptibility to such loss of structure, depending on the relative importance of other factors such as the inherent rigidity of the cell wall, the strength of the middle lamella bond and the degree of support from within the cell due to structures (such as starch granules and from other tissues such as fibres). Hence, a thin-walled, largely two-dimensional structure such as a lettuce leaf loses its characteristic texture much more readily than a potato, which has a greater three-dimensional structure of cell walls giving each other mutual support. In addition, each cell is to some extent supported internally by the starch granules within it, and the tissue is much less susceptible to water loss as a result of the much smaller surface-to-volume ratio.

Most animal and plant tissues used for food consist of 55–95% water, and the behaviour of this component is a vital factor in the effects of food processing on structure and texture. Water acts as a plasticiser in biological materials, reducing the glass-transition temperature (see chapter

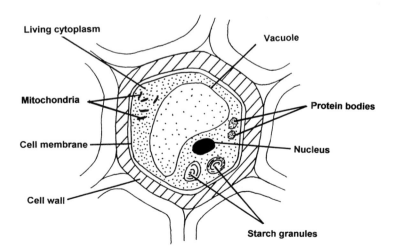

Figure 10.1 Diagram showing the main structural features of a plant cell.

2) from about 200°C (390°F) to about −10°C (14°F), without which they would be glassy. Using composite mechanics, Vincent [2] showed that the tensile and fracture properties of grass were governed by the longitudinal sclerenchyma fibres in the leaves. The stiffness of grass leaves changes with water content, increasing rapidly below a water content of 20% of dry weight, owing to the disappearance of 'free' (i.e. liquid not bound) water which acts as a plasticiser. A fibrous plant material such as grass is 'notch-insensitive', that is, its strength is directly proportional to the load-bearing cross-sectional area. Therefore, a small notch cut in the edge does not reduce its strength markedly. In contrast, 'notch-sensitive' materials such as fresh apples, carrots or potatoes lose a very large proportion of their strength if only a small notch is cut in the side. Notch insensitivity is due to the matrix material binding the continuous parallel elements together so that shear is not transmitted effectively; hence, if a single fibre breaks, stress is redistributed evenly across the other fibres, rather than being concentrated at any one point. At high moisture contents, fractures pass straight across laminae between notches. Below about 50% water content, the fracture takes a more complex course and is more readily deflected by fibre bundles.

Food processing methods, which have the effect of killing the cells, destroy the integrity of the plasmalemma and the ability of the cell to maintain turgor. Food processing by heat will also result in changes to the cell wall, particularly the middle lamella, and gelatinisation of the starch. The final texture will therefore depend on the relative importance of each factor contributing to texture, and the degree to which that factor has been changed by the processing method used.

10.2.1 Effects of moist heat on plant tissues

As the temperature rises from 20°C to 60°C (68°F to 140°F), thickening of cell walls occurs, together with disruption of the plasmalemma. Loss of cellular integrity has been seen at around 60°C (140°F) in carrots [3]. The rigidity of the tissue was closely related to the shear modulus, which declined rapidly at about 55–60°C (132–140°F), in contrast to the slow, steady decline in the modulus of elasticity (chapter 5 [4]). On prolonged heating, hemicellulose and cellulose both undergo some breakdown [5].

In mature cellulose walls, two principal layers, the primary and secondary cell walls, may be distinguished by microscopy. The more mature primary cell wall is less susceptible to breakdown than the younger secondary cell wall. Here, the typically reticular (net-like) microscopic structure of the raw material is lost and replaced by an amorphous structure embedded within a fibrillar network. The central zone of the secondary cell wall, seen as densely packed tubules (small tubes) in raw tissue, is transformed into a mass of dense amorphous matter after cooking

[6]. The cytoplasm of the cells forms dense clumps of granules, which are usually arranged along the cell walls [3].

Cell separation. As well as breakdown of the cell walls themselves, breakdown of the pectin in the middle lamellae of some cells begins, leading to cell separation [7]. This results in a change in the fracture properties of the tissue. In raw or lightly cooked plant tissue, the weakest plane, through which the tissue fractures, is across cell walls, and so a fractured surface reveals details of the internal structure of the cells (Figure 10.2). As middle lamella breakdown proceeds, the weakest plane becomes the line of the middle lamella so a fracture surface consists of the outer surfaces of the cells (Figure 10.3) [8–10]. Middle lamella breakdown occurs earlier in potato varieties giving a mealy texture than in those giving a firmer texture [5,11–15]. Cells in the fracture planes of firm potatoes have large intercellular contact areas; more of the cell walls are flat, and cell surfaces show folds and cracks. In mealy potatoes, cells in fracture planes have much smaller intercellular contact areas, and the cells are rounded and turgid (i.e. swollen by turgor pressure) with smooth surfaces. This suggests that final texture in potatoes is determined by a combination of the structure of the cell wall and middle lamella, combined with starch swelling distending the cell (see section 10.2.2) [8]. Release of amylose from ruptured cells gives rise to a sticky texture [5].

Middle lamella breakdown is believed to be due to the breaking of hydrogen bonds in pectins and other cell-wall polysaccharides [15]. High-

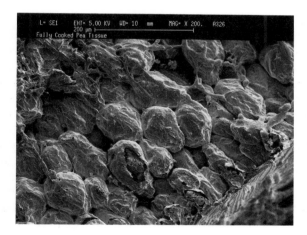

Figure 10.2 Scanning electron microscope (SEM) view of the fracture surface of blanched pea cotyledon tissue. Note that the tissue has fractured across the cells, revealing cell contents such as starch granules. Freeze-dried specimen; magnification ×200.

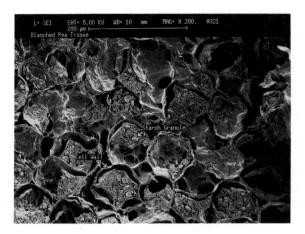

Figure 10.3 SEM view of the fracture surface of fully cooked pea cotyledon tissue. Note that the tissue has fractured across the middle lamellae, revealing the outside of each cell wall. Freeze-dried specimen; magnification ×200.

methoxyl pectins, with a high degree of substitution of hydroxyl groups with methoxyl groups, tend to be less water soluble and hence more resistant to this breakdown than low-methoxyl pectins. Therefore, the ratio of high and low methoxyl pectins in the tissue influences the rate at which cell separation occurs. Changes in these ratios are often responsible for increasing tenderness during ripening of fruits, cells becoming increasingly separated as the tissue matures. This is controlled by the native enzyme polygalacturonase; genetic engineering has been used to control the activity of this enzyme in order to produce a tomato which ripens very slowly and hence has an extended supermarket shelf-life [16,17]. Pectinase enzyme activity can also be manipulated to improve the texture of frozen vegetables such as green beans, which are normally given a hot-water blanch to de-activate enzymes before freezing (see chapter 14). Instead of a single blanch at about 93°C (200°F), the beans are given a preliminary 70°C (158°F) blanch, during which pectinase enzymes partially demethylate the pectins, leaving hydroxyl sites free on the pectin chain to cross-link with other pectin molecules via a calcium ion bridge, resulting in a firmer texture [18,19]. The beans are then given a second, high-temperature, blanch to destroy enzyme activity. This must be controlled very carefully; pectinase activity, which occurs in products kept warm for too long during processing, may destroy sufficient pectin to reduce the final product to a soft paste. Extraction of pectins into the liquor in canned apricots and peaches can result in a softer final fruit texture [20], although this phenomenon has also been related to fungal pectinases which are heat-resistant. In some canned vegetables, calcium and magnesium are often added to soft water used in canning, to give a firmer texture.

Hard-to-cook legumes. In navy beans (*Phaseolus vulgaris*), the so-called 'hard-to-cook' (HTC) phenomenon has been attributed to a failure of the middle lamella to dissolve during cooking, preventing cell separation [6]. A possible mechanism by which the middle lamella becomes insoluble was proposed by Jones and Boulter [21], who showed that during storage under conditions leading to HTC, phytate is partially hydrolysed and pectin is de-esterified. This results in the formation of calcium magnesium pectate, rendering the middle lamella pectins insoluble. Although Hincks and Stanley [22] provided tentative evidence for lignification of cell walls in HTC beans, Srisuma *et al.* [23] could detect no significant differences in lignin content between normal and HTC beans.

Keshun Liu *et al.* [24] found evidence of decreases in solubility and the thermal stability of intracellular proteins in HTC cowpeas during storage, and suggested that this prevented water reaching the starch granules. However, they also showed a lack of cell separation at the middle lamella, and their work failed to demonstrate clearly that the protein changes were the cause rather than a consequence of the phenomenon. Garcia and Lajolo [25] linked changes in the gelatinisation temperature of starch to starch alterations, resulting in increased resistance of starch granules to enzyme attack, and the hard-to-cook phenomenon in beans.

In a similar phenomenon, some samples of dry beans fail to take up water when soaked prior to heat processing, leading to dry, hard patches in the final product in affected peas. These are known as 'non-soakers'. Work at Campden (Edwards, Pither and Bedford, unpublished) and elsewhere [26] has shown that this is due to a failure of the middle lamella to soften and separate during soaking, thereby preventing cells from swelling fully as water is taken up. This was linked with differences in the relative contents of low- and high-methoxyl pectins. At the opposite end of the scale, some pea samples produce a gelled brine in the can after processing; this was related to relatively high proportions of water-soluble pectins in the peas leaching out during processing.

10.2.2 *Effects of processing on starch*

The two most important factors, other than changes in turgor pressure, are breakdown of the cell walls and the effect of the starch. The effect of these following heat processing is seen in starchy foods such as the pea [9] and the potato [8,10]. The first starch granules gelatinise as the temperature reaches about 60°C (140°F). However, some granules may remain ungelatinised until the temperature exceeds 70°C (158°F) [13,14].

Kaczynska *et al.* [27] studied heat-induced structural changes and starch swelling in faba bean starch paste. Amylose is released from the granules as they are heated and the amylopectin inside the granules is highly swollen.

In potatoes and other starch-based tissues, some workers have concluded that swelling of the gelatinised starch causes distension of the cell wall, resulting in splitting [5,11,28]. Reeve [5,11,29] has related the degree of starch swelling and cell-wall rupture to stickiness in mashed and processed potato products. However, other workers have suggested that, whilst gelatinised starch may swell to fill the space available, it does not cause cell separation and splitting [30,31].

The gelatinisation of starch is heavily dependent on water availability. Gelatinisation of potato granules in a solution of amylose or amylopectin results in less swelling of the granules than in water [32]. Much higher energy levels are required to produce proper gelatinisation of starch at low water content [33] than in high-moisture gelatinisation. In baked products, with relatively low water activity, swelling of granules is limited, and many granules do not gelatinise, even at quite high temperatures.

10.2.3 Effects of dry heat on plant tissues

The effects of dry heat on plant tissue are concerned principally with the manufacture of bakery products. The bulk of the grain or seed is made up of endosperm, a storage tissue in which the cells contain starch granules and protein bodies. Much of the structure of the grains used in this type of food is destroyed by milling, although fragments of cell structure are still clearly identifiable after milling [34]. Mixture of the flour with water to produce dough results in significant structural changes. The protein gluten (found in protein bodies within the cell cytoplasm) is responsible for much of the structural development in baking of bread; of the proteins comprising gluten, only glutenin swells in contact with water, whilst gliadin dissolves completely [35]. Individual gluten networks, already existing in flour particles, amalgamate to form a coherent whole during dough making, and protein films are formed by kneading. These structures form the gas cell walls in the final product (see chapter 12). After heating gluten at different moisture contents, the first major change at moisture contents <20% is a decrease in solubility of large glutenin aggregates and hydrophobicity of some proteins. At higher moisture contents, disulphide bonding and irreversible tertiary structure (the overall shape of the molecule) changes can also be observed [36]. After heat treatment, proteins are stretched in elongated sheets and starch granules are gelatinised [34]. Endosperm cell walls become highly fragmented during baking, and along with the starch, help to form a continuous matrix [37,38].

10.2.4 Oil frying of plant tissues

Cooking in oil is widely used in the production of snack foods of various kinds, and results in drastic changes to the structure of the food. Structural changes to the cytoplasm and protein bodies and cell separation all occur more rapidly in oil-cooked peanuts than when they are oven-cooked [39], and this is probably typical of most such products.

Penetration of the oil is particularly important for structure development in products such as potato crisps [40], tortilla chips [41] and potato chips [42]. In products which become totally hard and crisp, the oil first coats the surface and begins to move into the chip. As the heat builds up inside, moisture turns to steam and exits the chip, leaving a sponge-like network of tunnels which fill with oil. This process occurs within the first 20 seconds of frying. Starch granules gelatinise inside the chip, but not at the surface. The interior becomes smooth and plastic as protein, starch and lipids interact to form a continuous phase which hardens upon dehydration. In products with a soft interior, such as potato chips (French fries), the dehydrated oil-infused layer is confined to the peripheral four or five layers of cells [42]. The hard crisp surface pellicle (or crust) develops at the same time as the interior of the chip is changed to a structure comparable to that produced by moist heat.

10.2.5 Effects of microwaving on plant-cell structure

Although many acceptable food products are prepared using microwave energy, results are less than satisfactory with some food products, particularly starch-based foods. Relatively few studies have been carried out on the effects of microwave heating on plant-cell structures and food texture, and even fewer on animal tissues. The poor quality of some starch-based microwave-treated foods may be related to fast heating rates, difference in heat and mass-transfer mechanisms, or specific interactions of the components of the food with microwave radiation [43]. Microwave energy can produce comparable softening of potato tissue in about one-third of the time required by boiling water [44]. A more mealy texture results in microwave-treated potatoes. This may suggest that more complete cell separation occurs, or that there is less leakage of amylose from damaged cells, reducing apparent stickiness.

10.2.6 Effects of freezing on plant-cell structure

The principal damage caused to plant cell structure by freezing is the result of ice crystal formation, and hence the position and size of ice crystals is critical. Ice crystal formation is a phase change in which water molecules stop moving and form an ordered crystalline structure; in so doing, they give up energy as heat to their surroundings. This slows down the freezing of other water molecules around, so that during freezing the temperature of the water or the food drops only very slowly, but once freezing is complete there is a sudden fall in temperature (Figure 10.4).

Plant material is generally recognised as being more difficult to freeze satisfactorily than animal material because of the wide range of tissues in a fruit or vegetable; they also have a variety of enzyme systems and substrates, and so a wide range of enzyme reaction sequences is possible.

There are two processes which cause damage to cellular structure and lead directly to a loss of firmness in plant tissue. The first is when crystals puncture the cell membrane, leading to loss of turgor pressure. Secondly, there is damage to the cell wall structure caused by growing ice crystals. Fruits tend to have thin-walled cells, relying heavily on turgor pressure for firmness of texture, and are easily damaged. Vegetable tissues generally have somewhat thicker cell walls, and are more resistant. Ice crystals tend to form first outside the cells, in the intercellular spaces. When freezing is slow, these crystals grow very large, and water is withdrawn from within the cells to add to the crystals. Fast freezing, however, prevents the water from moving before it is frozen, and therefore gives smaller crystals both inside and outside the cells. This results in less damage to the tissue.

The formation of ice crystals also punctures internal cell membranes, allowing enzymes and substrates to mix which are normally separated. This results in a wide range of chemical reactions leading to breakdown of the cells and the development of off-flavours and colours. There is a number of potentially damaging enzyme reactions, listed in Table 10.1.

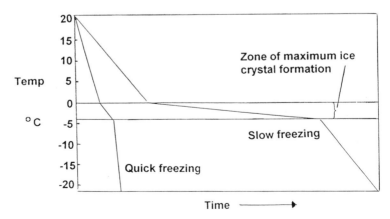

Figure 10.4 Diagram showing the effect of the water–ice phase change on the change in temperature of a food sample during freezing.

Table 10.1 Potentially damaging enzyme reactions in frozen fruit and vegetables

Enzyme	Substrate	Product
Catalase	Hydroperoxides	Oxygen
Lipase	Triglycerides	Monoglycerides, glycerol, fatty acids
Lipoxygenase	Unsaturated fatty acids	Fatty acid hydroperoxides, carbonyl compounds
Pectic enzymes	Pectins	Pectic acids, uronides, galacturonic acid
Peroxidase	Peroxides and other electron donors	Oxidised donors
Polyphenol oxidase	Phenols	Melanins

Where these enzyme reactions are important – particularly, for instance, in green vegetables – a brief heat treatment (blanching) can be used to inactivate the enzyme.

Freezing of plant tissue slows down chemical reactions. However, because withdrawal of water to form ice crystals concentrates the remaining solutions, reaction rates during storage may be increased in some cases. Proteins may be denatured by alteration of their tertiary structure or by breakage into smaller units which do not reassemble on thawing. Vitamin C (ascorbic acid) may be converted to dehydroascorbic acid, which is active as vitamin C, and to 2,3-diketogulonic acid, which is not active. Chlorophyll may be converted to pheophytin and other pigments may also be degraded.

The main physical damage occurring during frozen storage is dehydration as a result of fluctuating storage temperatures. As the temperature rises, water evaporates from the food. As the temperature drops again, water vapour condenses on the packaging, because this cools before the food. This results in noticeably dry and discoloured areas on the food, known as 'freeze burn'.

Starch granules in vegetables, such as potatoes, may be partly gelatinised during blanching, and hence absorb some water. Starch gelatinisation may be reversed to some degree during frozen storage due to progressive loss of water as a result of fluctuating temperature. This retrogradation of starch results in poor texture and a loss of water-holding capacity of the starch gel. Starch pastes can be used as protective systems for the solid elements in precooked frozen foods, by minimising dehydration and chemical changes in storage. Rapid freezing of starch pastes results in smaller ice crystals and the absence of retrogradation. Retrogradation and ice crystallisation during storage of frozen pastes contribute to degenerative changes such as spongy structure, decrease in apparent viscosity and increased syneresis (separation of liquid from the gel). Addition of xanthan gum minimises this deterioration, but affords no protective effect on ice crystal size and amylopectin retrogradation [45].

It is not always appreciated that the most damaging process in processing frozen food is often thawing. During thawing of plant tissue, the temperature rise moves inwards and can cause dehydration in the same way as fluctuations in storage temperature. As ice crystals melt, they undergo the same phase change as in freezing, but in reverse, and so they take in heat energy. This cools their surroundings and there is partial re-freezing, resulting in the growth of very large ice crystals during slow thawing, and causing further damage to cell walls and membranes. Due to puncturing of cell walls and membranes, much water is lost as 'drip', and any enzymic reactions which were retarded by freezing will speed up again. This may include any of the enzyme reactions mentioned earlier, and since more membranes will have been ruptured during freezing and thawing, more mixing of enzymes and substrates is possible.

10.3 Animal tissues

Meat and poultry consist principally of skeletal muscle tissue, which is basically similar in all species. The fact that a single tissue type is involved means that only a relatively small range of structures, enzymes and substrates have to considered, making the work of the food processor somewhat easier than for plant tissues. The muscles within an animal vary in shape and in their detailed micro-anatomy, but the structural hierarchy and the macromolecular units of which they are constructed are common to all. Muscle tissue is composed of bundles of cells known as 'myofibres'. The proteins in the myofibres are principally the myofibrillar proteins myosin, actin, actomyosin and troponin with sarcoplasmic proteins which are mainly albumins. Each myofibre is surrounded by a fine connective tissue membrane, the endomysium. The cell bundles are separated from each other and are collectively held together by sheaths or membranes of connective tissue comprised of collagen and elastin (the perimysium) (see Figure 10.5). Muscle firmness is related to the density and arrangement of collagen fibrils in the pericellular connective tissue [46]. A sheath of white connective tissue (the epimysium) surrounds the entire muscle; the terminal processes of this sheath form tendons by which the muscle is attached to the skeleton of the animal.

The myofibres comprising the fibre bundles are long, thin, multi-nucleated cells which may be >100 mm (4 in.) long, but only 0.1 mm (0.004 in.) in diameter. Each cell is bounded by a cell wall, the sarcolemma, consisting of an internal membrane or plasmalemma and an external basement membrane composed of collagen and mucopolysaccharide. Contained within each myofibre is a highly organised structure of myo-

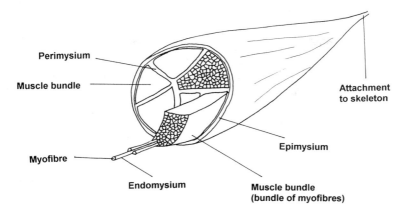

Figure 10.5 Diagrammatic representation of the overall structure of skeletal muscle.

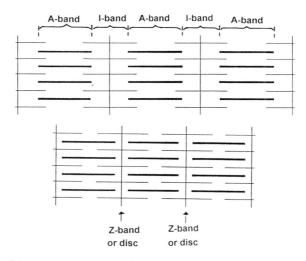

Figure 10.6 Diagrammatic representation of the microscopic structure of striated muscle.

fibrils, consisting of repeating, overlapping filaments which are able to slide across each other telescopically, so shortening the length of the cell and resulting in contraction. Each repeating unit is known as a 'sarcomere'. The repeating arrangement of these fibres results in banding or striations which are readily seen under the light microscope. Figure 10.6 gives a diagrammatic view of the bands. The energy for contraction comes from the breakdown of adenosine triphosphate (ATP), the presence of which also prevents cross-bridges from linking together the sarcomeres. Contraction is triggered by the release of calcium ions from the sarcoplasmic reticulum, a network of tubules surrounding the myofibrils. The energy for ATP production comes from metabolism of glycogen stored in the muscle. Lawrie [47] gives further information on muscle structure.

There are a number of natural post-mortem changes in muscle tissue which can profoundly affect the texture of the meat, and which interact with any food processing steps. Myofibres remain sensitive to a stimulus and capable of contraction for some time after death. Immediately after slaughter, meat has a tender texture; during this period, the ATP and glycogen in the muscle are used up in continuing metabolic processes. Because of the lack of a pumped blood supply, these processes have to continue without oxygen, and so glycogen is broken down to produce lactic acid instead of carbon dioxide, and the meat becomes more acid. This drop in pH causes the denaturation of some sarcoplasmic and myofibrillar proteins, and the sarcoplasmic proteins are precipitated onto the myofibrils in a more or less random pattern. This also results in loss of water-holding capacity, resulting in drip. Unfortunately, this drip carries with it some of the soluble flavour components of meat. Once the ATP supply is exhausted, it can no longer hold the actin and myosin filaments apart in a relaxed state and they become firmly attached to one another, resulting in rigor mortis. This gives a very tough meat until autolysis separates the fibres and causes the muscle to soften again. Post-rigor weakening of the rigor linkages formed between actin and myosin weakens the myofibrils, resulting in fading of the Z-band (Figure 10.6) [48,49] and increased tenderness. The endomysium has been shown to break down into separate collagen fibres, and the sheets of perimysium also resolve into separate collagen fibres in extended post-mortem conditioning [50]. When pre-rigor muscle is cooled to $<10°C$ ($<50°F$), the sarcoplasmic reticulum can no longer store calcium ions efficiently, so they are released, resulting in muscle contraction, known as 'cold-shortening'. Once the muscle pH has fallen to about pH 6, the myofibres are no longer excitable and chilling does not cause contraction. During cold-shortening, some fibres contract actively whilst others are passively contracted [51]; this may be related to the observation that there is more than one type of muscle cell. These are referred to as red, white and intermediate and have been found to be structurally and biochemically distinct [52].

Infusion of carcases with a solution of maltose, glycerin, dextrose and phosphate can be used to break down muscle structure [53]. The myofibrils are both fragmented and separated. Immediately post-mortem, the weakest plane (the fracture plane) is along the endomysial-sarcolemmal sheath, whilst 24 hours later the sheath is weakened enough for the fracture to occur along the surface of the myofibrils. Papain injections have also been used to tenderise meat in a similar way to breaking down meat proteins.

10.3.1 Effects of freezing on muscle tissue

The damage caused to muscle tissue by freezing, like that caused to plant tissue, depends on the rate of freezing. When freezing is rapid, large numbers of small, needle-like ice crystals are formed within individual muscle cells [54]. These cause little detectable structural damage, but mitochondria are damaged by ice formation, and meat can be shown to have been previously frozen by the detection of enzymes leaking from the mitochondria into the sarcoplasm [55–57]. When freezing is slow, large ice crystals are formed extracellularly, producing large spaces within the tissue [54] (compare Figures 10.7 and 10.8). The slower the freezing rate, the larger the crystals and the more extensive the damage. Large crystals may damage tissue components and cause fragmentation of the myofibres. Withdrawal of water from the cells to form ice increases the salt concentration in the remaining sarcoplasmic fluid, resulting in protein denaturation and a loss of water-binding capacity. The permeability of the cell membrane is also affected. On thawing, some of the water in the ice crystals returns to the cells, but some is also lost as 'drip'.

If muscle tissue is frozen before the onset of rigor mortis, rapid freezing results in massive 'cold-shortening', similar to, but more marked than that seen in chilled muscle tissue; very rapid freezing is required to overcome this. Massive contraction ('thaw-shortening') also occurs in pre-rigor muscle on thawing, unless it is thawed very slowly to use up all the available ATP before the support of the ice disappears. Both of these phenomena can result in tough meat and increased 'drip'. Muscle may be most easily frozen and thawed after rigor mortis has passed, when the size and position of ice crystals become the main factors.

10.3.2 Effects of heating on muscle tissue

Light microscopy shows that the most obvious effect of heat on muscle structure is the production of transverse breaks in the myofibres [58,59]. The breaks may occur right across the fibre, or they may be incomplete, giving a zig-zag appearance. Heat denaturation of the fibrillar proteins results in increased optical density of the myofibres in a stained section, also resulting in the dull appearance of cooked meat as compared with the translucent appearance of fresh raw meat. Sarcomeres are shortened and cell diameter reduced [60,61]. There is a loss of ultrastructural detail as seen by transmission electron microscopy; in particular the A-band regions become amorphous, detail is lost from the Z-discs and the I-bands are disrupted [61,62]. The degree of ultrastructural change caused by heat is temperature-dependent; for instance, chicken breast myosin heated at 55°C or 65°C (131°F or 149°F) gives a more elastic gel than that heated at

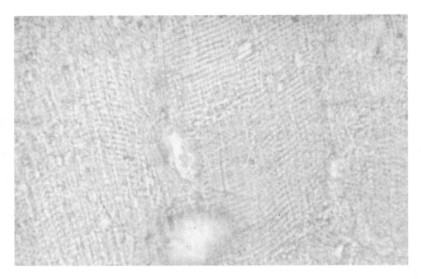

Figure 10.7 Light microscope view of the structure of muscle before freezing.

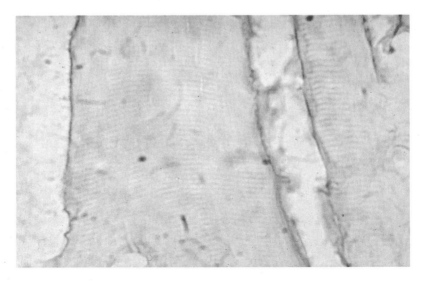

Figure 10.8 Light microscope view of the structure of muscle after freezing: note the large gaps left by ice crystals between the muscle cells.

75°C (167°F) [63]. The degree of damage is limited by the contractile state of the muscle and the consequent overlap between thick and thin filaments [64].

Collagen fibres undergo marked shrinkage in the temperature range 60–70°C (140–158°F); collagens from different sources have specific heat-shrink temperatures [65]. This shrinkage pulls the myofibres into a wavy configuration, and the entire piece of meat is reduced in size.

As heating proceeds, collagen is solubilised or converted to gelatine at about 80°C (176°F). In older animals, more inter- and intra-molecular cross-links are formed in the collagen fibres [66], and so the amount of solubilisation which can occur on heating is decreased. The amount of solubilisation occurring at a given temperature depends on the rate of heating and on the length of time at that temperature; hence, muscles with a high collagen content require longer, slower cooking than muscles of low collagen content.

The changes described above result in marked changes in the texture of the meat. Fragmentation of the myofibres tends to increase tenderness, since the tissue is more easily broken up by chewing. However, denaturation of myofibrillar proteins increases the hardness of the tissue, resulting in a loss of tenderness. Denaturation of the perimysial connective tissue may lead to a tendency to separation at the perimysium, so that the meat tends to separate into bundles of muscle fibres [67]. When meat is broken across the fibres, all the fibres must be broken to separate the pieces; cooked meat is ten times stronger along the fibre direction than perpendicular to it. Elastic fibres to not contribute significantly to texture. The conventional instrumental method of assessing meat texture employs the Warner–Bratzler shear press, which uses a shear action to imitate the chewing action. However, most studies of the structural basis of meat tenderness have used tensile tests to detect the preferential fracture planes. Perimysial connective tissue determines breaking strength when meat is pulled apart transversly; in longitudinal tensile tests, the initial event is detachment of fibre bundles from the perimysium, so that they contribute independently to load-bearing. At greater extensions, it seems likely that the fibre bundles fail progressively, leaving the perimysial strands as the last structure to break [68]. The final texture, therefore, is the net result of all of these contributing factors together with the effects of post-mortem treatment, chilling and freezing.

10.3.3 Effects of chemical additives on meat structure

The microstructure of meat, such as ham, may be significantly altered by the addition of salt and phosphates [69] and the chopping or tumbling of meat pieces to incorporate the additives. The injection of brine, together with the high injection pressure, separates the sarcolemma from the muscle cell and its myofibrils, and subsarcolemmic spaces are created. Myofibrils swell due to salt and water absorption; myofibrillar proteins are released. In heat processed ham, the shape of the myofibrils remains unaffected, but the typical structure of the sarcomere is no longer recognizable. Polyphosphate treatment alone results in a complete loss of the Z-disc on heating, whilst salt alone induces the loss of material from the actin-rich I-band. Heating of meat treated with both salt and polyphosphates results in the replacement of the Z-disc and the A- and I-bands with thinner bands of apparently coagulated material. These bands appear to be due to actomyosin, which is resistant to dispersion on processing; the water-retaining gel is likely to be formed by the action of polyphosphate in dispersing free actin and myosin by mimicking the action of ATP [70]. Interfibrillar spaces are created which are partially filled with granulated protein. This protein skeleton provides structural stability, and absorbs and holds water by capillary forces. Impact and friction energy generated during ham tumbling loosen the tissue and muscle cells are disintegrated. Cell disintegration occurs particularly in the external area of the meat, and together with the influence of brine, it increases the amount of swollen fibrils and promotes the release of protein from the muscle cells in the form of a slimy exudate. Continuous tumbling results in the production of more exudate, and this is distributed all around the meat, resulting in improved binding.

10.3.4 Effects of processing on fish cell structure

At the molecular level, the structure of fish muscle is similar to that of mammals and birds, but the muscle fibres are organised into blocks, or myotomes, lying between sheets of connective tissue or myocommata, giving fish flesh its flaky structure.

The body muscle of fish is of two distinct types, normally referred to as 'white' and 'dark' (or 'red') muscle. There are differences in the chemistry and physiology of the two types. White muscle constitutes the bulk of the musculature of most fish, particularly the white-fleshed demersal species (the so-called 'white fish', such as cod). It does not contain significant stores of glycogen, and relies on anaerobic glycolysis to supply energy for contraction. Dark muscle is present to a larger extent in pelagic or fatty fish, e.g. mackerel than in demersal fish. It is better supplied with blood vessels and stores of glycogen, and its metabolism is basically aerobic.

Fish muscle is much more susceptible than that of mammals or poultry to changes during freezing and storage because of the high unsaturated fat content and rather unstable protein composition.

Post-mortem glycolysis follows a similar path in fish to that in mammals and birds. However, because fish are generally hunted from the wild, they are usually stressed at death, resulting in a much smaller fall in muscle pH. Although the biochemistry of rigor mortis in fish is somewhat different from that in other animals, the physical phenomena follow a similar sequence. In most food-processing operations, fish are chilled or frozen after death. As fish muscle goes into rigor mortis, it contracts with a force which increases with temperature. The innate strength of the muscle tissue, however, decreases with rising temperature, and above a certain point rigor contractions progressively weaken the tissue so that separation of the flakes (gaping) occurs; this is most noticeable in fish which has been frozen, thawed and then filleted. If fillets are removed in a pre-rigor state, rigor tensions are unopposed; so gaps are not produced.

Post-mortem tenderisation of fish muscle is caused, not by weakening at the Z-discs, but by disintegration of the plasmalemma (cell membrane) allowing cell contents to be shed into the extracellular space. The basement membrane becomes more diffuse, but collagen fibres may still be found, although collagen fibres in pericellular connective tissue break down in rainbow trout [46,71]. Sarcoplasmic enzymes, including glycolytic and hydrolytic enzymes, are principally responsible for the quality deterioration of fish muscle after death [72].

Thaw rigor in fish appears to be due to damage to sarcoplasmic reticulum membranes caused by ice crystals during freezing and thawing. The resultant calcium ion release from the sarcoplasmic reticulum probably causes an increase in ATP consumption and an acceleration of thaw rigor by activating myofibrillar Mg^{2+} ATPase [73].

Storage of fish under unsuitable frozen conditions gives rise to a tough, chewy, dry and stringy texture, due at least in part to changes in the myofibrillar proteins [74]. As with the freezing of meat, the size, shape and position of ice crystals depends on the rate of freezing and whether the fish is frozen pre- or post-rigor [75]. On thawing, the water formed as ice crystals is largely resorbed, although fissures are left in the flesh where crystals were present. In poorly stored fish, water is only partially resorbed, and the fibres retain the structure of frozen muscle. This appears to be due to compression of the sarcoplasmic reticulum into a thin layer [76], which degrades during frozen storage and acts as a glue to cement the myofibrils together, although other work has failed to demonstrate the expected reductions in filament spacing in such material [77,78].

The main structural changes associated with heating fish muscle occur at relatively low temperatures, in the range 5–40°C (41–104°F) [79]. Water loss at these temperatures is probably due to denaturation and melting of collagen. Maximum water loss is attained when the muscle cells shrink due to denaturation of myosin. The reduced water loss at higher temperatures (50–70°C, 122–158°F) is probably caused by aggregates of sarcoplasmic proteins stabilising the aqueous phase.

10.4 Conclusions

The structural changes to animal and plant cells caused by food processing are not only dependent upon the overall structure of the cells, but are also the result of complex interactions between the response of their chemical components to heat, cold and chemical treatments such as brining. In plant tissues, the more important factors in texture and structure development are cell separation and starch gelatinisation. Animal materials for food use, in contrast, are principally one type of tissue – muscle. Here changes are strongly influenced by the interaction of food processing treatments with natural post-mortem changes in the tissue derived from both natural muscle function and tissue autolysis. Further improvements in the control of food quality during commercial processing are likely to be derived from a better understanding of these interactions.

References

1. Christensen, C.M. Food texture perception. *Adv. Food Res.* **29** (1984), 159–199.
2. Vincent, J.F.V. *Structural Biomaterials*. Basingstoke: Macmillan (1982).
3. Grote, M. and Fromme, H.G. Electron microscopic investigations of the cell structure in fresh and processed vegetables (carrots and green bean pods). *Food Microstruct.* **3** (1984), 55–64.
4. Ramana, S.V., Wright, C.J. and Taylor, A.J. Measurement of firmness in carrot tissue during cooking using dynamic, static and sensory tests. *J. Sci. Food Agric.* **60** (1992), 369–375.
5. Reeve, R.M. Histological survey of conditions influencing texture in potatoes. I. Effects of heat treatments on structure. *Food Res.* **19** (1954), 323–332.
6. Shomer, I., Paster, N., Lindner, P. and Vasiliver, R. The role of cell wall structure in the hard-to-cook phenomenon in beans (*Phaseolus vulgaris* L.). *Food Struct.* **9** (1990), 139–149.
7. Burton, W.G. *The Potato*, 3rd edn. New York: Longman (1989).
8. van Marle, J.T., Clerkx, A.C.M. and Boekestein, A. Cryo-scanning electron microscopy investigation of the texture of cooked potatoes. *Food Struct.* **11** (1992), 209–216.
9. Edwards, M.C. The measurement of pea texture and its interpretation. *Proceedings of the Symposium on The Measurement of Maturity in Vining Peas*, Ed. Arthey, V.D. Chipping Campden: Campden Food and Drink Research Association (1994).
10. Rose, D.J., Edwards, M.C., Rendell, E. and Roberts, C. *The effect of process variables on the structure and texture of vegetables using the potato as a model*. Technical Memorandum No. 550. Chipping Campden: Campden Food and Drink Research Association (1989).
11. Reeve, R.M. Histological survey of conditions influencing texture in potatoes. II. Observations on starch in treated cells. *Food Res.* **19** (1954), 333–339.
12. Reeve, R.M. Estimation of extra-cellular starch of dehydrated potatoes. *J. Food Sci.* **28** (1963), 198–206.
13. Reeve, R.M. A review of cellular structure and texture qualities of processed potatoes. *Economic Botany* **21** (1967), 294–308.
14. Reeve, R.M. *Proceedings of the 16th Annual Potato Util. Conference*, Ed. Collins, F.E. ARS 74-40, U.S. Department of Agriculture (1967).
15. Sterling, C. Anatomy and histology of the tuber with respect to processed quality. In *Proceedings of the Plant Science Symposium, Camden, N.J.* Campbell Institute of Agricultural Research (1966), 11–25.

16. Smith, C.J.S., Watson, C.F., Ray, J., Bird, C.R., Morris, P.C., Schuch, W. and Grierson, D. Antisense RNA inhibition of polygalacturonase gene expression in transgenic tomatoes. *Nature* **334** (1988), 724–726.
17. Schuch, W. Improving fruit quality through biotechnology. *AgBiotech News Information* **3** (1991), 249–252.
18. Moledina, K.H., Haydar, M., Ooraikul, B. and Hadziyev, D. Pectin changes in the precooking step of dehydrated mashed potato production. *J. Sci. Food Agric.* **32** (1981), 1091–1102.
19. Adams, J.B. and Robertson, A. *Instrumental methods of quality assessment: texture.* Technical Memorandum No. 449. Chipping Campden: Campden Food Preservation Research Association (1987).
20. Dobias, J., Curda D., Vana, V. and Zakova, P. Influence of pectic substance extraction into the brine on the texture of canned fruit. *Potravinarske Vedy* **11** (1993), 233–242.
21. Jones, P.M.B. and Boulter, D. The cause of reduced cooking rate in *Phaseolus vulgaris* following adverse storage conditions. *J. Food Sci.* **48** (1983), 623–626, 649.
22. Hincks, M.J. and Stanley, D.W. Lignification: evidence for a role in hard-to-cook beans. *J. Food Biochem.* **11** (1987), 41–58.
23. Srisuma, N., Hammerschmidt, R., Uebersax, M.A., Ruengsakulrach, S., Bennink, M.R. and Hosfield, G.L. Storage induced changes of phenolic acids and the development of hard-to-cook in dry beans (*Phaseolus vulgaris* var. Seafarer). *J. Food Sci.* **54** (1989), 311–314, 318.
24. Keshun Liu, Yen-Con Hung and Phillips, R.D. Mechanism of hard-to-cook defect in cowpeas: verification via microstructure examination. *Food Struct.* **12** (1993), 51–58.
25. Garcia, E. and Lajolo, F.M. Starch alterations in hard-to-cook beans (*Phaseolus vulgaris*). *J. Agric. Food Chem.* **42** (1994), 612–615.
26. Stanley, D.W. *Tuff beans – textural problems in beans. The emerging bean,* Winter 1988. Ontario Bean Producers' Marketing Board (1988).
27. Kaczynska, B., Autio, K. and Fornal, J. Heat-induced structural changes in faba bean starch paste: the effect of steaming faba bean seeds. *Food Struct.* **12** (1993), 217–224.
28. Jarvis, M.C., Mackenzie, E. and Duncan, H.J. The textural analysis of cooked potato. 2. Swelling pressure of starch during gelatinisation. *Potato Res.* **35** (1992), 93–102.
29. Reeve, R.M. Histological survey of conditions influencing texture in potatoes. III. Structure and texture in dehydrated potatoes. *Food Res.* **19** (1954), 340–349.
30. Bretzloff, C.W. Some aspects of cooked potato texture and appearance. II. Potato cell size stability during cooking and freezing. *Am. Potato J.* **47** (1970), 176–182.
31. Warren, D.S. and Woodman, J.S. The texture of cooked potatoes: a review. *J. Sci. Food Agric.* **25** (1974), 129–138.
32. Svegmark, K. and Hermansson, A.M. Microstructure and rheological properties of composites of potato starch granules and amylose: a comparison of observed and predicted structures. *Food Struct.* **12** (1993), 181–193.
33. Vainionpaeae, J., Forssell, P. and Virtanen, T. High-pressure gelatinisation of barley starch at low moisture levels and elevated temperature. *Starch/Staerke* **45** (1993), 19–24.
34. Gallant, J.D.J., de Monredon, F., Bouchet, B., Tacon, P. and Delort-Laval, J. Cytochemical study of intact and processed barley grain. *Ferment* **6** (1993), 111–114.
35. Eckert, B., Amend, T. and Belitz, H.D. The course of the SDS and Zeleny sedimentation tests for gluten quality and related phenomena studied using the light microscope. *Zeitschr. Lebensmittel-Untersuch. Forsch.* **196** (1993), 122–125.
36. Weegels, P.L., Verhoek, J.A., de Groot, A.M.G. and Hamer, R.J. Effects on gluten of heating at different moisture contents. II. Changes in physico-chemical properties and secondary structure. *J. Cereal Sci.* **19** (1994), 39–47.
37. Parkkonen, T., Haerkoenen, H. and Autio, K. Effect of baking on the microstructure of rye cell walls and protein. *Cereal Chem.* **71** (1994), 58–63.
38. Virtanen, T. and Autio, K. The microscopic structure of rye kernel and dough. *Carbohydrate Polym.* **21** (1993), 97–98.
39. Young, C.T. and Schadel, W.E. A comparison of the effects of oven roasting and oil cooking on the microstructure of peanut (*Arachis hypogaea* L. cv. 'Florigiant') cotyledon. *Food Struct.* **12** (1993), 59–66.
40. Gamble, M.H. and Rice, P. Effect of pre-fry drying on oil uptake and distribution in

potato crisp manufacture. *Int. J. Food Sci. Technol.* **22** (1987), 535–548.
41. McDonough, C., Gomez, M.H., Lee, J.K., Waniska, R.D. and Rooney, L.W. Environmental scanning electron microscopy evaluation of tortilla chip manufacture during deep-fat frying. *J. Food Sci.* **58** (1993), 199–203.
42. Selman, J.D. and Hopkins, M. *Factors affecting oil uptake during the production of fried potato products.* Technical Memorandum No. 475. Chipping Campden: Campden Food and Drink Research Association (1989).
43. Goebel, N.K., Grider, J., Davis, E.A. and Gordon, J. The effects of microwave energy and convection heating on wheat starch granule transformations. *Food Microstruct.* **3** (1984), 73–82.
44. Collins, J.L. and McCarty, I.E. Comparison of microwave energy with boiling water for blanching white potatoes. *Food Technol.* **23**(3) (1969), 63–66.
45. Ferrero, C., Martino, M.N. and Zaritzky, N.E. Stability of frozen starch pastes: effect of freezing, storage and xanthan gum addition. *J. Food Proc. Preserv.* **17** (1993), 191–211.
46. Ando, M., Toyohara, H. and Sakaguchi, M. Post-mortem tenderization of rainbow trout muscle caused by the disintegration of collagen fibers in the pericellular connective tissue. *Bull. Japanese Soc. Scient. Fisheries (Nihon Suisan Gakkai-shi)* **58** (1992), 567–570.
47. Lawrie, R.A. *Meat Science*, 5th edn. Oxford: Pergamon Press (1991).
48. Davey, C.L. and Dickson, M.R. Studies in meat tenderness. 8. Ultrastructural changes in meat during ageing. *J. Food Sci.* **35** (1970), 56–60.
49. Penny, I.F., Voyle, C.A. and Dransfield, E. The tenderising effect of a muscle protease on beef. *J. Sci. Food Agric.* **25** (1974), 703–708.
50. Nishimura, T., Hattori, A. and Takahushi, K. Structural weakening of intramuscular connective tissue during conditioning of beef. *Meat Sci.* **39** (1995), 127–133.
51. Voyle, C.A. Some observations on the histology of cold-shortened muscle. *J. Food Technol.* **4** (1969), 275–281.
52. Gauthier, G.F. In *The Physiology and Biochemistry of Muscle as a Food*, Vol. 2, Eds Briskey, E.J., Cassens, R.G. and Marsh, B.B. Madison: University of Wisconsin Press (1970), 103.
53. Farouk, M.M., Price, J.F. and Salih, A.M. Post-exsanguination infusion of ovine carcasses: effect on tenderness indicators and muscle microstructure. *J. Food Sci.* **57** (1992), 1311–1315.
54. Grujic, R., Petrovic, L., Pikula, B. and Amidzic L. Definition of the optimum freezing rate. I. Investigation of structure and ultrastructure of beef M. longissimus dorsi frozen at different freezing rates. *Meat Sci.* **33** (1993), 301–318.
55. Hamm, R. and Kormendy, L. Biochemisches Routineverfahren zur Unterscheidung zwischen Frischfleisch und aufgetautem Gefrierfleisch. *Fleischwirtschaft* **46** (1966), 615–616.
56. Gottesmann, P. and Hamm, R. New biochemical methods of differentiating between fresh meat and thawed, frozen meat. *Fleischwirtschaft* **63** (1983), 219–221.
57. Toldra, F., Torrero, Y. and Flores, J. Simple test for differentiation between fresh pork and frozen/thawed pork. *Meat Sci.* **29** (1991), 177–181.
58. Paul, P.C. Influence of methods of cooking on meat tenderness. In *Proceedings of the Meat Tenderness Symposium, Camden, N.J.* Campbell Soup Company (1963), 225–241.
59. Reid, H.C. and Harrison, D.L. Effect of dry and moist heat on selected histological characteristics of beef semimembranosus muscle. *J. Food Sci.* **36** (1971), 206–208.
60. Paul, P.C. Storage and heat-induced changes in microscopic appearance of rabbit muscle. *J. Food Sci.* **30** (1965), 960–968.
61. Schmidt, J.D. and Parrish, F.C., Jr. Molecular properties of postmortem muscle. 10. Effect of internal temperature and carcass maturity on structure of bovine longissimus. *J. Food Sci.* **36** (1971), 110–119.
62. Lewis, D.F. An Electron Microscope Study of the Factors Affecting the Structure of Meat. PhD Thesis, University of Leeds (1974).
63. Shue Fung Wang and Smith, D.M. Dynamic rheological properties and secondary structure of chicken breast myosin as influenced by isothermal heating. *J. Agric. Food Chem.* **42** (1994), 1434–1439.

64. Voyle, C.A. *Proceedings of the MRI Symposium No. 3 Meat Freezing: Why and How*, 6.1–6.6. Langford, Bristol: Meat Research Institute (1974).
65. Mohr, V. and Bendall, J.R. Constitution and physical chemical properties of intramuscular connective tissue. *Nature* **223** (1969), 404.
66. Shimokomaki, M., Elsden, D.F. and Bailey, A.J. Meat tenderness: age related changes in bovine intramuscular collagen. *J. Food Sci.* **37** (1972), 892–896.
67. Purslow, P.P. The fracture behaviour of meat – a case study. In *Food Structure and Behaviour*, Eds Blanshard, J.M.V. and Lillford, P. London: Academic Press (1987), 177–179.
68. Offer, G., Knight, P., Jeacocke, R., Almond, R., Cousins, T., Elsey, J., Parsons, N., Sharp, A., Starr, R. and Purslow, P. The structural basis of the water-holding capacity, appearance and toughness of meat and meat products. *Food Microstruct.* **8**, (1989), 151–170.
69. Katsaras, K. and Budras, K.D. The relationship of the microstructure of cooked ham to its properties and quality. *Lebensmittel-Wissenschaft Technologie* **26**, (1993), 229–234.
70. Kotter, L. *Zur Wirkung Kondensierkers Phosphate und anderen Salze auf Tierisches Eiweiss*. Hanover: Verlag M. and H. Schaper (1960).
71. Ando, M., Toyohara, H., Shimizu, Y. and Sakaguchi, M. Post-mortem tenderization of fish muscle due to weakening of pericellular tissue. *Bull. Japanese Soc. Scient. Fisheries (Nihon Suisan Gakkai-shi)* **59** (1993), 1073–1076.
72. Shahidi, F. Seafood proteins and preparation of protein concentrates. In *Seafoods: Chemistry, Processing Technology and Quality*, Eds Shahidi, F. and Botta, J.R. London: Blackie Academic and Professional (1994), 3–9.
73. Long Bin Ma, Yamanaka, H., Ushio, H. and Watabe, S. Studies on the mechanism of thaw rigor in carp. *Bull. Japanese Soc. Scient. Fisheries (Nihon Suisan Gakkai-shi)* **58** (1992), 1535–1540.
74. Sikorski, Z., Olley, J. and Kostuch, S. Protein changes in frozen fish. *CRC Crit. Rev. Food Sci. Nutr.* **8** (1976), 97–129.
75. Love, R.M. Ice formation in frozen muscle. In *Low Temperature Biology of Foodstuffs*, Eds Hawthorn, J. and Rolfe, E.J. Oxford: Pergamon Press (1968).
76. Connell, J.J. and Howgate, P. In *Freezing and Irradiation of Fish*, Ed. Kreuzer, R. London: Fishing News (Books) (1969), 145–146.
77. Jarenback, L. and Liljemark, A. Ultrastructural changes during frozen storage of cod (*Gadus morhua* L.). I. Structure of myofibrils as revealed by freeze etching preparation. *J. Food Technol.* **10** (1975), 229–239.
78. Aitken, A. and Connell, J.J. In *Freezing, Frozen Storage and Freeze-drying*. Paris: International Institute of Refrigeration (1977), 187.
79. Ofstad, R., Kidman, S., Myklebust, R. and Hermansson, A.M. Liquid holding capacity and structural changes during heating of fish muscle: cod (*Gadus morhua* L.). *Food Struct.* **12** (1993), 163–174.

11 Dairy products
J.E. HOLDSWORTH and S.J. HAYLOCK

11.1 Introduction

A wide variety of consumer and industrial products is manufactured today by processing raw milk into dairy products or though combining products derived from milk with other ingredients. Discussion of all of them is not possible in the space available. Therefore this chapter concentrates on the various physico-chemical processes associated with the manufacture of the following:

- Milks
- Butters
- Creams
- Cheese
- Cultured products
- Ice cream
- Powdered consumer products

11.2 Milks

Milk is a complex mixture of water, lipids, carbohydrates, proteins, salts and a long list of miscellaneous constituents. The lipids are present in emulsified, globular form, and most of the proteins are colloidally dispersed and can easily be separated from the lactose and various salts that are in true solution. This section concentrates on the major effects of milk processing on the key components. However, a fuller description of the effects can be found in reference [1].

The composition of raw milk is influenced by the breed of cow(s), the stage of lactation and the time of the dairy season, as well as by various other on-farm factors such as nutrition, disease and the milking procedure.

A schematic diagram of the processing options for town milk is shown in Figure 11.1, in which the key processes important to the final product are pumping, standardisation, heating, cooling and homogenisation. The major changes occurring during each stage are discussed below.

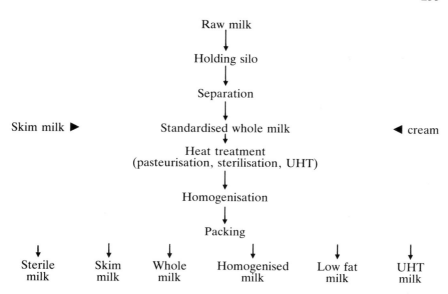

Figure 11.1 Processing options for town milk.

11.2.1 Pumping

Dairy plants typically process milk at high flow rates which result in a high degree of turbulence within the product, and can result in increased air incorporation and partial homogenisation. If the turbulence during pumping is very high, it can result in disruption of some of the fat-globule membranes, particularly if it occurs while the fat in the globule is partially liquefied.

11.2.2 Standardisation

Because the fat content of milk varies, it is often standardised to a set milkfat value (usually 3.0–3.8% for whole milk and 0.5–2.5% for low-fat milk). This is done either by separating the milk into a cream stream and a skim milk stream by centrifugation and then recombining them to the desired concentration, or by simply adding additional skim milk or cream to whole milk.

11.2.3 Homogenisation

Homogenisation is principally used in the dairy industry to prevent or delay the formation of a cream layer in full cream milk, by reducing the diameter of the fat globules (see chapter 3). It is effectively a process by which the raw milk is subjected to high turbulent flow while the fat in the fat globules is fully liquid. The process has a number of major physico-chemical effects on the resulting milk, which include imparting a whiter colour (due to an increased number of fat globules and hence greater light scattering), a slightly greater viscosity (due to an increased volume of suspended matter and because more casein adsorbs on to the surface of the greater number of fat globules) and a slightly greater surface tension (due to slight removal of surface-active proteins from the skim-milk phase).

11.2.4 Heat processing

Various heat processes are applied commercially to milks and the severity of the physico-chemical effects on the finished product is governed by the intensity and duration of the heating step.

Commercial pasteurisation is relatively mild and is carried out for the purpose of killing pathogenic bacteria. The process typically involves holding milk at 75–80°C (167–176°F) for 15 s and causes only minor changes to the milk components, including up to 25% whey protein denaturation, and the development of some cooked flavour.

Ultra-high-temperature (UHT) milk undergoes a more rigorous heating process than pasteurisation. Typically milk is heated to 135–150°C (275–300°F) for 2–8 s followed by aseptic packaging. The most marked difference between UHT and pasteurised milk lies in the flavour of the final product, with UHT milk having a noticeably cooked flavour after processing, which generally dissipates after about a week. This is replaced by a stale flavour after 3–4 weeks, which becomes progressively more intense. The many factors that influence the development of this flavour are complex and not universally agreed on; a full account of the subject has, however, been documented [2]. Some other chemical/biochemical changes also accompany UHT processing, such as an increase in protein digestibility due to the partial break down of protein tertiary structure.

Sterilisation of milk is carried out by holding at 109–115°C (228–240°F) for 20–40 min and has the effect of destroying all microflora and spore-formers and also causes complete enzyme denaturation. However, at such extreme temperatures, adverse physico-chemical reactions such as browning and caramelisation of the lactose also begin to occur. These reactions cause the sterilised milk to have an intensely cooked flavour.

11.3 Butters

Butter is a complex water-in-oil emulsion, in which the continuous fat phase is partially crystalline so as to provide the desired plasticity, and hence functionality. These days, butter is used in a wide variety of applications, from spreading to baking and pastry-making. The manufacturing process aims to achieve the correct composition (particularly moisture (16%), fat (80–82%) and salt (0–2%) contents), as well as the optimum functional properties.

The fat phase has the greatest influence on the functional properties of butter, and varies considerably in composition and physical properties due to seasonal, regional, feed and breed effects. In each end-use application, butter has its imitators, many of which have been formulated to overcome some of the functional limitations imposed by the melting properties of standard milkfat. Examples of competitor products include vegetable oil-based margarines (spreading, baking, pastry-making), low-fat spreads (spreading), blended spreads (spreading) and liquid vegetable oils (frying). However, none can match the desirable flavour and mouth-feel that butter imparts to food, and in many cases butter has secured high value markets based on flavour and natural dairy identity.

Butter is generally manufactured from fresh milk using four physical operations:

- Concentration
- Crystallisation
- Phase inversion
- Dispersion

Traditionally, butter was made by batch churning. This process is too restricted in throughput for a highly mechanised industry, and thus continuous butter churns were developed. However, the same key physico-chemical processes are involved in both, namely fat crystallisation, emulsion destabilisation (surface-tension and foaming effects) and rheological operations on a thixotropic fluid.

11.3.1 Concentration

Milk, which contains approximately 4% fat, is concentrated using centrifugal separation to produce a cream of about 40% fat. A concentration temperature of 55°C (131°F) gives optimal separation efficiency and a low-fat content in the resulting skim milk.

The cream is then pasteurised by heating to 72–77°C (162–171°F) for up to 30s to destroy pathogenic bacteria and other food-spoilage organisms.

11.3.2 Crystallisation

The cream is shock-cooled to a final temperature of 12–16°C (54–61°F) and carefully maintained at this temperature for several hours, or overnight, in a crystallising silo, during which time it undergoes a crystallisation process which is crucial to the quality of the finished butter.

Cream crystallisation can be more strictly defined as crystallisation of the fat in the creams which is present in small globules approximately $0.5-20\,\mu m$ ($0.02-0.75 \times 10^{-3}$ in.) in diameter, enclosed within a globule membrane. The aims of cream crystallisation are firstly to achieve an optimum solid fat content within each globule (perhaps 45%), which will give rapid phase inversion during churning, and secondly to produce the appropriate crystal forms that will confer the desired functional properties (particularly spreadability and hardness) on the final butter. These aims are achieved by carefully controlling the temperature–time treatment of the cream after pasteurisation.

A number of cooling regimes are used around the world to manipulate the crystallisation, to compensate for the seasonal variation in milkfat composition and thus to ensure that the butter has consistent functional properties. Space does not permit discussion of them here, but a full account can be found in [3].

11.3.3 Phase inversion

The cream resulting from the crystallisation stage is an oil-in-water emulsion containing about 40% fat. During the phase inversion, or churning process, this is converted to a water-in-oil emulsion by destabilising the cream emulsion through incorporation of air and mechanical agitation.

Churning commences when air is beaten into the cream and dispersed as small bubbles on which proteins then form an interfacial film, resulting in an unstable foam. Fat globules touching the air bubbles appear to suffer damage to their membranes (probably by surface-tension effects), releasing some liquid fat which can provide oil 'bridges' to encourage the agglomeration of fat globules.

As the air bubbles collide, coalesce or burst at the surface, the interface between the air and the plasma diminishes and the attached fat globules are drawn towards one another, forming small clumps with the oil bridges acting as 'glue'.

As the process continues, larger clumps of fat are formed with liquid fat and membrane substances finely dispersed throughout. The volume of air, which increased rapidly at the outset of churning, is dispersed rapidly as the volume of clumped fat increases and acts as a foam suppressant. At this stage, mechanical agitation effects predominate and the fat clumps aggregate, with the liquid fat from the globules providing the clumps with sufficient coherence to withstand disruption.

The point at which the foam subsides is known as the 'break point' and is when butter granules and residual liquid (known as 'buttermilk') are clearly visible in the mixture. Churning is continued to allow the clumps of butter to reach about 0.5–1.0 cm (0.2–0.4 in.) in diameter and the buttermilk is then drained off.

11.3.4 Dispersion

In batch churning, the butter may be washed with cooled water so that the temperature rise that has occurred during churning is reversed, allowing some of the fat in the granules to resolidify and producing firm granules that can be worked effectively. Sufficient wastewater is retained in the churn to give a moisture content of 15.2–15.6% and salt is added if desired.

The churn contents are then worked together to form a homogeneous mass butter with a fine moisture distribution (moisture droplets $<10\,\mu m$; 0.4×10^{-3} in.), which is necessary to prevent microbiological growth. The product is then packed and allowed to set.

The setting process is observed as the hardness of the butter increasing with time, reaching about 95% of the final hardness after 30 days at 5°C (41°F). This is thought to be a thixotropic effect (reversible, isothermal, sol–gel transformation; chapter 6), rather than a continued crystallisation effect [4], and appears to involve the formation of a scaffolding structure by the dispersed particles. If the set butter is subsequently reworked, then the thixotropic effects are partially reversed; however, setting occurs again after reworking, although the final hardness at 5°C (41°F) is lower than that of the original butter by 20–60%.

11.3.5 Continuous butter-making

The continuous butter-making process was developed in 1940 by W. Fritz and is often termed the 'Fritz process'. The principles behind the process are as for batch churning. The process involves three basic elements, notably:

1. A high-speed churning beater.
2. A separating section to separate butter granules and buttermilk.
3. A working system.

The continuous process is extremely rapid relative to its batch counterpart, typically producing butter within 4 min at up to 15 t/h on one machine. Recent developments in Fritz butter-making equipment have broadened the range of products, particularly spreads, that can be made with this technology. If the plastic ribbon of butter is pumped through a mixing device prior to packing, then other liquid ingredients can be added to give homogeneous products. High-moisture (low-fat) products and spreads containing milk fat fractions or vegetable oils can also be made in this way.

11.3.6 Scraped-surface processing

The continuous butter-making process has the drawbacks that cream must be held for a period of time during the crystallisation stage and that moisture control of the resultant product is difficult. Alternative processes that do not suffer from these problems were adapted from margarine processing technology in the 1950s and employ the principle of shock chilling a liquid emulsion that has the composition of butter, using a scraped-surface heat exchanger.

A schematic representation of such a process for the production of butter is shown in Figure 11.2.

Fresh cream is firstly converted to anhydrous milkfat. Cream is concentrated to a high fat content (75%) by centrifugation, and phase inversion is achieved by subjecting this cream to high shear rates when it is passed through a homogenising valve under high back-pressure (100-150 bar; 1450-2180 psi). The anhydrous milkfat is then blended with a serum source and pumped to the scraped-surface plant. Nucleation and initial crystallisation occur in the scraped-surface heat exchangers, which are jacketed tubes each containing a central rotating shaft on which scraper blades are mounted. Product is pumped through the central annulus, and liquid refrigerant (direct expansion or glycol-water) circulates through the jacket. The scraper blades provide high shear and high rates of heat exchange at the tube walls, which together control the crystal growth rate, size and network formation.

The majority of crystallisation and dispersion occurs in pinworkers which have a much larger volume and hence residence time than the scraped-surface units. A pinworker consists of a hollow tube with a series of pins spaced around the inner wall at regular intervals and a rotating shaft, with pins radiating out from it, running through the centre. As the shaft rotates, the product is sheared between the stationary and moving sets of pins which prevents the build-up of large crystal networks and also ensures a fine moisture dispersion, resulting in a product with a smooth, plastic texture.

Resting tubes may also be included, usually just prior to packing, so as to provide further time for crystallisation within the plant, but under less shear than in the pinworkers. Resting tubes may contain one or more sieve plates to increase shear in the product, and are sized to provide the holding time required for the desired degree of crystallisation. The configuration is chosen to provide the firmness at packing and the final product consistency needed.

Scraped-surface plants may contain a variety of scraped-surface heat exchangers and pinworkers in series to provide the desired texture in the final product. Products with texture ranging from brittle to plastic and hard to soft can be produced from the same fat source, and different fat

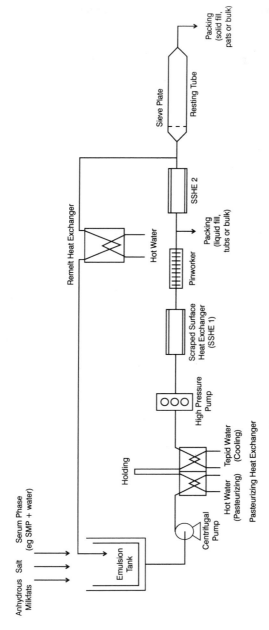

Figure 11.2 Typical recombined butter plant layout.

sources offer even further scope to optimise product properties to suit a given end use. In general, the more crystallisation that occurs under static conditions in the final pack, the more brittle the product will be.

Scraped-surface processing is today being applied to produce a wide range of consumer dairy spreads and functional butters for use in industrial applications, particularly bakery. In the manufacture of spread products, the incorporation of non-dairy fats that have a softening effect on the milkfat (e.g. soya bean oil) is more easily done by a scraped-surface process than by using Fritz technology. The former also allows the manufacture of products with a lower fat content as it is difficult to achieve moisture contents much above 20% using the Fritz process. However, scraped-surface plants can produce products with moisture contents of >60%.

The use of scraped-surface processing in the manufacture of specialised bakery butters for use in the manufacture of laminated pastry products such as croissants, Danish pastries and puff pastries is increasing. Historically, churned or Fritz butter used in these applications has performed poorly, partly because it tends to be too soft at typical bakery temperatures, if air conditioning is not available, and partly because this butter work softens far more than bakery margarine. These two effects combine so that it is difficult to keep the dough layers apart during the laminating process (the formation of consecutive layers of dough and fat) using standard butter. Work softening is largely due to the destruction of the intercrystal networks and can be accompanied by moisture leakage from the butter, particularly at temperatures below 20°C (68°F). Obviously this is undesirable in pastry products, where the integrity of the dough and fat layers must be preserved to achieve a high-quality pastry product. The use of scraped-surface processing has meant that high-quality functional butters suitable for laminated pastry-making can now be manufactured. With this technology, a fine moisture dispersion and a uniform lattice of crystals in the final butter can easily be achieved, which has the effect of producing a butter with extremely pliable plastic properties that withstand the work-softening process.

11.4 Creams

A variety of fluid creams is sold in consumer packs around the globe. They tend to differ from each other both in their mode of processing and in their fat content. A wide variety of non-dairy toppings and whips is also available, but they are outside the scope of this chapter.

All commercial creams, apart from those made by recombining, are produced by centrifugation as discussed in sections 11.2 and 11.3. Typical fat contents may vary from 10 to 55% milkfat and the creams are often classified on this basis.

11.4.1 Single creams

Creams with a low-fat content are often referred to as 'single creams' and careful processing is required to produce a product with the desired viscosity and thermal stability properties. Untreated milk is first separated, pasteurised and standardised as previously described. After fat analysis, the cream undergoes homogenisation, which is the key unit process in providing a product of the desired thickness. The cream is heated to homogenisation temperature (typically 40–60°C; 104–140°F) and is homogenised within the range 10 000–24 000 kPa. Control of the temperature and pressure of this process has a strong bearing on the end-product consistency, with an increase in homogenisation pressure increasing the viscosity of the cream. However, this effect is reduced by an increase in temperature, which is generally kept as low as possible, but must be maintained at least 40°C (104°F), as below this temperature the fat in the milkfat globules will not be liquid.

Low-fat creams are subjected to a single-stage homogenisation to make them thick, and less prone to fat separation, and the pressures used are generally at the high end of the range indicated. In contrast, creams specifically for use in coffee tend to be produced using homogenisation pressures at the low end of the scale or employ a two-stage homogenisation process. For UHT coffee creams, two homogenisations, both upstream and downstream of the UHT process, are favoured.

The temperature at which homogenisation occurs also has a bearing on the heat stability of the cream, which is an important consideration, particularly if the cream is to be utilised in hot coffee. Basically, the lower the temperature, the poorer is the stability.

Careful control of the homogenisation technique is therefore required to produce a product with the desired balance of attributes.

11.4.2 Whipping creams

Good-quality whipping creams must be easy to whip and produce a fine foam and a good increase in volume ('overrun') on whipping. The resultant foam must be long lasting and must not leak fluid on storage.

Generally, to produce such whipping qualities, the cream needs to have a fat content of 30–40%; below this level, other ingredients would need to be added to the formulation to induce good whipping characteristics.

Air must be excluded from the process at all points as its presence could lead to milkfat globule damage, formation of foam and granulation of the fat.

To produce whipping cream, untreated milk undergoes the same separation, pasteurisation and standardisation processes as single cream. Pasteurisation is generally fairly light (80–85°C (176–185°F) for 15–20 s) as severe pasteurisation would reduce the viscosity of the cream and could result in it having a cooked flavour. A low temperature of separation, compared with that used for single cream, may also be favoured because phospholipids impart good whipping properties to the cream and this would lead to more of the phospholipid being retained in the cream. After pasteurisation, the cream is chilled to 4°C (39°F) as rapidly as possible and left to crystallise (sometimes referred to as 'ripening') for at least 24 h. During these processes, the phospholipids migrate from the serum to the fat globule surface and assist in subsequent whipping by making the membrane more shear-sensitive and promoting interaction of the globules at the air–water interface during whipping. Other physico-chemical changes associated with this process are the same as those during the crystallisation phase of butter-making, which were described in section 11.3.2.

If the crystallisation process is not carried out prior to packaging, it is possible that adhesion and lining of the carton with a thick, sticky layer of cream can occur due to, amongst other things, the heat produced during crystallisation of the fat.

It is generally recommended that whipping creams are not homogenised as this leads to a more stable emulsion and thereby makes the cream more difficult to whip. However, for long-life whipping cream, some homogenisation is required to prevent creaming and surface agglomeration of the globules during the long storage life. Such cream takes longer to whip than non-homogenised cream, although additives (so-called 'emulsifiers') can assist by promoting globule interactions and emulsion destabilisation during whipping.

11.5 Cheese

Cheese manufacture is considered to have originated perhaps 6000–7000 years ago. The reason for making cheese then was to preserve milk. Preservation is still important; however, the sensory appeal of cheese has undoubtedly contributed to its popularity today. Cheese manufacture uses several key preservation techniques, some of which differ markedly from those for other dairy products. It is estimated that perhaps 1000 different cheese varieties exist. These varieties use different processing or raw materials to achieve different properties. Thus, it is difficult to generalise about the physico-chemical changes that occur during cheese manufacture. A detailed review of the processes for major natural cheese varieties and processed cheese has been published by Fox [5]. Simplified processes for both natural and processed cheese are summarised in Figure 11.3.

Clearly, the processes for the manufacture of natural and processed cheese differ markedly. The key difference between the two products is that natural cheese can be considered as dynamic. The flavour and texture of natural cheese are constantly changing, as residual micro-organisms and enzymes remain active in the product during storage. Processed cheese, however, is more inert. Here, all enzyme activity and most microbial activity are inactivated during processing.

Different natural-cheese types require the whole milk to be standardised to different component levels and to be pasteurised under different conditions. These initial differences provide a key means of differentiation between cheese types. The components present in the cheese milk greatly affect the final texture. Differences in pasteurisation conditions can result in changes to residual protease and lipase activity and consequent changes in the flavour and texture of the cheese. Some of the major physico-chemical changes in cheese-making are initiated with the addition of lactic acid starter bacteria (LAB), rennet enzyme and calcium salts. LAB are

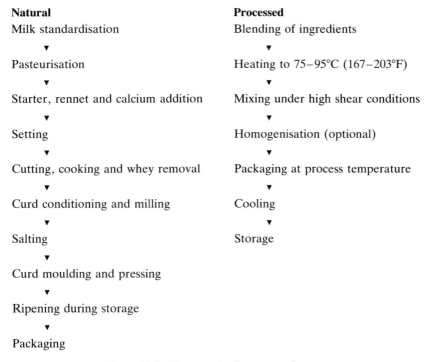

Figure 11.3 Key steps in cheese manufacture.

added to metabolise lactose to lactic acid. The generation of lactic acid during cheese manufacture results in a reduction of 1–2 pH units. Rennet enzyme or chymosin addition results in the cleavage of a hydrophilic peptide from k-casein on the casein micelle (an arrangement of casein proteins, calcium and phosphorus with colloidal properties). The net effect is to promote hydrophobic interaction between casein micelles and the formation of a protein network. At a macro level, this is seen as the gelation of the milk. Calcium ions are often added to promote the formation of the protein network. At this point, the gel formed contains all the moisture in the milk; however, it is necessary to increase the solids-to-liquid ratio during the cheese-making process. To achieve this, whey removal is important. Release of this moisture from the milk gel is achieved by first disrupting the gel by cutting into centimetre-sized (half-inch) cubes and applying heat in the cooking process. This results in a tightening of the protein structure, and the expulsion of water or curd syneresis. This moisture is then separated physically from the cheese curd. The curd conditioning (in Cheddar cheese known as 'cheddaring') and milling processes result in further tightening of the structure and further moisture loss. The starter bacteria are active throughout this process and continue to cause a gradual reduction in pH. This reduction in pH causes further syneresis.

The salting process is carried out by different means depending on the cheese variety. The addition of salt has several functions: control of the final pH by limiting micro-organism growth, promotion of further syneresis through tightening of the protein matrix, microbiological safety by limiting growth of undesirable bacteria and limiting enzymatic proteolysis (chapter 9).

Curd moulding and pressing are further physical processes that increase the structural integrity of the cheese and cause further syneresis. The ripening or maturation process used varies depending on the cheese type. Some cheese varieties can be matured for several years before being released for sale. Considerable changes occur in the product texture and flavour during this time. Proteolysis occurs, resulting in changes to the protein network. In Cheddar cheese, extensive proteolysis results in a high level of bitterness, decreased elasticity and increased hardness. In Swiss varieties, propionic acid bacteria remain active during maturation. These organisms ferment lactose to propionate and carbon dioxide, resulting in the characteristic flavour and 'eye' formation. Different microbiological floras play a significant role in the development of characteristics in other cheeses such as Camembert, Roquefort and Stilton. In these varieties, the process of lipolysis is also important in the development of the final cheese flavour.

Natural cheese is one of the main ingredients used in processed cheese manufacture. The selection of the correct blend of natural cheese ingredients greatly affects the texture and flavour of the final product. Besides natural cheese, a variety of other dairy ingredients can be used such as: milk proteins, whey powders, cheese powders, skim milk powder and anhydrous milkfat. A further important ingredient group is the emulsifying salts. These are mixtures of calcium chelating salts, usually phosphates, polyphosphates or citrates. Although these salts are termed 'emulsifying salts', they are not surface-active components in the strict sense. Various combinations of emulsifying salts are used to modify the texture and melting properties of the finished processed cheese product.

Several important physico-chemical processes occur during the heating and shearing processes. The emulsifying salts sequeste calcium ions bound to the protein. Sodium ions then replace the calcium ions on the protein. The removal of calcium ions from the protein in this way has the effect of reducing ionic interactions between proteins and causing proteins and protein fragments to become more soluble. These proteins are then more able to participate in the emulsification of the milkfat and the production of a stable emulsion. The mixture begins to thicken on prolonged mixing of the newly formed emulsion at high temperature; this is known as the 'creaming reaction'. This thickening is a clear sign of reformation of the protein matrix. It is postulated that this is not a re-establishment of the matrix originally found in the natural cheese.

It is considered that the matrix reformation is affected by the following process. During the shearing process, fat globules are formed and these are stabilised by the protein that has been partially solubilised by the emulsifying salts. The protein bound to the fat globule membrane has hydrophilic strands that protrude from the fat-globule membrane and these are thought to link to other protein strands on other globules through ionic interactions. The re-release of calcium from the emulsifying salts is considered to play some part in this. The creaming reaction, that is responsible for the transformation from cheese ingredients through to processed cheese product, is not well understood and further study in this area is required.

From a consumer point of view, cheese products, both natural and processed, are characterised by their flavour, texture and appearance. These features can be linked to several physico-chemical properties. Important physico-chemical properties of cheese, both natural and processed, are water activity and rheology. Cheese has water activity values between 0.9 and 1. Water activity values in this range can sustain bacteria growth and enzyme activity, and thus are critical in promoting the important reactions that occur during the maturation process. The water activity tends to decrease as natural cheeses age. This is a consequence of the generation of proteolysis products that can bind water more effectively than their substrates. The texture of cheese products is one of the most important factors for consumers. It can be measured by a range of different rheological techniques. Several methods have been reviewed [6].

11.6 Cultured products

The earliest forms of cultured foods, as we know them today, are thought to have originated in the Middle East. Here, nomadic peoples first recognised the benefits that the controlled culturing of milk could provide in preserving their very precious and limited milk resource.

Today, an extremely broad range of cultured foods is produced in different parts of the world. These products range from the more simple cultured milk drinks, where there is little or no concentration of the milk solids, to more complex products, where the milk components are varied in both concentration and ratio to achieve specific textures. Additives such as sweeteners, fruits, cereals and candy are commonly used in these products.

The manufacture of yoghurt demonstrates a lot of the features common to many cultured foods. Yoghurt processing is described in detail by Tamime and Robinson [7]. A summary of a typical process is shown in Figure 11.4.

Several physico-chemical changes occur during the transformation of liquid milk into yoghurt. In the mix standardisation, the concentration of the milk solids is typically increased from that in milk by up to 50% by evaporation or the addition of dried milk ingredients. Concentration in this way, particularly with the addition of isolated whey protein, affects the texture of the final yoghurt product. Addition of stabilisers such as gelatin and modified starches is often used to achieve a similar texture more economically. Heat treatment of the milk to near 90°C (194°F) is a

Mix standardisation
▼
Homogenisation at 200/30 bar
▼
Heat treatment of the milk at 90°C/5 min
▼
Cooling to incubation temperature (~40°C)
▼
Starter inoculation
▼
Holding until pH reaches ~4.4
▼
Packaging

Figure 11.4 Key steps in yoghurt manufacture.

further process step that affects the final texture of the product. This results from the promotion of both hydrophobic reactions between proteins and the chemical reaction between disulphide bonds on κ-casein and β-lactoglobulin, two of the major milk protein components. Both interactions contribute to the formation of a protein network in the yoghurt, which results in the characteristic texture of the product.

The fermentation process in early cultured foods would have developed haphazardly. In modern cultured foods, the key elements of the fermentation process are to select lactic acid bacteria strains that can provide desirable flavours and to achieve the desired final product pH. Commercial yoghurt products are differentiated into stirred and set varieties. Set yoghurt has a more rigid gel structure. The key processing differences for set yoghurt is that the fermentation is carried out in the consumer yoghurt pot. Modern processing technology such as the UHT treatment of liquid milks and milk powder drying has eliminated the need to use fermentation merely for milk preservation; however, cultured products remain immensely popular throughout the world, because consumers now derive additional benefits from these products. They are consumed for their pleasant taste and texture and often for health-providing benefits. Some benefits attributed to cultured foods containing lactic cultures are as follows: high calcium content and assistance with calcium metabolism, reduced lactose intolerance, improved gastrointestinal metabolism, inhibition of pathogenic bacteria and perhaps reduced accumulation of blood serum cholesterol.

Two of the major factors that affect texture in cultured foods, i.e. formulation and process, have been discussed above. However, fermentation and the type of fermentation organism also have a significant effect. As the fermentation of lactose to lactic acid proceeds, calcium migrates from the casein micelle. When this process occurs slowly, hydrophobic interactions between proteins in the micelles increase and thus a protein matrix or network is established. If acidification is carried out rapidly or at high temperature (as in casein manufacture), complete coagulation of the protein occurs and the structure is unable to retain moisture within the gel matrix. Traditionally, fermentation in yoghurt is carried out using organisms such as *Lactobacillus bulgaricus* or *Streptococcus thermophilus*. Some yoghurt manufacturers use mucogenic bacteria that produce extracellular polysaccharides. These polysaccharides are present as fine filaments extending from the starter organism. They become part of the three-dimensional structure of the yoghurt and thus contribute to the overall texture of the product. The starter organism also has an effect on the gel structure through its tendency to cause void spaces in the yoghurt.

The texture of cultured product gels can be characterised by a variety of rheological techniques. These can range from very simple 'cup-and-funnel' systems, that can be used to ensure consistency during production, through to more elaborate techniques that can provide quantification of viscoelastic properties for semi-solid materials (chapter 5). Clearly, the properties and the rheological methods used will be dependent on the type of cultured product. Yoghurt has been the most studied cultured food to date. For stirred yoghurt, rotational viscometry can be used. Here a constant shearing action is applied and this can provide information on the viscosity of the product. However, in carrying out this measurement, the shearing action causes irreversible structural changes in the product. Using this rheological technique, these products show thixotropic behaviour; that is, as the shear rate is increased, the viscosity of the product decreases. Oscillatory rheometry is a technique that is applied to determine the viscous and elastic components of foods. This technique is particularly suitable for application to systems with a weak gel network such as yoghurt. For set yoghurt, it is common to measure the texture of the product using a penetration probe or an oscillatory rheometer.

Apart from being a pleasurable eating experience, consumers see cultured foods as nutritious and health-providing and part of this is the use of selected strains of lactic acid-producing bacteria. Bio yoghurt using Bifidobacteria has become popular in some markets, as have yoghurts using *Lactobacillus acidophilus* strains. *Lactobacillus casei* is used in cultured beverages. More recently, MD Foods have released the Gaio product which uses a starter strain combination of *Enterococcus faecium* and *Streptococcus thermophilus* bacteria. The cultured food products produced using these bacteria are all promoted on the basis of different health-providing claims. In the case of the Gaio product, a significant reduction in total blood serum cholesterol has been claimed.

11.7 Ice cream

Ice cream is thought to have originated in China perhaps as long ago as 1000 BC. Marco Polo is reputed to have brought the idea back from China to Italy in the 13th century. However, it was not until 1851 that the first commercial ice-cream factory was opened in Baltimore in the USA.

As with many traditional dairy foods, the modern marketplace has required the differentiation of ice cream to the point where there is now a broad range of products. Ice cream is commonly categorised on the basis of fat content, with low or no fat products containing <4% fat, standard ice cream containing 10% fat and premium and super premium ice-cream formulations containing ≤22% fat. Standard ice cream typically has an overrun of 100%; that is, for every kilogram of solids, one litre of air is dispersed through the product. Premium ice-cream formulations that contain higher levels of milkfat usually have lower overrun levels.

Mix standardisation
▼
Homogenisation at 115 bar (1668 psi)
▼
Pasteurisation at 80°C for 20 s
▼
Aging at 4°C
▼
Air injection
▼
Scraped surface cooling to −5°C
▼
Packaging
▼
Hardening between −20°C and −30°C

Figure 11.5 Key steps in ice-cream manufacture.

The key steps used in the processing of ice cream have been covered in detail by Arbuckle [8], and summarised in Figure 11.5.

The components in an ice-cream mix are varied depending on the type of ice cream to be manufactured. A typical formulation will contain skim-milk solids, fat (milkfat or vegetable fat), sugar, stabiliser and emulsifier. Having a fine dispersion of fat globules is critical to achieving stability and smoothness in the final ice cream product. Homogenisation has the major influence on the emulsion properties in the early stages of the process. Processing on a single-stage homogeniser at about 115 bar (1668 psi) reduces the median fat globule size to about $0.7\,\mu m$ (0.03×10^{-3} in.).

Generally, pasteurisation is carried out at higher temperatures than used for standard milks. Manufacturers have noted improvements in product smoothness and uniformity and it is likely that this arises from the interactions between the casein and whey proteins.

Fat crystallisation also plays an important part in achieving a stable foam. Following pasteurisation, all the fat is in a liquid form. By cooling the mix to 4°C (39°F) and holding for a minimum of 3 hours during the ageing process, crystallisation of triglycerides occurs within the fat globules. This process, where the liquid fat is crystallised in the fat globule, is important subsequently in achieving good foam stability in the ice cream.

Air injection and scraped surface cooling to −5°C (23°F) are carried out in an ice-cream churn. These processes are important in developing the correct texture and structure in the product. This is achieved by generating air vacuoles ranging between 10 and $60\,\mu m$ (0.4–2.5×10^{-3} in.). These vacuoles are initially stabilised by hydrophobic proteins adhering to the vacuole wall. Added emulsifiers are important at this stage of the process. It is thought that these components cause a controlled destabilisation of the fat-globule membrane that subsequently enables the fat globules to adhere to the vacuole wall and provide additional stability to the foam. The inclusion of emulsifiers in ice-cream formulations causes a large reduction in the processing time required to reach optimum overrun during churning and a large reduction in meltdown rate of the ice cream product. As the globules adhere to the surface of the vacuole, it is important that the level of liquid fat is limited. The ageing process, referred to previously, ensures that about 75% of the milkfat is crystalline at the time of churning. In this way, the amount of free fat that could spread onto the surface of the air vacuole is limited. It is thought that fat in this form causes collapse of the air vacuole by reducing the thickness of the stabilising protein film [9]. From a consumer's point of view, these effects are seen as defective product which melts rapidly and has a wet or icy texture.

The final step in ice-cream production is hardening. This process is usually carried out in a blast freezer to achieve a rapid reduction in temperature and thus limit the size of ice crystal growth. It is important that the ice crystal size be $<55\,\mu m$ (2×10^{-3} in.) to eliminate a coarse texture in the final product. Thermal abuse of ice cream results in the growth of large ice crystals. This defect can be minimised by using polysaccharide stabilisers.

Figure 11.6 is a scanning electron micrograph showing the structure of a 10% fat ice cream. Small fat globules of about $1\,\mu m$ (0.4×10^{-3} in.) can be seen as white dots on the surface of the air vacuoles. Dispersed between the air vacuoles is the continuous aqueous phase that contains hydrated protein and carbohydrate material. The structure, and hence the final textural properties, of the ice cream are very dependent on achieving stable air vacuoles.

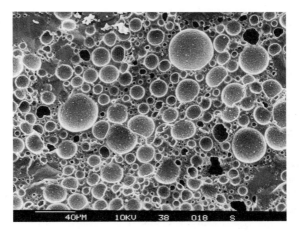

Figure 11.6 Scanning electron micrograph of a 10% fat ice cream.

Rheological and textural analysis of ice cream is difficult. These products are generally stored at about −20°C (−4°F), but are consumed at warm ambient temperatures. Thus, the temperature profile of an ice cream changes during consumption. The measurement of texture and the relationship of this to a consumer's preferences is a difficult area of study. However, rheological properties of the ice-cream mix can be measured during processing. These data can provide information on the meltdown rate and the tendency to resist thermal abuse. However, the results tend to be specific to individual formulations and process equipment.

11.8 Powdered consumer products

Drying of milk products is a further technique that is used to achieve the preservation of milk. Dried-milk products may date back as far as the 7th century. However, milk-powder products, as we know them today, had their origins in 1872 when the patent for spray drying was granted. Although many dried-milk products are commonly used as ingredients in a range of food products, some are important as consumer dairy products in their own right.

Dried-milk products generally contain <4% water to ensure prolonged stability to microbial spoilage. The most common method of drying milk is evaporation of water and removal of the vapour from the milk solids using spray-drying techniques. This is the basic principle used in most forms of dried-milk products today. A further important innovation, particularly for consumer powders, was the development of highly dispersible or 'instant' powders in 1955. Here, spray-dried milk powders are agglomerated, causing the powder particles to 'wet' more readily when added to water. This rapid uptake of moisture into the powder particle ensures that the subsequent processes of dispersion and solubilisation in the reconstitution process are not restricted.

Several important dairy-based consumer powders are instant milk powder, infant formula, coffee whitener and various health and sports products. All products are similar in that they are reconstituted by adding to water before consumption. Thus, wettability, dispersibility and solubility are important properties for these powders.

The basic manufacturing processes for consumer powders have been reviewed by Carié [10], and a summary of the key unit operations is shown in Figure 11.7.

Mix standardisation is often achieved by milk separation or the addition of components. The effects of pasteurisation, heat treatment and homogenisation on the physico-chemical properties of milk have been covered in other parts of this chapter. Just as in other milk products, these processes

Mix standardisation
▼
Pasteurisation
▼
Heat treatment
▼
Homogenisation
▼
Evaporation to about 50% solids
▼
Drying
▼
Packaging

Figure 11.7 Key steps in the processing of consumer milk powder.

have a major influence on the emulsion state, the microbiological stability and the levels of interactions between protein components.

Concentration of milk solids by evaporation to about 50% total solids before spray drying is important from an economical standpoint, as well as from the point of view of powder properties. Providing a feedstream to the drier at about 50% total solids results in powders with larger particles and smaller amounts of occluded air than feedstreams at lower total solids. Roller drying and spray drying are the two most common commercial means of drying milk concentrate; however, consumer milk powders are almost entirely produced using spray drying. In principle, this process involves atomisation of the fluid milk stream into hot air in a drying chamber. There is considerable scope to change the properties of powders by varying the atomisation system, the hot air temperature, the chamber design and the recycle of fines, and using secondary drying chambers.

Several physico-chemical properties of powders are important in ensuring their acceptability. Water activity is known to have a major influence on the growth of mould, yeast and bacteria, on enzyme activity, on Maillard reaction activity and on fat oxidation (chapter 1 and [11]). Uncontrolled activity in any of these factors will cause significant defects in the powder product. Thus, powder manufacturers attempt to optimise the water activity of powders to minimise the activity of each of the above factors. This is clearly a compromise; however, moisture contents of about 3.0% giving rise to water activity between 0.20 and 0.30, are commonly found in powders.

Emulsion and surface properties are also important physico-chemical properties of consumer milk powders. The state of the lipid material is a key factor in determining the functional performance of a powder. In a coffee whitener, whitening is achieved through having the fat finely dispersed and in a stable emulsion. Thus, emulsion stability is critical in this product. Fat oxidation is a prime cause of off-flavours in high-fat products. Poor emulsification of fat during processing can lead to high levels of free fat. This migrates to the surface of powder particles and is thus more susceptible to oxidation. Advances in inert atmosphere packaging have reduced this risk of oxidation. A requirement of some high-fat consumer powders is good dispersibility. It is often necessary to coat the surface of these powders with a surface active material such as lecithin.

Powdered milk products have an ample level of lactose and protein to participate in Maillard reactions. Generally, powder manufacturers attempt to minimise Maillard reactions because they result in darker powders with a cooked, burnt or caramel flavour. Manufacturers thus attempt to avoid processing conditions that enhance these reactions. This often involves minimising the length of time for high-temperature processing steps and ensuring that the product is not stored at a high ambient temperature (>25°C; 77°F). Extreme care must be taken with the processing of low pH powders as these show an even greater potential to participate in Maillard reactions.

One major defect in some consumer powders is their tendency to 'cake' on storage. This term describes the manner in which individual powder particles clump together on storage, thus reducing the flowability of the powder. Caking is more common in powders with high lactose contents and it has been postulated that the transition from amorphous lactose to crystalline lactose on storage is an important factor in the caking process. Glass-transition theory (chapter 2) has provided an explanation of how the key factors of moisture content, temperature and time affect the tendency of powders become sticky and cake [12].

In addition to instant milk powders, several other milk powders are designed to meet specific consumer needs. Infant formula has been designed as a substitute for human milk. The trend is to further 'humanise' infant formulae by attempting to replicate the composition of human milk with isolated dairy components. To achieve this, the whey-to-casein ratio, the level of lactose and the calcium-to-phosphorus ratio have been increased over what is typically found in cows' milk. Further development has seen the use of α-lactalbumin in preference to the major whey protein found in cows' milk, β-lactoglobulin, which is not found in human milk. Modification of the triglyceride content is also required as cows' milk has higher levels of saturated fatty acids than human milk. More recent advances in infant formulae have seen the addition of components with antimicrobial activity, such as immunoglobulins, lactoferrin and lactoperoxidase.

Coffee whitener is a further specialist consumer powder. These products are commonly manufactured with the only dairy component being the protein emulsifier, sodium caseinate. Thus, vegetable fat is substituted for milkfat and additional emulsifiers are used to increase the stability of the emulsion in hot acid conditions. Good emulsion stability under these conditions is critical to achieving good whitening in the hot beverage.

Dietetic, health and sports powders are usually manufactured with a lower fat content and higher protein and carbohydrate contents than standard whole-milk powder. In a simple form, these products can be simply vitamin-enriched milk powders. For consumers suffering from lactose intolerance, lactose levels can be reduced through enzyme hydrolysis.

11.9 Conclusion

Although dairy products are amongst the oldest foods known to man, a knowledge of the science of composition and processing has considerably extended their range of use. This has not only led to traditional products having new textures, greater stability and longer shelf-life, but also to entirely new ones, such as infant formula, which uses cows' milk ingredients to copy another product.

The development of products with a beneficial health factor is a field which is likely to extend considerably in the not too distant future. A knowledge of the physico-chemical aspects of their processing will be essential to their production.

Acknowledgements

The following staff of the New Zealand Dairy Research Institute are acknowledged for their contribution and helpful discussions during the writing of this chapter: Mr C. Towler, Dr J.B. Smart, Miss S.E. Croft, Mr F. Dunlop, Mr A.M. Fayerman, Dr R. Norris, Dr P.A. Munro, Dr P.A.E. Cant, Mr A.B. McKenna, Dr D.F. Newstead, and Mr K. Palfreyman.

References

1. Jenness, R. Composition of milk. In *Fundamentals of Dairy Chemistry* 3rd edn. Eds Wong, N.P., Jenness, R., Keeney, M. and Marth, E.H. Van Nostrand Reinhold, New York (1988), 1, 38.
2. Mehta, R.S. Milk processed at ultra-high temperatures – a review. *Journal of Food Protection*, **43** (1980), 212–225.
3. Frede, E. and Buchheim, W. Buttermaking and the churning of blended fat emulsions. *Journal of the Society of Dairy Technology*, **47** (1994), 17–27.
4. De Man, J.M. and Wood F.W. Hardness of butter. II. Influence of setting. *Journal of Dairy Science*, **42** (1959), 56–61.
5. Fox, P.F. *Cheese; Chemistry, Physics and Microbiology*, 2nd edn. Chapman and Hall, London (1993).
6. International Dairy Federation. *Rheological and Fracture Properties of Cheese*. International Dairy Federation Bulletin No. 268. International Dairy Federation, Brussels (1991).

7. Tamime, A.Y. and Robinson, R.K. *Yoghurt: Science and Technology*. Pergamon Press, Oxford (1985).
8. Arbuckle, W.S. *Ice Cream*, 3rd edn. Avi Publishing Company, Westport, Connecticut (1984).
9. Brooker, B.E. The stabilisation of air in foods containing fat–a review. *Food Structure*, **12** (1993), 115–122.
10. Carié, M. *Concentrated and Dried Dairy Products*. VCH Publishers, New York (1994).
11. Rockland, L.B. and Beuchat, L.R. *Water Activity: Theory and Applications to Food*. Marcel Dekker, New York (1987).
12. Roos, Y. and Karel, M. Effects of glass transitions on dynamic phenomena in sugar containing food systems. In *Glassy State in Foods*, Eds Blanshard, J.M.V. and Lillford, P.J. Nottingham University Press, Leicester (1993).

12 Cereal processing: The baking of bread, cakes and pastries, and pasta production
R.C.E. GUY

12.1 Introduction

Foods that contain wheat flour as their major ingredient represent a large proportion of the food market in most countries of the world [1,2]. In this chapter, two of the largest sectors are considered, bakery products and pasta. These products have little in common, other than that they may be formed from a dough, and therefore they will be treated separately. In the bakery sector, the doughs or batters form a continuum stretching from high-moisture systems such as wafer biscuits and cake batters, through the different types of bread doughs, to the low-moisture biscuit and pastry doughs. All these products are based on cereal flours, usually wheat flour, with appropriate additions of egg, sugar and fats to give the traditional recipes for the products. There is a common order to the processing methods employed, with ingredients being mixed with water to form a dough or batter, then subdivided into portions for the second stage of processing by the application of heat. Most of these products are heated in an oven to form their structures, except for wafer biscuits and pasta, which have their own unique thermal processes, with the heated wafer plate and boiling water, respectively.

Physical changes taking place in the doughs or batters may vary according to the technological process being used, but normally they will include the hydration of the raw materials, mixing and other mechanical handling procedures, which modify the physical forms of the natural ingredients in the doughs or batters, and determine the initial bulk rheological properties of the systems. A more detailed view of the physical changes taking place within the doughs or batters as they are mixed will reveal the entrapment of bubbles, decreasing density, changes to the physical form of biopolymers, some heat input and changes to the biopolymers controlling their bulk rheology. In the heating stages, there are further changes, such as the expansion of occluded gases into bubbles or fissures, the vaporisation of dissolved gases and the melting of crystalline structures. Major changes to the rheological characteristics of the doughs occur, which are related to phenomena such as starch gelatinisation, protein denaturation and variations in moisture levels, due either to migration or loss by evaporation. Enzymic and chemical reaction rates change with temperature and are also affected by subtle effects of the physico-chemistry within the doughs and batters. The main chemical reactions are non-enzymic browning caused by Maillard-type reactions and caramelisation. They occur at the edges of the doughs or batters during baking, where particular phenomena related to moisture migration occur. The structures, colours and flavours formed in the baked products are metastable and change with time during storage. Such changes are referred to as the 'staling phenomena'. In the products with high-moisture contents, microbiological deterioration may occur to reduce the shelf-life of the products. Their stability is closely related to physical factors, such as the water activity within their structures and the diffusion of water from fillings, the addition of preservative chemicals in the recipe and the packaging technologies used to protect them during their shelf-life in retail outlets and in the home.

In this review of the cereal processing within the bakery sector, the main physical changes taking place during manufacture will be reviewed for three typical examples of the products, pastry, cake and bread. A brief examination will be made of the very high-moisture system used to form wafer batters to complete the bakery range. The chemical and shelf-life aspects of bakery products will be considered together, because they have common phenomena. Finally, pasta products will be described separately as unique examples of dough products, which are de-aerated, dried and then cooked in boiling water.

12.2 Bakery products, pastry, cakes and bread

12.2.1 Physical changes during the manufacture of baked products; pastries

If the full range of all the types of products available in the bakery industry is examined in detail, it may be seen to consist of a range, extending from very low- to high-moisture products (Figure 12.1). At the low-moisture end of the range, the first products encountered will be biscuits and pastries, with moisture levels of 2–5% w/w and water activities <0.35. Some of these products are extremely simple in their formulations. For example, a short pastry may contain the ingredients shown in Table 12.1 [3]. The dough will contain ca. 59.5% w/w flour solids (moisture content 13–14% and starch 75–80%) and 8.9% w/w water, at the start of the mixing process. Therefore, the amount of water available to plasticise the starch polymers is of the order of 30–35% w/w of the starch. During dough-making the wheat-flour proteins will tend to hydrate and swell, thereby binding some of the water within a protein gel phase. The full development of the protein phase gel phase, commonly known as

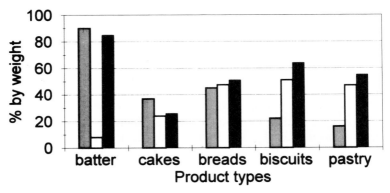

Figure 12.1 Relationships of starch to water in bakery doughs and products.

Table 12.1 Recipe for a short pastry

Ingredients	Baker's weight per 100 of flour	Weight (%)
Wheat flour	100	59.5
Bakery fats	50	29.5
Water	15	8.9
Salt	3	1.8

'gluten', is inhibited by the use of high fat and relatively low water levels. The coating of fat on the flour particles reduces the number of elastic strands or sheets of protein gel, which form between the flour particles. If such a network is too well developed, problems arise due to elastic recoil effects during the sheeting and cutting of the dough. It might also trap air during mixing, which could expand into unsightly blisters during baking.

Approximately two-thirds of the dough water should be available to plasticise the amorphous regions of the starch granules. Therefore, from observations of the relationship between the glass-transition temperature (T_g; see chapter 2) for starch or wheat proteins and water (Figure 12.2), it can be determined that at the beginning of the baking process at temperatures of 30–35°C (86–95°F), the starch granules are held in their rigid structure by the crystalline junction zones, whereas their amorphous regions and the gluten particles are present in the rubbery elastic form [4]. During the baking process, as the temperature rises within the dough, the water vapour pressure increases and causes some expansion in the fissures within the pastry. However, there are very few bubbles with well-developed gluten films to trap the vapours within simple pastry doughs. Therefore, the vapour is lost to the atmosphere through the fissures

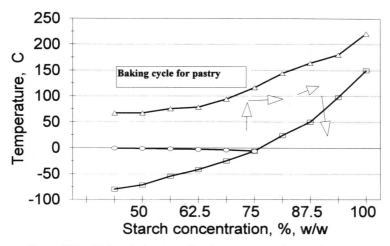

Figure 12.2 Melt and glass-transition temperatures of starch in pastry.

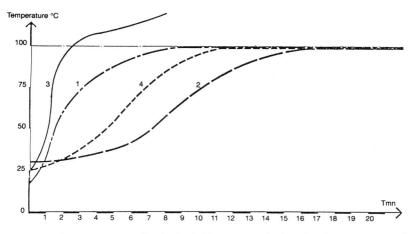

Figure 12.3 Temperature profiles in the baking process for bread, cakes and biscuits [25]. 1 = French bread ('baguette'); 2 = pan bread; 3 = hard biscuit; 4 = sponge cake.

between the flour particles and although several litres of water vapour are evaporated from the dough, the expansion remains very small.

Moisture is lost from the dough more rapidly as the temperature rises and eventually the dough temperature becomes constant at ca. 100–103°C (212–217°F) (depending on the dissolved solute), because the heat loss during evaporation prevents any further temperature rise, while the water is being boiled off. The measurements, shown in Figure 12.3 for various bakery products, indicate the plateau effect in the larger products at the boiling point of the aqueous phase. It has been shown that T_m (melt temperature) value for wheat starch increases as the moisture falls within the dough pieces [5] and therefore it is not possible to exceed T_m, within the bulk of the dough, for the pastry product illustrated in Figure 12.2. In this type of pastry dough, it is not possible to melt the crystalline regions within the starch granules, if their T_m value rises above the boiling point of the aqueous phase, except in small regions near the surface or where the pastry is in contact with a moist filling. In the bulk of the pastry, the temperature remains at the boiling point of the aqueous phase until the water has been evaporated to a very low level. Then it rises in the same manner as shown in the graph of the hard biscuit in Figure 12.3.

The pastry leaves the oven with a structure formed from ungelatinised starch granules and plastic gluten particles, which have been dehydrated to 2–5% moisture level. At the surface where the dehydration occurred early in the bake, allowing temperatures to rise to 130–150°C (266–302°F), some coloration of the crumb occurs and flavour compounds are created. On cooling, the polymer phases tend to stiffen and may become glassy in large areas of the pastry (Figure 12.2). This change results in formation of a texture within the products, which in terms of its components is hard, brittle and crisp. However, the overall structure has many lines of weakness and breaks down easily in the mouth. In addition, there are oils and fats present in the structure of the pastry, which serve to lubricate the particles and to give excellent sensory qualities.

The microbiological stability of the pastry is normally very high because its water activity is very low, <0.2–0.3, unless it becomes moist due to the transfer of water from other food components. Physically, the crumb structures are also very stable because the biopolymers are in their glassy forms at ambient temperatures. However, exposure to moisture, leading to an increase in moisture within the pastry, causes the loss of the crisp brittle texture at moisture levels of 10–12%. In this region, the proteins linking the structures together become rubbery, changing the nature of the product.

In many forms of biscuits, the recipes are very similar to pastry, but with a significant addition of sugar. They tend to have the same type of structure as the pastry, but with coarse cavities expanded by the escaping water vapour [6]. The sugar forms a viscous solution during the dough formation and links the flour particles together. As the water is removed from the dough, the sugar solution is concentrated and on cooling forms a sugar glass or a mixture of crystalline and glass phases, between the flour particles [4]. Biscuit doughs also tend to be in the low-moisture range, where it is not possible to melt the crystalline structures within the starch granules during a normal baking process. An example of this is given in a study at FMBRA on Lincoln biscuits [7]. It was found that all the starch granules retained their birefringent properties (i.e. appear bright under crossed Polaroids on a microscope), indicating the retention of their crystallinity, except for a narrow band at the surface of the biscuits.

As the biscuit doughs enter an oven, a phenomenon of 'moisture condensation' occurs [6]. The relatively cool doughs encounter an atmosphere, at the entrance to the tunnel of the travelling oven, in which water vapour pressure is high, due the temperature, and on contacting the relatively cold biscuit dough, condenses to form a surface film. This extra moisture might help to lower the T_m for starch and assist the melting of some starch granules on the surface, although it will also be swept away very quickly by the heat flux over the surface of the dough piece. It was also interesting to note that it was observed [7] that biscuits have larger cavities at the central regions than at the crust. Thus, although the water was escaping by evaporation through the crevices in the dough without any bubble structure to retain them, there was sufficient restriction to cause the cavities to expanded a little before the water vapour escaped. In the central region, the restriction was greater due to the greater length of the passage out of the dough, which might account for the greater expansion.

12.2.2 Cakes and flour confectionery

Cakes and other flour confectionery products have a wide range of recipes, because the amounts of sugar, fat and fluids, such as egg and milk, have been varied widely in the past [3]. There are many versions of cakes ranging from those with dense pudding-like textures, to other products which are light, aerated foams, such as sponge cakes and meringues. The sponge cake might be considered to be a representative choice to illustrate the physico-chemical phenomena occurring in the cake field. A typical recipe for a sponge cake, shown in Table 12.2, may be used as an example. If the fat phase is discounted, the flour represents ca. 29% w/w and the water (including that in the flour and liquid egg) 36% w/w of the batter. The proteins may bind ca. 25% w/w of the water, leaving a substantial amount available for the starch, which is almost equal to its

Table 12.2 Recipe for high ratio sponge cake

Ingredients	Baker's weight per 100 of flour	Weight (%)
Soft wheat flour	100	27.5
Castor sugar	105	28.9
Liquid egg	95	26.1
Water	45	12.4
Fats	15	4.1
Milk soldis	3.5	1.0

own weight. In addition, it would appear from the graph of T_m versus moisture content (Figure 12.4) that in the baking process, the starch granules should be gelatinised and, indeed, this is the practical finding [8]. There is sufficient water to permit the starch to be plasticised and melted as the batter is heated to 65–75°C (149–167°F). However, it should be noted that the castor sugar, sucrose, is present at a concentration of ca. 29% w/w of the batter or >50% w/w of the aqueous phase. It forms a concentrated sucrose solution as the batter is mixed and aerated. The increase in viscosity, caused by formation of the sugar syrup, helps to retain the bubbles of air within the batter, slowing their upward flotation according to the Stokes law which relates the velocity of the bubble to gravity and the viscosity of the batter. At this early stage in the process, the sugar syrup diffuses into the amorphous regions of the starch granules. The presence of sucrose in the aqueous phase within the starch granules has been shown to increase the T_m of the crystalline regions [4], thereby raising the gelatinisation point of the starch in a cake batter to 85–90°C (186–194°F).

The initial bubble population in the cake batter is extremely important in the production of cake structure. Only those bubbles which are present in the batter after mixing, are available for the expansion of the batter. Bubbles cannot be formed spontaneously during the baking process due to the very high excess pressure needed to initiate new bubbles [9]. Therefore, the initial mixing and aeration process is very important [10]. It was noted that the presence of the aqueous sugar syrup in the batter helps to retain the bubble population by its contribution to batter viscosity.

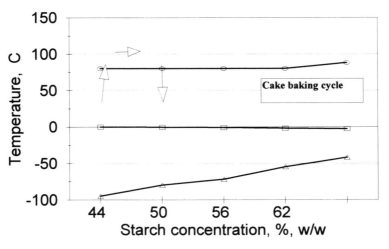

Figure 12.4 Melt and glass-transition temperatures of starch in cakes.

Other components of the recipe also contribute to the batter viscosity, notably the flour proteins. Some components interact more directly with the bubbles by forming layers at the air–water interface, thereby preventing bubble coalescence. In this role, the emulsifiers are chosen for their surface activity and the fats and egg proteins may also play a significant role. Crystalline fats, which have been aerated during their cooling process, may retain air bubbles in the batter, which will be stable in the fat phase until the crystal structure of the fat melts. The bubble system in the batter will be most stable when the bubble population is of a uniform size, and in the initial stages made up of small bubbles. If the population is uneven in its size distribution, the small bubbles will tend to coalesce with the larger ones, and there will be a greater tendency, according to the Stokes law, for the large bubbles to float upwards. The tendency for the bubbles to coalesce will be accelerated at higher temperature, because the batter viscosity falls with temperature. As the temperature of the batter rises from 30 to 70°C (86 to 158°F), the starch granules and egg proteins remain almost inert. Other reactions take place, such as the release of gases from baking powders and of dissolved air from the aqueous syrup. The bubbles which were trapped in the partial crystalline fats are also released at ca. 35–37°C (95–99°F), as the fat crystals melt, and as these gases enter the existing bubbles, the batter expands. It continues to expand as the temperature rises, due to the increase in gas pressure within the bubbles. It was stated earlier that no bubbles are created in the batter, but some may be lost by coalescence and escape from the batter. A model [10] was derived for the expansion of the bubbles in cake systems, showing the importance of the initial volume of the air bubbles, the gas from the baking powder and the water vapour. Initially, the batter viscosity falls as the temperature rises up to 75–80°C (167–176°F). There is a region of rapid change in the bubble size, as the gas pressure increases rapidly and the resistance of the fluid to expansion decreases (Figure 12.5). At the lowest viscosity value reached by the batter in the heating profile, two changes take place very quickly, which increase the viscosity to a very high level that is sufficient to cause a complete resistance to flow in the cell walls. The first major factor affecting viscosity is the swelling of starch granules to occupy much of the free volume in the fluid system. This change occurs rapidly after the T_m of the starch granules is exceeded. At that time, the starch polymers within the granules become amorphous, allowing the granular aggregates to swell rapidly in the aqueous phase [8]. Egg albumen proteins are denatured at similar temperatures and undergo a polymerisation of their molecules into linear chains and eventually a cross-linked gel phase [11]. The egg proteins, being too large to enter the granular aggregates, are concentrated in the intergranular space and form a concentrated gel phase between the hydrated starch granules. The whole structure is a composite of two

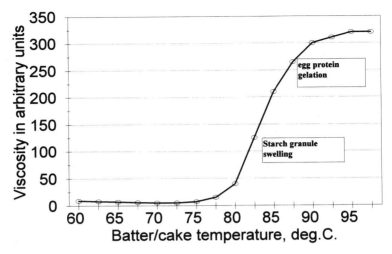

Figure 12.5 Changes in the viscosity of cake batter as it sets.

gelled polymer systems [13]. During the formation of structure within a cake, the batter is heated from the outside and the temperature rise, within the fluid mass of batter, occurs mainly from the outside towards the centre. The application of heat from the base of the batter in the early stages may give rise to convection currents and movement in the batter, but in a short period, crusts form at the top surface and outer edges of the batter. This means that the outer crumb is formed from the batter and set while the central crumb is still in a fluid state. This may cause splits in the crusts as the centre pushes outward in certain cakes. The massive increase in viscosity during starch gelatinisation and egg-protein gelation, inhibits the expansion of the gas bubbles and causes the internal pressure within bubbles to rise until the cell walls rupture at their weakest points and release the excess pressure. The phenomenon of 'bubble rupture' occurs in the outer layers of the batter and moves progressively towards the centre of the cake following the temperature profiles. The last bubbles to rupture are those in the central regions of the cake. If these bubbles are not ruptured, due to inadequate baking, they will collapse when the temperature of the cake falls, as it is removed from the oven, and form an unaerated wedge of dense cake crumb.

The physical changes in the batter which lead to the formation of the ruptured gas cells are very important and have unwittingly been used by bakers in their formulation of cake recipes. In a cake-baking process, the batter rises until the viscosity increase and cell-rupture phenomenon stops the expansion. It is generally observed that the later the large viscosity increase occurs in the bake, the larger the cake and the greater its specific volume. There are two ways in which the baker may create a larger lighter textured cake. If the sugar level is increased relative to the water used in the recipe, the T_m value of the starch granules will be raised and the rupture point will be later in the bake. This type of change is used in recipes which set at temperatures as high as 95°C (203°F) in certain types of cakes. An alternative method to increase cake volume is to reduce the amount of flour used in a recipe, while keeping the water and sugar levels constant. In this example, the T_m value for the starch will not increase, but the viscosity rise in the batter will be slower. The viscous rise is due to the contacts between the swollen aggregates of the starch granules with interactions between the starch and protein phases. In this system, the viscosity is related to an exponential function of the starch concentration. A reduction in the flour level requires an increase in the starch swelling and protein interaction to achieve the required viscosity values for the rupture of cell walls. The delay, caused by the reduction in swelling process, gives an increase in specific volume. The reduced starch level in the cooled cake crumb causes it to have a softer texture than the crumb with the higher starch content, and this type of cake is usually preferred by the consumer. However, there is a point at which the reduction in flour may lead to a collapse in the cake structure during the final stages of baking, or immediately on removal from the oven. It was discovered in the 1930s that the addition of small amounts of chlorine gas (0.1–0.2% w/w) improved the performance of cake flours, enabling their batter levels to be reduced from 29 to 24% by weight. These cakes, known as 'high-ratio cakes', because of the sugar-to-flour ratios used, were superior in eating quality, being lighter and more melting in their textural qualities than the heavier traditional recipes [13]. For many years the secret of the chlorination process defied investigation, but in recent years it has been shown that it changes the starch granules so that their swollen form [12,14] increase the viscosity of the systems more effectively and therefore the viscosity of the lower concentration of chlorinated starch is equal to that of the higher level of untreated starch. Thus, the simple change in the starch, caused by chlorine, can be related to the stabilisation of cake structure. Other methods of improving the flour have been discovered, such as heat treatment of flours or their wheats and semolina derivatives [15–17].

12.2.3 Shelf-life of cakes

Cakes may be regarded as intermediate moisture products because their water activity is in the range 0.7–0.75 and when packed in a moisture impermeable film, they have a microbiological shelf-life of 2–3 weeks. In the crust regions of cakes, the moisture levels may be as low as 4–5% w/w during the last few minutes of the bake. In these dry regions the temperature during baking rises to sufficiently high levels to give the browning reactions. However, after the cakes have cooled to ambient temperatures, a moisture redistribution occurs. It was shown in Madeira cakes that the diffusion of moisture proceeds until an equilibrium is reached and that the diffusion process can be slowed by lowering the storage temperature until it is completely inhibited at −18°C (0°F) [18]. The diffusion process in an open foam structure takes place largely through the vapour phase and is driven by the differences in water activity in the crust and crumb regions of the cakes. Many cake products are made with icings, butter creams or fruit gels as their fillings. In each case the filling must be balanced in its formulation so that its water activity is close to that of the cake crumb or the equilibrium relative humidity (ERH) of the cake within in packaging. An expert system called ERH Calc has been developed at CCFRA, Chorleywood, UK, for the rapid calculation of the ERH in any recipe or filling to achieve maximum shelf-life in bakery products [19]. Often some migration from the filling is allowed to keep the cake moist, but too much leads to the growth of surface moulds and shrinkage of the filling layer. A filling which picks up moisture would lead to a dry cake, which would appear staler than normal to the consumer.

The physical structures which are formed in cake crumb after baking are formed from starch and protein polymers. It can be seen that, at ambient temperatures, the amorphous starch granules are in a fluid state well above their T_g values of −10 to −20°C (14 to −4°F). Consequently, the polymers can rearrange their physical forms into helical complexes and crystalline structures. There are rapid changes immediately on cooling, in which the linear component forms a crystalline complex with lipids. In the first few hours of storage, the linear polymer of starch, amylose, is thought to crystallise and to stiffen the cake structure. Later, it is probable that the crystallisation of the branched polymer of starch, amylopectin, takes place over a period of up to 4–6 days, causing the firmness of the crumb to increase markedly. These changes are recognised as staling by the consumer [18].

12.2.4 Bread and morning goods

Bread and related yeast-raised products have simpler formulations than cakes. A standard recipe for bread in the UK is shown in Table 12.3. The recipe contains ca. 50% w/w of dry flour solids and a total moisture level of ca. 43% w/w. In the initial stages of processing, the flour is mixed with the water to form a viscoelastic dough. This is achieved in two stages – a rapid hydration process, followed by changes in the protein phase caused by the intensive mixing of the dough. In the first stage, the flour components, such as the wheat proteins, hydrate to their full capacity and the starch granules hydrate within their amorphous regions. There is a rough

Table 12.3 Recipe for bread and morning goods

Ingredients	Baker's weight per 100 of flour	Weight (%)
Strong bread flour	100	57.6
Water	63.5	36.6
Wheat gluten	2.5	1.4
Fats	2.0	1.2
Soya improver	1	0.6
Baker's yeast	2.5	1.4
Salt	1.5	0.9
Emulsifier	0.5	0.3

distribution of the water between the continous protein gel phase, which also contains the pentosan hydrocolloids, and the dispersed phase of native starch and those starch granules which had been damaged during milling. Damaged starch granules lose their crystallinity during milling [20] and their polymers may be fully plasticised by the available water, so that they swell and form a viscoelastic gel. Within the hydrated flour, the native granules remain rigid, whereas the remaining polymer phases, which are above their glass-transition temperatures at the mixing temperature range of 25–30°C (77–86°F), are present as viscoelastic forms in the rubbery state. Continued mixing of the dough transforms the physical structures of the wheat flour proteins to form a continuous gluten phase, which entraps air bubbles and has both the ability to flow under low stresses and to maintain good gas-holding properties.

The rheology of the bread doughs has been studied in great detail for many decades [21]. It has been shown that the essential features of breadmaking are the incorporation of a good bubble population within the dough [9] and the development of the native proteins, particularly the large glutenin molecules into a suitable gelled matrix [22]. The quality of the dough's protein system depends both on the selection of wheats with the appropriate protein compositions and the application of mechanical energy and chemical-improving agents during the mixing process. If large glutenin molecules are present, they can be manipulated by physical shearing and chemical additives to form a complex structure with the other constituents such as gliadins and pentosans. This viscoelastic matrix helps to control the gas bubbles during the initial handling procedures, proving (leaving it to rise) and baking processes. During the shaping and moulding of doughs, the aerated dough containing its air bubble population, is rolled out and shaped with considerable deformation. At this stage, the protein gel must have rheological characteristics which permit it to be stretched and compressed without losing its bubble structures. If it is too stiff and elastic, the air may be squeezed out, leaving sections of unaerated dough which cannot be expanded during the baking process.

Doughs are proved, after moulding into the product form, at 35–37°C (95–99°F) for up to 60 min, to allow the yeast to produce a good gas supply for raising the loaf and also to gently expand the bubbles, before the more rapid expansion occurs in the baking process.

On entering the oven, the dough is heated and its temperature gradient begins to vary as the heating occurs from the outside towards the centre. It is thought that the heat-transfer mechanism is a form of heat pipe. Water is vaporised in the outer layers and may condense on the cooler inner regions of dough, transferring the latent heat and raising the dough temperature, thus carrying on the mechanism towards the centre of the dough piece. This means that the final moisture at the centre may be as high as the initial dough. There is less opportunity for convection and flow in bread doughs than cake batters because of their higher viscosity. The outer layers expand under the increasing gas pressure and are heated to exceed the T_m for the starch granules, which is ca. 65–70°C (149–158°F) for the moisture level of a bread dough. At this point, the viscosity of the cell walls increases sharply and the cells rupture, releasing the gas pressure. This phenomenon occurs sequentially from the edge of the loaf towards the centre region, with successive layers of aerated dough setting and releasing their excess gas pressure through the outer layers of the bread structure. Eventually, the centre of the dough piece is set and releases its gas pressure, so that an open foam structure is formed throughout the loaf. The T_m values for bread doughs are lower than for cakes, because little or no sucrose is used in the recipes. Therefore, it is easier to ensure that the structures are set during the bake and there is less likelihood of collapse. A loaf may reach 95°C (203°F) at the centre during the normal baking period. The ERH for breadcrumb, at 95%, is higher than for cakes. Therefore, the shelf-life of bread is only a few days compared to several weeks for cakes. The physical stability of breadcrumb is similar to that of cakes, because the starch polymer systems are also in a fluid state at ambient temperatures, being well above the T_g values of −5°C (41°F) for the normal breadcrumb. This means that the formation of helical complexes between starch polymers takes place, leading to their partial crystallisation at junction points within the granules. Such changes proceed at rates roughly similar to those observed for cake crumb. Today, there is a much clearer view of the staling of bread and cakes due to the introduction of a food polymer science approach to this subject [4].

12.2.5 Wafer products

Wafers are made from liquid batters which have the highest water levels used in baked products (see Figure 12.1). After forming the dilute batter, a layer of the liquid is heated very quickly on the hot-metal surfaces of the wafer plates, where the temperatures may reach 110°C (230°F) for a few seconds. It has been shown that all the structure of the starch granules is destroyed [23] and a continuous starch phase is produced in the wafer. During the baking process on the wafer plate, most of the water is evaporated from the batter, leaving a highly concentrated mass of starch polymers. At high temperatures, this has a fluid-plastic rheological characteristic, but on cooling, it become a brittle glassy structure for moisture levels of <5% w/w. If normal wafers are exposed to the atmosphere, they absorb water and lose their crisp texture at levels of moisture >8–12%.

12.3 Pasta products

Pasta products are sold in the form of a dough or dried dough which must be cooked before it is ready for eating. The dough, which is prepared by mixing and shaping on an extruder, is often dried for long-term storage before sale through the retail outlets. Other methods of preservation are also used, such as canning, refrigeration or chill storage. In each case, the method of preservation affects the pasta product and the sensory characteristics may be significantly different for the various types of pasta. The moist products are also susceptible to change during storage in chilled or frozen conditions.

If a pasta dough is made in the traditional manner, it will be made from a simple recipe of durum wheat semolina, water and probably some egg [24,25]. The durum wheat has a hard vitreous endosperm texture and forms a gritty semolina, with a particle size range 200–350 µm (7.5–13 × 10^{-3} in.). In the mixing stage, the semolina particles must be hydrated so that they soften and stick together, but do not break down and form a homogeneous dough, as in a bread dough. The rheology of the pasta dough is different to the more intensely mixed bread doughs. Therefore, the particle size is very important and should have as narrow a range as possible in order to obtain an even hydration pattern across the dough. Otherwise, the finer particles would absorb water too quickly and become soft while the larger particles were still hard. If the large particles were allowed time to soften, the fine ones would be mixed into a bread-type dough.

The mixing action for pasta is gentle, allowing hydration and mixing without too much shear to develop the protein phase. At the end of the mixing period, a dough formed of hydrated particles is shaped by extrusion or by sheeting between rolls. Normally, it is forced into a single screw extruder and de-aerated. During extrusion, the hydrated particles are pressed and sheared against each other to form a continuous cohesive dough and the air is removed from the crevices to form a dense extrudate. The use of a vacuum during extrusion produces a translucent extrudate, with no obvious signs of air bubbles.

The moist extrudate may be dried to produce the stable pasta, which is the most widely sold product. However, the water must be removed from the moist extrudate without weakening the integrity of the extrudate, or introducing any distortions in its appearance such as blisters or cracks. Therefore, the drying process relies on the careful removal of water by evaporation at temperatures of 40–70°C (104–158°F) under controlled humidities [23]. Any air bubbles present in the extrudate might detract from the smooth translucent appearance, particularly if they are expanded during the drying process. The water has to be removed at a rate which allows the diffusion within the extrudate to keep pace with the losses from the surface. In the process, it is necessary to allow the air flowing over the surface to have a relatively high humidity, while the slower process of moisture diffusion from the centre is taking place. If such steps are not followed, changes in the rheology of the pasta with the moisture gradient across the product may create strain within the structure. This may not be obvious when the product is dried and all the polymers are in the stable glassy states, but when the pasta is cooked in boiling water, the products may distort or split as the weaknesses are revealed. The cooking process for pasta requires the rehydration of the dried product from 2–4% w/w to the level of 40–45% w/w which must be achieved without any sign of disintegration of the products. The presence of any air bubbles would obviously lead to expansion and distortion unless they were eliminated in the process. This is not so important for fresh pasta because the cooking time is shorter and textures are usually softer. As the water penetrates the pasta, the level becomes high enough to permit the T_m value for the wheat starch to be exceeded in boiling water. The pasta softens as the starch granules lose their crystallinity and gradually, from the surface to the centre to the core, the pasta becomes soft. The controlled development of the protein phase binds the original semolina particles together and prevents the starch dispersing into the cooking water.

12.4 Colour formation in cereal products

12.4.1 Bakery products, pastries, cakes and bread

The appearance of baked products is highly attractive due to the golden brown colours which form at the crust regions during baking. These colours are accompanied by delicious odours from the bakery and leave subtle flavours in the crusts and crumb of the products. It is generally recognised that they are formed by the combination of chemical reactions known as 'non-enzymic browning' (see chapter 4). These may encompass Maillard reactions, caramelisation and breakdown of carbohydrates and lipids. In the baking processes for products such as pastry, biscuits, cakes and bread, the common feature is that colour formation occurs at the surface of the products. Little or no colour is formed with microwaved products which are cooked at constant moisture levels of 30–45% w/w. There is normally a surface region, known as the 'crust', where most of the colour appears. This is similar to colour development in fried products, such as doughnuts, where the hot oil surrounds the bottom half of the floating dough piece and colour formation takes place only in the area of contact.

In the hot-air ovens used for baking, the surface is swept by a stream of air which removes the layer of moisture and serves to dehydrate the product. For a short time, the moisture migration from the neighbouring layers and from deeper within the structure may keep the moisture content at the surface fairly high, so that evaporative cooling may occur. This phenomenon will prevent the temperature of the dough rising above the boiling point of the aqueous phase. Thus, for a short time, the surface regions remain at 100–105°C (212–221°F). At this temperature, the main reactions which form the colours are slow and little reaction occurs, even with precursors present. As the moisture level falls to <5%, the evaporative cooling fails to restrict the temperature rise and the contact with the hot air stream causes the temperature to rise to 130–170°C (266–338°F). Browning of the crust is an important feature of bread baking, both because of the attractive appearance and the flavours which develop within the coloured regions of the crust due to Maillard browning reactions. As in other baked products, Maillard reactions occur in a thin layer at the surface, where the moisture level is reduced by evaporation to a low level, allowing the temperature to rise well above 100°C (212°F). The direct contact with the hot air in the oven raises the temperature to 130–150°C (266–303°F) and accelerates the browning reactions.

The reactions taking place are complex and clearly involve small molecules such as reducing sugars, amino acids and peptides. They may also involve larger molecules such as proteins with free amino groups and even large carbohydrates such as starch, by dehydration reactions. In the three classes of products discussed in this chapter, the recipes described vary in the amounts of small precursors which are added. Pastries are usually fairly simple recipes, as illustrated in Table 12.1, and may not colour very much unless they are sprayed or coated with a glazing solution prior to baking. A glaze may contain skimmed milk or whey solids, glucose and maltodextrins. Such a mixture supplies the reducing sugars, amino acids and peptides for the browning reactions.

In cakes, the recipes are more complex and usually contain some skim-milk solids or whey. The main sugar, sucrose, is a non-reducing sugar and does not take part in browning reactions until heated to high temperatures when it may break down to reducing compounds. In the bulk of the cake, the sucrose remains in its normal form and takes no part in browning reactions. If it is replaced by glucose or fructose, these sugars are reactive and darken the crusts and outer regions of the cakes. Therefore, the recipe must be chosen carefully to ensure that the level of reactants for colour formation is well balanced.

In bread products, recipes may vary from simple systems of flour, yeast, salt and water to richer recipes with soya, milk solids and sugars. There may be some hydrolytic enzymes present, such as amylases, proteases and hemicellulases which may release precursors from the polymers during the mixing, proving and the early stages of the bake.

12.4.2 Colour in pasta products

The colour of pasta is a translucent golden yellow in high-quality products, but may be more opaque and less yellow in low-cost products. Natural colours are derived from the durum wheat and may be preserved in the products by good processing methods. It is possible for some less-attractive colours to be developed in the drying process by the non-enzymic browning reactions. These are caused by precursors, such as those mentioned in section 12.4.1, i.e. basic amino acids, peptides and reducing sugars. The rates of non-enzymic browning are slow at the temperatures of the dryers, but their reaction rates reach a maximum at 0.3–0.7 water activity.

12.5 Conclusions

In looking at the manufacture of these cereal products, it is obvious that a knowledge of the science, described in the first ten chapters, is important in understanding the changes which are taking place. This in turn enables products and processes to be optimised and possible pitfalls avoided. For example, water activity and T_g measurements may enable a product's shelf-life to be increased, whereas changes to rheological properties will have a profound effect upon texture. Thus, although our current knowledge is incomplete, scientific principles are enabling food scientists and technologists to have much better control over their production processes.

References

1. Bushuk, W. and Scanlon, M.G. Wheat and wheat flour. In *Advances in Breadmaking Technology*, Eds Kamel, B.S. and Staffer, C.E. (1993), Blackie Academic and Professional, Glasgow, 1–19.
2. Bushuk, W. Wheat: chemistry and uses. *Cereal Foods World* **31** (1986), 218–226.

3. Street, C.A. Basic products and finished goods. In *Flour Confectionery Manufacture*, Chapter 6 (1991), Blackie, Glasgow.
4. Slade, L. and Levine, H. Structure–function relationships of cookie and cracker ingredients. In *The Science of Cookie and Cracker Products*, Ed Faridi, H. (1994), Chapman and Hall, London, 86–91.
5. Donovan, J.W. A study of the baking process by differential scanning calorimetry. *J. Sci. Food Agric.* **28**(6) (1977), 571. Also: Donovan, J.W. Phase transitions of the starch water system. *Biopolymers* (2) (1979), 263.
6. Lawson, R.L. Mathematical modelling of cookie and cracker ovens. In *The Science of Cookie and Cracker Products*, Ed. Faridi, H. (1994), Chapman and Hall, London, 387–438.
7. Burt, D.J. and Fearn, T.E. A quantitative study of biscuit microstructure. *Starch* **35** (1983), 351–354.
8. Bean, M.M. and Yamazaki, W.T. Wheat starch gelatinisation in sugar. I. Sucrose microscopy and viscosity. *Cereal Chem.* **55** (1978), 936. Also: Bean, M.M., Yamazaki and Donelson, D.H., II. Fructose, glucose and sucrose: cake performance. *Cereal Chem.* **55**, 945.
9. Baker, J.C. and Mize, M.D. The origin of the gas cell in bread doughs. *Cereal Chem.* **18** (1941), 19.
10. Davies, A.P. Protein functionality in bakery products. In *The Chemistry and Physics of Baking: Materials, Processes and Products*, Eds Blanshard, J.M.V., Frazier, P.J. and Galliard, T., chapter 7, pp. 89–104 (1986), RSC Press, London.
11. Mizukoshi, M., Maeda, H. and Amano, H. Model studies on cake baking. II. Expansion and heat setting of cake batter during baking. *Cereal Chem.* **57** (1980), 352–355.
12. Guy R.C.E. and Pithawala, H. Rheological studies of high ratio cake batters to investigate the mechanism of improvement of flours by chlorination or heat treatment. *J. Food Technol.* **16** (1981), 153.
13. Allen, J.E., Sherbon, J.W., Lewis, B.A. and Hood, L.A. Effect of chlorine treatment of wheat flour and starch: measurement of thermal properties by differential scanning calorimetry. *J. Food Sci.* September/October (1982), 1508–1511.
14. Frazier, P., Brimblecome, F.A. and Daniels, N.W.R. Rheological testing of high-ratio cake flours. *Chemy Ind.* **21**, December (1974), 1008.
15. Russo, J.V. and Doe, C.A.F. *J. Food Technol.* **5** (1970), 563.
16. Hodge, D.G. Alternatives to chlorination in high-ratio cake flours. *Baking Ind. J.* **8**(1) (1975), 12–19.
17. Cauvain, S.P., Hodge, D.G., Muir, D. and Dodds, N.J.H. Improvements related to the treatment of grain. UK Patent 1 444 173 (1976), July.
18. Guy, R.C.E. Factors affecting the staling of Madeira slab cake. *J. Sci. Food Agric.* **34** (1982), 477–491.
19. Young, L.S. Expert systems in baking technology. *Food Technol. Int. Europe* (1992), 91–93.
20. Kent, N.L. and Evers, A.E. Chemical components. *Technology of Cereals*, Pergamon, Oxford (1994), 62.
21. Bloksma, A.H. Rheological aspects of structural changes. In *The Chemistry and Physics of Baking: Materials, Processes and Products*, Eds Blanshard, J.M.V., Frazier, P.J. and Galliard, T. (1986), RSC Press, London, 170–178.
22. MacRitchie, F. Physico-chemical processes in mixing. *The Chemistry and Physics of Baking: Materials, Processes and Products*, Eds Blanshard, J.M.V., Frazier, P.J. and Galliard, T. (1986), RSC Press, London, 133.
23. Blanshard, J.M.V. The significance of the structure and function of starch in baked products. In *The Chemistry and Physics of Baking: Materials, Processes and Products*, Eds Blanshard, J.M.V., Frazier, P.J. and Galliard, T. (1986), RSC Press, London, 2.
24. Feillet, P. Present knowledge of the biochemical basis of pasta cooking quality; consequences for wheat breeders. *Sci. Aliment.* **4** (1984), 551–556.
25. Antognelli, C. The manufacture and applications of pasta as a food ingredient. *J. Food Technol.* **15** (1980), 125.
26. Audidier, Y. Effects of thermal kinetics and weight loss kinetics on biochemical reactions in dough. *The Baker's Digest*, October (1968), 36–42.

13 Freezing and cooking of meat and fish
S.J. JAMES

13.1 Introduction

The main foods of animal origin, fish and meat, are very perishable raw materials. If stored under ambient conditions, 16–30°C (61–86°F), the shelf-life of fish can be measured in hours and meat, tens of hours to a few days. Under the best conditions of chilled storage, close to the initial freezing point of the material, the storage life can be extended to a few weeks for fish and approaching 6 weeks for some red meats. Even under the best commercial practice, strictly hygienic slaughtering, rapid cooling, vacuum packing and storage at super chill (-1 ± 0.5°C, 30.0 ± 1.0°F) the maximum life that can be achieved in red meat is approximately 20 weeks. Freezing will extend the storage life of meat to a number of years. With fish, Morrison [1] wrote that "freezing is the only preservation technique that delivers fish in essentially the same quality as freshly caught and processed fish".

Over a hundred years ago, the first large frozen food store was built in the USA [2]. Since then, a gargantuan amount of data has been generated dealing with freezing and frozen storage of fish and meat but there is little agreement on frozen storage times or conditions. In 1947, Ramsbottom [3] stated that a "review of the literature indicated little agreement amongst investigators as to the frozen storage life of fresh meat" and little seems to have changed. Twenty-five years later, Bengtsson et al. [4] found that if normal 95% confidence limits were applied, the high-quality storage life of lean meat could lie between 8 months and 3 years.

One main reason for lack of agreement between data is the wide range of rather confusing definitions used to define storage life and the use of sensory panels to assess it. Usually, storage is defined by practical storage life (PSL) which is described in the International Institute of Refrigeration (IIR) recommendations [5] as "the period of frozen storage after freezing during which the product retains its characteristic properties and remains suitable for consumption or the intended process". Panels are often small and are really only a measure of product acceptability within that select group and assessments cannot necessarily be related to consumer preferences.

The two prime purposes of cooking are (1) to render the food safe to eat and (2) to facilitate chemical changes within the food that will improve its organoleptic properties. In industrial cooking, there is also considerable pressure to maximise the yield from the cooked product and minimise cooking time.

13.2 Freezing

The factors that influence the storage life of frozen fish and meat may act in any one of three stages: (1) Prior to freezing, (2) during the actual freezing process and (3) post-freezing in the storage period itself.

13.2.1 Pre-freezing treatment

There are a number of pre-freezing factors that have been considered to affect the storage life of fish and meat. Some variation (i.e. species differences) between individuals or differences between cuts/portions, are inherent in the animal or fish. With meat, there are also other factors, including feeding and transport, which may have an effect on frozen storage.

Species differences. The main pre-freezing factor that is commonly believed to influence the frozen storage life is the species of animal or fish that produced the product. Fish are commonly divided into two categories: fatty (pelagic) and non-fatty (white or demersal). Pelagic fish contain large amounts of dark muscle and are capable of swimming at high speeds for long periods of time. Demersal fish contain predominantly white muscle and, on the whole, forage for food at, or near, the sea bed at slow speeds.

The polyunsaturated fat in the pelagic fish is prone to oxidation resulting in rancidity during frozen storage. The oil content of pelagic fish differs between species, hence the rate of deterioration will also differ. In white fish, the main quality deterioration during frozen storage is often in texture. During storage, chemical changes result in the water-holding capacity being reduced and the cooked flesh becoming tough, fibrous and dry. Different species deteriorate at different rates.

In meat the industry accepts the frozen storage life of pork to be less than that of the other species. Recommended storage lives from the IIR [5] reflect this view (Table 13.1). However, when all the available data found in the literature are considered the picture becomes confusing. Average values for the storage lives of the different species show differences between the species (Table 13.1) but with a ranking that is different from that which would normally be expected (e.g. pork actually has the longest shelf-life!). In all species, the range of storage lives found in the literature is very large and indicates that factors other than species have a pronounced effect on storage life.

Table 13.1 Frozen storage life (months) of meat from different species at −18°C (0°F)

Meat	IIR [5]	Average
Beef	18	10.2
Pork	12	17.4
Lamb	24	7.8
Chicken	18	13.6

Animal-to-animal variation. Animal-to-animal variation is believed to cause wide variations in storage times. Winger [6,7] has stated that animal-to-animal variation can lead to storage-life differences as great as 50% in lamb, but does not give any definite reason for this variation. Differences would appear to be caused by genetic, seasonal or nutritional variation between animals, but there is little reported work to confirm this view. Crystall and Winger [8] found that variations exist between the fatty acids and ratio of saturated:unsaturated fatty acids in lamb from New Zealand, America and England. Differences related to area, sex and cut were mainly a reflection of fatness, with ewes having a greater percentage of body fat than rams. Differences between areas were, however, found to produce larger variations between animals than sex differences.

There appears to be potentially large variations between animals, which cause changes in the storage life of meat, but why these differences exist is not completely understood.

Feeding. The way in which an animal is fed can influence its frozen storage life. It has been reported that chops from pigs fed on household refuse have half the storage life of those fed on a milk/barley ration [9,10]. Again, pork from pigs that had been fed materials containing offal had half the PSL and higher iodine numbers in the fat than that of pigs which had not been fed this type of diet [11]. Conversely, Bailey et al. [12] did not find any differences between meal and swill fed pigs after 4 and 9 months at −20°C (−4°F). Rations with large amounts of highly unsaturated fatty acids tend to produce more unstable meat and fat [13].

Many fish, for example herring and mackerel, may have individual fat contents as low as 1% after winter starvation, rising to 30% after intense feeding. After spawning, protein and fat reserves become depleted and the flesh becomes watery [14]. Generally, the keeping quality of well-fed fish is improved due to higher glycogen contents (see chapter 10).

Variation within an animal. Reports of variations in the storage life of different cuts of meat are scarce and primarily deal with dark and light meat. Both Ristic [15] and Keshinel et al. [16] have found that breast meat stores better than thigh meat. Ristic [15] states that breast meat will store for 16 months while thigh meat can only be stored for 12 months due to its higher fat content. Judge and Aberle [17] also found that light pork meat stored for a longer time than dark meat. This was thought to be due to either higher quantities of haem pigments (from blood) in the dark muscle (which may act as major catalysts of lipid oxidation), or to higher quantities of phospholipids which are major contributors to oxidised flavour in cooked meat.

Handling and transport. Since most fish are hunted, they will have struggled significantly during catching, depleting muscular glycogen reserves, initially increasing the lactic acid content. Immediately after death, their pH of 6.4–6.8 is higher than that of rested meat and the onset of rigor is accelerated.

The way animals are handled and transported before slaughter is thought to affect meat quality and its storage life. Increased stress or exhaustion can produce PSE (pale, soft and exudative) or DFD (dark, firm and dry) meat which is not recommended for storage, mainly due to its unattractive nature and appearance. Jeremiah and Wilson [18] found that frozen storage may increase shrinkage especially in PSE muscle. PSE muscle was found to give low yields after curing and it was concluded that PSE meat was unsuitable for freezing or further processing.

Chilling and ageing. Meat is generally not frozen until rigor is complete and a degree of conditioning has taken place, otherwise toughening and increased drip can occur ([2]; see also chapter 10). In red meat, there is little evidence of any relationship between chilling rates and frozen storage life. However, there is evidence for a relationship between storage life and the length of time that elapses before freezing occurs. Chilled storage of lamb for 1 day at 0°C (32°F) prior to freezing can reduce the subsequent storage life by as much as 25% when compared to lamb which had undergone accelerated conditioning and 2 hours storage at 0°C (32°F) [7]. Harrison et al. [19] have also shown that pork which had been held for 7 days deteriorated at a faster rate during storage than carcasses chilled for 1 and 3 days. Ageing for periods greater than 7 days was found by Zeigler et al. [20] to produce meat with high peroxide and free fatty acid values when stored at −18°C or −29°C (0°F or −36°F). However, Pool et al. [21] have shown that there were no detectable flavour differences over an 18-month period between turkey that had been frozen immediately and turkey that had been held at 2°C (36°F) for 30 hours. Although shorter ageing times appear to have a beneficial effect on storage life, there is obviously a necessity for it to be coupled with accelerated conditioning to prevent any toughening effects.

In poultry, chilling method does have an effect on storage life. Grey et al. [22] found that air-chilled broilers had significant flavour changes after 3 months at −12°C and −20°C (10°F and −4°F), whereas immersion-chilled birds only exhibited changes at −12°C (10°F) after 6 months and were stable at −20°C (−4°F). The air-chilled birds had a better texture initially, but became tougher than the immersion-chilled birds after 6 months. Ristic [23] also found that water chilling of broilers produced a more favourable taste in the leg and breast meat than air chilling. Significant differences were found by a taste panel after 3 months in the thigh meat and 4 months in the breast.

In fish, rigor mortis, where muscles go into a state of contraction, occurs between 1 and 12 hours post-mortem and they can remain in rigor for up to 24 hours. Delays in cooling fish from temperate waters can result in the rapid onset of rigor and lead to extreme muscular contractions, reaching a maximum at 17°C (63°F) for cod [1]. Separation of white-fish muscle flakes, or segments, known as 'gaping', will occur in this situation. Any attempts to straighten the fish before rigor is complete and the muscles relaxed, will cause severe gaping in the fillets [24].

13.2.2 The freezing process

Meat for industrial processing is usually frozen in the form of carcasses, quarters [25] or boned-out primals in 25-kg (55-lb) cartons [26,27]. Most bulk meat, consumer portions and meat products are frozen in air-blast freezers. Some small individual items, e.g. beefburgers, may be frozen in cryogenic tunnels, and a small amount of offal and other meat is frozen in plate freezers.

With fish, the preferred treatment is to plate-freeze the fish at sea. White fish such as cod are usually gutted and often deheaded before being plate frozen in 45–65-kg (100–140-lb) blocks. Pelagic fish are often frozen whole inside plastic film with sea water to minimise oxidative reactions.

It is not unusual for both fish and meat to be frozen twice before it reaches the consumer. As already stated, meat and fish are often frozen in bulk. During industrial processing the raw material is then thawed or tempered before being turned into meat/fish-based products, pies, convenience meals, burgers, etc. or consumer portions, fillets, steaks, etc. These consumer-sized portions are often refrozen before storage, distribution and sale.

There are little data in the literature to suggest that, in general, the method of freezing or the rate of freezing has any substantial influence on a food's subsequent storage life. There is some disagreement in the literature as to whether fast or slow freezing is advantageous. Slightly superior chemical and sensory attributes have been found in food cryogenically frozen in a few trials [28–30], but other trials did not show any appreciable advantage [31], especially during short-term storage [32]. Jackobsson and Bengtson [33] indicated that there is an interaction between freezing rate and cooking method. Meat that had been cooked from frozen was found to show a favourable effect of faster freezing rates.

The method of freezing clearly affects the ultrastructure of muscle. Slow freezing (1–2 mm/hour (0.05–0.1 in./hour) for example [34]) tends to produce large ice crystals extracellularly, whilst quick freezing (e.g. 50 mm/hour (2 in./hour)) gives smaller crystals in and outside cells [34,35]. Obviously, a temperature gradient will occur in large pieces of meat and result in a non-uniform ice morphology [35]. However, there are no data to suggest that the ultrastructure influences storage life.

13.2.3 During frozen storage

Three factors during storage: (1) the storage temperature, (2) the degree of fluctuation in the storage temperature and (3) the type of wrapping/packaging. These are commonly believed to have the main influence on frozen storage life.

Storage temperature. To quote from the IIR 'Red book' [5], "storage life of nearly all frozen food is dependent on the temperature of storage . . ." and in the book, a table is provided of PSL of different foods at three storage temperatures. Few would disagree that storage temperature influences frozen storage life and that, in general, the lower the temperature the longer the PSL.

For fish reducing the storage temperature from −12°C to −24°C (10°F to −11°F) produces an extension of at least threefold in the PSL (Table 13.2).

Experimental data from many different publications have been plotted against the temperature of storage for beef (Figure 13.1) and similar data are available for pork and lamb [36]. There is a clear effect of temperature on storage life, with lower temperatures resulting in extended storage, but there is considerable scatter between results at any one temperature.

Table 13.2 Practical storage lives (months) of fish [5] at various temperatures

Product	−12°C (10°F)	−18°C (0°F)	−24°C (11°F)
Fatty fish, glazed	3	5	>12
Lean fish	4	9	>12
Cooked lobster, crab	4	6	>12
Shrimps (cooked/peeled)	2	5	>9

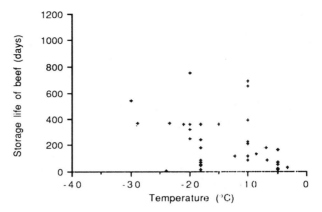

Figure 13.1 Experimental data on storage life of beef at different temperatures.

Some products do not store better at lower temperatures. Lindelov and Poulson [37] have shown that free fatty acids accumulate more readily in bacon stored at −30°C (−22°F) than bacon stored at −12°C (10°F). The highest rate of free fatty acid accumulation in bacon was found to be at −62°C (−80°F). Cured-pork products are known to have an abnormal temperature profile between −5°C and −60°C (23°F and −76°F) and store less well between −30°C and −40°C [86°F and −40°F] [38].

Temperature fluctuations. Generally, fluctuating temperatures in storage are considered to be detrimental to the product. However, it has been reported that repeated freeze–thaw cycles do not cause any essential change in the muscle ultrastructure [39] and that several freeze–thaw cycles during a product's life cause only a small quality damage (in terms of juice loss) or possibly no damage at all. Jul [40] found a slight, but significant, improvement in samples that had been frozen and unfrozen several times when tested by a taste panel.

Minor temperature fluctuations in a stored product are generally considered to be unimportant, especially if they are below −18°C (0°F) and are only of the magnitude of 1–2°C (~1°F). Well-packed products and those that are tightly packed in palletised cartons are also less likely to show quality loss.

There is disagreement on how much effect temperature fluctuations have on a product. Some authors consider temperature fluctuations have the same effect on quality of the product as storage at an average constant temperature [41,42].

Packaging. Packaging has a large direct effect on storage life, especially in fatty foods and, in extreme cases, indirectly due to substantially increasing the freezing time. As already discussed, freezing rate does not normally influence frozen storage life. In practice, a number of examples have occurred where large pallet loads of warm boxed meat have been frozen in storage rooms. In these cases, freezing times for product in the centre of the pallet can be so great that bacterial and enzymic activity results in a reduction of storage life.

Klose et al. [42] found that storage life of chicken pieces at $-18°C$ (0°F) could be quadrupled by the careful choice of packing material. Waterproof packing also helps to prevent freezer burn (see chapter 14) and tight packing helps to prevent an ice build-up in the pack. Packing may also delay the onset of rancidity if the wrapping is impermeable to air and moisture. Packaging can be effective, in some cases, in reducing discoloration by decreasing oxygen penetration into the meat.

13.3 Cooking

The main objectives of caterers and manufacturers of cooked foods are to produce economically products that are microbiologically safe and of high organoleptic quality. For safety, a minimum temperature–time treatment should be achieved during cooking, followed by sufficiently rapid cooling to minimise growth of surviving organisms. These processes should incorporate good handling procedures designed to eliminate bacterial contamination. For a cooking operation to be economic, it should achieve high throughputs (i.e. short cooking and cooling processes), incur low weight losses, use the lowest cost raw materials and require both minimal capital investment and kitchen/factory floor space.

In practice, these requirements are often in conflict because weight loss tends to increase with product temperature and heating time. Furthermore, the relationship between cooking temperature, time and tenderness is complex and it is generally accepted that poor-quality meats, rich in collagen, should be cooked at low temperatures for long times. Fish, in comparison, require a far less severe heat treatment and many products rely on the final heating by the consumer or caterer to produce the cooked product.

It has been shown that there is a small but significant decrease in tenderness when meat joints are cooked directly from the frozen state without an intermediate thawing stage [43]. A similar decrease was also found when moist rather than dry heat was employed in the cooking oven. Studies have also shown that, as an aid to the consumer, it is possible to pre-slice meat before cooking [44] and a small amount of frozen, pre-sliced, cooked meat is now being sold.

The above studies are typical of most studies of cooking methods and conditions that have generally been carried out with small joints and high operating temperatures typical of those used in domestic situations. Few studies have compared the heating times and weight losses of large joints processed by different commercial systems. Currently, there is no UK legislation that specifies a minimum temperature that has to be achieved during a cooking process. The Food Hygiene (Amendment) Regulations 1990 specify a minimum temperature of 63°C (145°F) for certain foods, if they are to be sold hot. The Chilled Food Association Chilled Food Guidelines and the DHSS Cook–Chill and Cook–Freeze Guidelines recommend a minimum temperature of 70°C (158°F) for at least 2 minutes during cooking.

The problems in specifying the optimum cooking conditions for different types of meat are best illustrated using data from three different studies carried out at Langford, UK.

13.3.1 Cooking of small meat joints

Small cylinders (80-mm diameter × 160-mm long; 3 in. × 6 in.) of lean beef were cooked in air at temperatures of 65.5, 117.5 and 175°C (150, 244 and 347°F) and air velocities of 0.15, 0.3, 0.45 and 0.6 m/s (0.5, 1, 1.5 and 2 ft/s) to represent the range of conditions found in commercial ovens [45]. In most cooking operations joints are placed in ovens for a set time and the relationship between time and minimum temperature and percentage weight loss are shown in Figures 13.2 and 13.3, respectively.

As expected, minimum temperature increases with increase in time, air temperature and air velocity. Weight loss increases with time, and the rate of weight loss increases both with increased air temperature and in general with increased air velocity. In commercial practice, it is quite possible for air velocities in different parts of a large oven to span the range used in these investigations. The results show that differences in air velocity can have a substantial effect on the final temperature of the meat and the percentage weight loss. For example, after heating for 100 minutes in air at 175°C (347°F), the minimum meat temperature is 67°C (153°F) in air at 0.15 m/s (0.5 ft/s) and 75°C (167°F) at 0.6 m/s (2 ft/s). Corresponding weight losses are 18% and 30%. Variations in weight loss of this magnitude would lead to substantial differences in the quality of the meat cooked in an oven with this degree of variation in air velocity. To the process operator relying solely on air-temperature control, all the meat joints would appear to have received the same cooking treatment.

Although meat is often cooked for a set time, this time has usually been chosen so that a known minimum internal temperature has been exceeded. When these results were analysed in terms of weight loss to a minimum internal temperature, no relationship between weight loss and air velocity was revealed. However, air temperature did have a significant effect on weight loss (Figure 13.4). Cooking to a lowest internal tem-

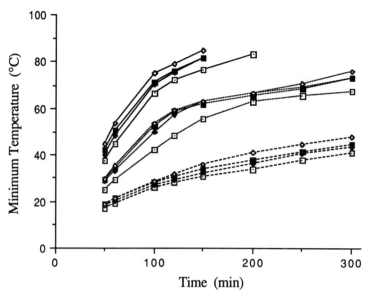

Figure 13.2 Minimum temperature in an 80-mm diameter × 160-mm long (3 in. × 6 in.) meat cylinder cooked in air at 65.5, 117.5 and 175°C (150, 244 and 347°F) and air velocities of 0.15, 0.3, 0.45 and 0.6 m/s (0.5, 1, 1.5 and 2 ft/s).

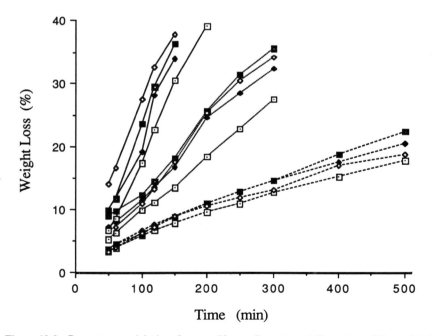

Figure 13.3 Percentage weight loss from an 80-mm diameter × 160-mm long (3 in. × 6 in.) meat cylinder cooked in air at 65.5, 117.5 and 175°C (150, 244 and 347°F) and air velocities of 0.15, 0.3, 0.45 and 0.6 m/s (0.5, 1, 1.5 and 2 ft/s).

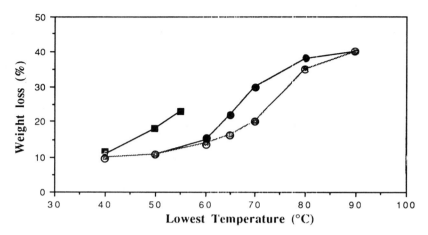

Figure 13.4 Percentage weight loss versus lowest temperature from an 80-mm diameter × 160-mm long (3 in. × 6 in.) meat cylinder.

perature of 75°C (167°F) the lowest weight loss (mean 28%) achieved in air at 175°C (347°F) was approximately 7% less than in air at 117.5°C (244°F).

Weight loss can be related to juiciness and consequently eating quality. These results, therefore, support the view that with small pieces of meat, steaks for example, a fast, severe cooking treatment should be recommended.

13.3.2 Cooking of whole poultry

The influence of cooking conditions on weight loss and time when cooking larger pieces of meat was demonstrated by cooking whole chicken carcasses (dressed weight approximately 1 kg (2 lb) to a minimum internal temperature of 80°C (176°F) [46]. Cooking was carried out in forced convection wet (dew point 75°C (167°F)) and dry air at air temperatures of 120, 150, 180 and 210°C (248, 302, 356 and 410°F) and air velocities of 0.5, 1.0 and 2.0 m/s (1.7, 3.3 and 6.6 ft/s).

Raising the oven temperature from 120 to 210°C (248 to 410°F) reduces the cooking time by a factor of approximately two under all conditions. At least 70% of this reduction is achieved during a 30°C (54°F) increase in oven temperature from 120 to 150°C (248 to 302°F).

At a set oven temperature, the use of wet air produces a substantial reduction in cooking time over that produced using dry air. The reduction is between 15 and 25 minutes at 120°C (248°F) and approximately 10 minutes 210°C (410°F).

Increasing the air velocity by a factor of four, from 0.5 to 2 m/s (1.7 to

6.6 ft/s), produces a maximum reduction of 30 minutes in the cooking temperature at 120°C (248°F). At 210°C (410°F), the maximum reduction is <10 minutes.

Weight loss increases with increase in temperature with a minimum of approximately 15% in wet and 20% in dry air at 120°C (248°F) to a maximum of 24% and 34%, respectively, at 210°C (410°F). In most conditions, weight loss increased with increase in air velocity. The effect was most pronounced at low temperatures and in wet air. The use of wet rather than dry air substantially reduced weight loss.

It would therefore appear that with whole poultry, a slower cooking process, using low temperatures and moist air, is required to minimise weight loss. These are the opposite recommendations to those for small pieces of meat.

13.3.3 Cooking and cooling of large meat joints

In industrial operations, both the cooking and cooling process need to be considered together. Three cooking and cooling methods (convective air cooking and cooling, water immersion cooking and cooling, and pressure cooking and vacuum cooling) have been compared when applied to three types of beef joint (slabs from m. semitendinosus, silversides and forequarters), boned-out ham and boned-out turkey [47].

In the convection method, each joint was cooked in air at 120°C (248°F), 0.5 m/s (1.7 ft/s) and 75°C (167°F) dew point followed by cooling in air at 0°C (32°F) and 1.2 m/s (4 ft/s). Before immersion processing, joints were vacuum-packed and then heated in agitated water at 95°C (203°F) before being cooled in water at 0°C (32°F). Pressure cooking was carried out in a vessel operating at 1.03 bar (15 psi) followed by vacuum cooling. All the joints were cooked from 5°C (41°F) to an internal temperature of 75°C (167°F) for beef, 80°C (176°F) for ham and 85°C (185°F) for turkey then cooled to a maximum internal temperature of 10°C (50°F).

Pressure cooking was always substantially faster than either of the other two cooking methods and was always twice as fast as the slowest cooking method (Table 13.3). With the small m. semitendinosus joints, immersion cooking was faster than convection cooking, but there was no significant difference between immersion and convection cooking times for the larger silverside and forequarter joints. With the large turkey and ham joints, convection cooking was significantly faster than immersion.

Weight losses after convection cooking were always lower than after pressure cooking with a mean difference of 8.5%.

Overall, weight losses after convection or immersion cooking and cooling were not significantly different when processing any of the beef joints. The pressure/vacuum technique produced significantly greater

Table 13.3 Processing times and weight losses from meat under different cooking and cooling regimes

Processing conditions	Cooking		Cooling		Overall	
	Time (min)	Weight loss (%)	Time (min)	Weight loss (%)	Time (min)	Weight loss (%)
Beef m. semitendinosus						
Convection	120	26.6	150	2.7	270	29.3
Immersion	71	–	100	–	171	32.7
Pressure/vacuum	50	32.4	52	9.5	102	41.8
Beef forequarter						
Convection	225	26.7	390	2.3	615	28.9
Immersion	208	–	282	–	489	25.5
Pressure/vacuum	110	38.0	43	7.9	153	45.9
Beef silverside						
Convection	219	28.4	338	3.3	557	31.8
Immersion	227	–	240	–	467	30.0
Pressure/vacuum	104	39.2	61	7.4	165	46.6
Boneless turkey						
Convection	360	32.7	526	1.8	886	34.5
Immersion	457	–	411	–	867	25.9
Pressure/vacuum	207	37.8	36	7.8	243	46.7
Boneless ham						
Convection	389	29.9	761	2.2	1150	32.1
Immersion	429	–	456	–	885	23.6
Pressure/vacuum	203	39.6	57	8.6	260	48.2

overall weight losses than either convection or immersion processing. Mean weight losses for all treatments for convection, immersion and pressure/vacuum were 31.3, 27.5 and 45.8% respectively.

Although higher throughputs can be achieved using the pressure/vacuum process, the large weight losses may be prohibitive for commercial use with thick joints. Increased losses during pressure cooking are probably due to the high surface temperatures achieved during the process coupled with evaporative loss during vacuum cooling, which is an inherent part of the process. However, it would be interesting to see if the addition of intermittent water sprays could reduce weight loss. It is possible that a hybrid system, possibly using convection cooking in bags followed by immersion cooling, could optimise yields with large joints.

13.4 Conclusions

Although a great deal has been written on the frozen storage life of different meats, the underlying data have probably been gathered from practical experience, backed up by a relatively small number of controlled scientific experiments. Much has been written about the basic freezing process, the equipment required and recommendation for different foods [5,48]. A lot of the scientific data on frozen storage dates back to the time when fish and meat were either stored unwrapped or in wrapping materials that are no longer used. It is not surprising, when we consider the changes in packaging and handling methods over the last century, that there is a considerable scatter in data on storage lives for similar products.

The importance of good cooking practice and oven design is demonstrated by the work previously reported. Very small changes in process control variables can make significant differences to the yield and hence profitability of the cooking operation and can also result in large variations in the quality of the final product. It is clear that the designer and manufacturer of commercial cooking equipment should pay particular attention to the variation in air velocity within the working section of cookers and write the instructions for their use so as to minimise these differences.

The data presented have shown that there is no single process that will meet the often conflicting aims of the commercial cooked meat producer, for example, to achieve a minimum final temperature needed for microbiological safety and a low weight loss for economic reasons. The importance of considering the effect of the combined process of cooking and subsequent cooling on overall weight loss has also been demonstrated. There is no advantage in optimising the cooking process if any gains made are subsequently lost in the cooling phase.

There appears to be tremendous opportunity and scope for industry to improve the efficiency and quality of meat cooking through better monitoring and control of the process. Without close monitoring of weight loss after cooking, there can be no measure of what effect changes in control settings are having and how economic the operation is. This may seem a fairly obvious statement but few producers of cooked meat routinely measure anything other than a few temperatures. However, the cooking of meat is considered by some to be an art and consequently in practice operations tend to be evolved rather than designed.

References

1. Morrison, C.R. Fish and shellfish. In *Frozen Food Technology*, Ed. Mallet, C.P. Blackie Academic and Professional, Glasgow (1993), 196–236.
2. Wirth, F. Chilling, freezing, storage and handling of meat: present state of our knowledge. *Fleischwirtschaft,* **59**(12) (1979), 1857–1861.
3. Ramsbottom, J.M. Freezer storage effect on fresh meat quality. *Refrigeration Engineering,* **53** (1947), 19–22.
4. Bengtsson, N., Lilejemark, A., Olsson, P. and Nillsson, B. An attempt to systemise time–temperature-tolerance (TTT) data as a basis for the development of time–

temperature indicators. *Bulletin of the International Institute of Refrigeration.* Commissions C2 and D1 Warsaw (Annex 1972) (2), 303–310.
5. *Recommendations for the processing and handling of frozen foods*, 3rd edn. International Institute of Refrigeration, 177, Boulevard Malesherbes, F-75017 Paris (1986).
6. Winger, R.J. Storage life of frozen foods. New approaches to an old problem. *Food Technology in New Zealand*, **19** (1984), 75, 77, 81, 84.
7. Winger, R.J. Storage life and eating related quality of New Zealand frozen lamb: A compendium of irrepressible longevity. In *Thermal Processing and Quality of Foods*, Eds Zeuthen, P., Cheftel, J.C., Eriksson, C., Jul, M., Leniger, H., Linko, P., Varela, G. and Vos, G. Elsevier Applied Science, Barking (1984), 541–552.
8. Crystall, B.B. and Winger, R.J. Composition of New Zealand lamb as influenced by geographical area, sex of animal and cut. *Meat Industry Research Institute of New Zealand (MIRINZ) Publication no. 842* (1986).
9. Wismer-Pederson, J. and Sivesgaard, A. Some observations on the keeping quality of frozen pork. *Kulde* **5** (1957), 54.
10. Palmer, A.Z., Brady, D.E., Nauman, H.D. and Tucker, L.N. Deterioration in freezing pork as related to fat composition and storage treatments. *Food Technology* **7** (1953), 90–95.
11. Bogh-Sorensen, L. and Hojmark Jensen, J. Factors affecting the storage life of frozen meat products. *International Journal of Refrigeration*, **4**(3) (1981), 139–142.
12. Bailey, C., Cutting, C.L., Enser, M.B. and Rhodes, D.N. The influence of slaughter weight on the stability of pork sides in frozen storage. *Journal of Science Food and Agriculture* **24** (1973), 1299–1304.
13. Klose, A.A., Hanson, H.L., Mecchi, E.P., Anderson, J.H., Streeter, I.V. and Lineweaver, H. Quality and stability of turkeys as a function of dietary fat. *Poultry Science* **32** (1953), 83–88.
14. Connell, J.J. *Control of Fish Quality*, 3rd edn., Blackwell Scientific Publications, Oxford (1990).
15. Ristic, M. Influence of the water cooling of fresh broilers on the shelf life poultry parts at −15°C and −21°C. *Lebensmittel* **15** (1982), 113–116.
16. Keskinel, A., Ayres, J.C. and Snyder, H.E. Determination of oxidative changes in raw materials by the 2-thiobarbituric acid method. *Food Technology* (1964), 101–104.
17. Judge, M.D. and Aberele, E.D. Effect of pre-rigor processing on the oxidative rancidity of ground light and dark porcine muscles. *Journal of Food Science* **45** (1980), 1736–1739.
18. Jeremiah, L.E. and Wilson, R. The effects of PSE/DFD conditions and frozen storage upon the processing yields of pork cuts. *Canadian Institute of Food Science and Technology Journal* **20** (1987), 25–30.
19. Harrison, D.L., Hall, J.L., Mackintosh, D.L. and Vail, G.E. Effect of post-mortem chilling on the keeping quality of frozen pork. *Food Technology* **10** (1956), 104–108.
20. Zeigler, P.T., Miller, R.C. and Christian, J.A. Preservation of meat and products in frozen storage. *The Pennsylvania State College, School of Agriculture, State College, Bulletin no. 530* (1950).
21. Pool, M.F., Hanson, H.L. and Klose, A.A. Effect of pre-freezing hold time and antioxidant spray on storage stability of frozen eviscerated turkeys. *Poultry Science* **29** (1950), 347–350.
22. Grey, T.C., Griffiths, N.M., Jones, J.M. and Robinson, D. The effect of chilling procedures and storage temperatures on the quality of chicken carcasses. *Lebensmittel Wissenschaft und -Technologie* **15** (1982), 362–365.
23. Ristic, M. Thawing and re-freezing of cut up broilers after short or long term storage. *Kaelte* **41** (1982), 71–74.
24. Aitken, A., Mackie, I.M., Merritt, J.H. and Windsor, M.L. *Fish Handling and Processing*, 2nd edn., HMSO, London (1980).
25. James, S.J. and Bailey, C.J. The Freezing of Beef Quarters. XVIIth International Congress on Refrigeration *1987, Vienna*, C2–4.
26. James, S.J., Creed, P.G. and Bailey, C. The determination of the freezing time of boxed meat blocks. *Proceedings of the Institute of Refrigeration* **75** (1979), 74–83.
27. Creed, P.G. and James, S.J. A survey of commercial meat block freezing in the United Kingdom and Eire. *International Journal of Refrigeration* **4** (1981), 348–354.

28. Sebranek, J.G., Sang, P.N., Rust, R.E., Topei, D.G. and Kraft, A.A. Influence of liquid nitrogen, liquid carbon dioxide and mechanical freezing on sensory properties of ground beef patties. *Journal of Food Science* **43** (1978), 842–844.
29. Dobryzcki, J., Pietrzak, E. and Hoser, A. Rheological characteristics of fresh and frozen chicken muscles and their relation to tenderness measured by sensory methods. *Acta Alimentaria* **6**(2) (1977), 107–111.
30. Sebranek, J.G. Cryogenic freezing of ground beef patties shows superior organoleptic effects. *Quick Freezing* (1980), (August) 50–53.
31. Lampitt, L.H. and Moran, T. The palatability of rapidly frozen meat. *Journal of the Society of Chemical Industry* L11 **21** (1933), 143t–146t.
32. Hill, M.A. and Glew, G. Organoleptic assessment of products frozen two methods: Liquid nitrogen and blast freezing. *Journal of Food Technology* **8** (1973), 205–210.
33. Jackobsson, B. and Bengtson, N. Freezing of raw beef: Influence of ageing, freezing rate and cooking method on quality and yield. *Journal of Food Science* **38** (1973), 560–565.
34. Buchmuller, J. Chilling and freezing meat products with liquid nitrogen. *Fleischwirtschaft* **66**(4) (1986), 568–570.
35. Bevilacqua, A., Zaritzky, N.E. and Cavelo, A. Histological measurements of ice in frozen beef. *Journal of Food Technology* **14** (1979), 237–251.
36. Evans, J.A. and James, S.J. Factors influencing the frozen storage life of meats. *Proceedings of the Institution of Mechanical Engineers Future Meat Manufacturing Processes*, London (1990).
37. Lindelov, F. and Poulsen, W.D. Time–temperature tolerance of foods, some exceptions to the general rule. *Proceedings of the XIV International Congress of Refrigeration* **3** (1975), 759–767.
38. Lindelov, F. Reactions in frozen foods; the reaction of myosin and single amino acids with some aldehydes. *International Journal of Refrigeration* **1**(2) (1978), 92–98.
39. Carrol, R.J., Cavanaugh, R.J. and Rorer, F.P. Effects of frozen storage on the ultrastructure of bovine muscle. *Journal of Food Science* **46** (1981), 1091–1094, 1102.
40. Jul, M. The intricacies of the freezer chain. *International Journal of Refrigeration* **5**(4) (1982), 226–230.
41. Dawson, L.E. Stability of frozen poultry meat and eggs. In *Quality and Stability of Frozen Foods – Time–temperature Tolerance and its Significance*, Eds Van Arsdel, W.B., Copley, M.J. and Olson, R.L. (1969), 143–167.
42. Klose, A.A., Pool, M.F. and Lineweaver, H. Effect of fluctuating temperature on frozen turkeys. *Food Technology* **9** (1955), 372–376.
43. James, S.J. and Rhodes, D.N. Cooking beef joints from the frozen or thawed state. *Journal of Science, Food Agriculture* **29** (1978), 187–192.
44. Burfoot, D., Swain, M.J. and James, S.J. Cooking of pre-sliced meat. *Journal of Food Technology* **20** (1985), 155–161.
45. Burfoot, D. and Griffin, W.J. Effect of dimensions on the heating time and weight losses of cylindrical beef joints. *International Journal Food Science and Technology* **23** (1987), 487–494.
46. Swain, M.J. and James, S.J. Cooking times and weight losses in meat cooking. *Institution of Chemical Engineers, Food Engineering in a Computer Climate*, Cambridge (1992).
47. Burfoot, D., Self, K.P., Hudson, W.R., Wilkins, T.J. and James, S.J. Effect of cooking and cooling method on the processing times, mass losses and bacteria condition of large meat joints. *International Journal of Food Science and Technology* **25** (1990), 657–667.
48. Mallet, C.P. *Frozen Food Technology*, Blackie Academic and Professional, Glasgow (1993).

14 Fruits and vegetables
H.F. JONES and S.T. BECKETT

14.1 Introduction

Fruits and vegetables are the parts of plants, except for grain, that are consumed as food. These parts are shown schematically in Figure 14.1.

Fruits may be regarded as the reproductive organs or seed-containing portions of flowering plants, which normally have a high sugar and acidic content together with aromatic flavours. Vegetables, on the other hand, include many roots, tubers, bulbs, stems and leaves, etc. and tend to be lower in sugar and aroma. The distinction between a fruit and a vegetable is often unclear. For example rhubarb, a petiole, is usually classified in the food industry as a fruit and the strawberry is a 'false fruit' that arises from the flower receptacle. Although the tomato is a fruiting body, it is classified in the industry as a vegetable. All fruits and vegetables however have a high water content (75–95%) and tend to be low in fat and protein, and although some may need special treatment, perhaps because of their delicate nature, many processing methods can be applied to both. This chapter reviews the factors governing their storage, preparation, preservation, cooking and reconstitution. In response to the increasing importance of fruit juices in the marketplace, a section has been included on their manufacture and packaging.

Not all the factors described in the first ten chapters are of equal importance with respect to fruit and vegetable processing. Because of their high moisture content, water activity is not of significance with respect to storage and preservation of most fruits and vegetables and their products. Water activity and the glassy state are, however, of great importance in dried products and the glassy state is of crucial significance in understanding the quality and shelf-life of frozen foods. The disruption of cell structure during processing can lead to the release of enzymes which can cause deleterious reactions. Such reactions are stopped or slowed by cooking and blanching, and a keen understanding of the relative rates of physical processes and biochemical and chemical reactions is required to achieve optimum results. The Maillard reaction is generally of more importance in processing fruit because of the higher sugar content, whereas rheology finds a role in the processing and packaging of both fruit and vegetable products. Special rheological methods have been developed to monitor and control the flow of vegetable pieces in water down pipes to try to ensure that the correct mixture is packed at the far

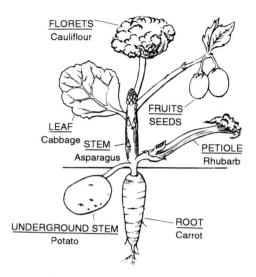

Figure 14.1 Schematic diagram of edible parts of the plant (reproduced from [1]).

end and the consumer does not get all peas or all brine, etc. Where fruits or vegetables are dehydrated, their wettability and other physical properties affecting rehydration become very important factors for the convenience of the consumer.

Fermentation is often used to prepare products such as alcoholic drinks or cocoa; see chapters 17 and 20. Finally, the change in cell structure upon heating or freezing is perhaps the most important factor governing most processing techniques.

14.2 Storage

Fruits and vegetables can deteriorate in many ways, e.g. sprouting, changes in metabolic respiration, loss of moisture, bruising and enzyme break down, etc. Fresh or minimally processed fruit and vegetable products (e.g., washed or sliced) require correct storage for the customer to receive them in a satisfactory condition.

Unprocessed produce can be divided into four types.

1. Those that should be marketed immediately if sold fresh – most salad crops, watermelon, asparagus and spinach.
2. Those requiring ripening before marketing – bananas, apricots and tomatoes.
3. Those that can be stored for short periods – peaches and citrus.

4. Those with good keeping properties under controlled conditions – potatoes, pears, apples and onions.

Temperature of storage is critical in determining the fresh shelf-life of all these types of produce. Lower temperatures will reduce the respiration and transpiration rates, inhibit enzymatic reactions, reduce the growth of micro-organisms and slow down ethylene production (which is a key process in the ripening of climacteric fruits). The optimum temperature of storage will depend upon the fruit or vegetable itself, and is critical as even short exposures to lower temperatures may result in freezing or chilling injury. This damage may not be evident while the produce remains cool, but may become apparent once the temperature is raised and is due to cells breaking and releasing enzymatic and non-enzymatic reaction products (see also chapter 10). The effects of it include loss of flavour and texture, discoloration, lack of ripening ability and increased susceptibility to fungal spoilage. The optimum storage temperature for fresh fruit and vegetables can be broadly classified into four groups and examples of some common products are given in Table 14.1.

When fruits are harvested ripe, it is good practice to 'pre-cool' them. This process is the removal of field heat as quickly as possible and often involves cool-air tunnels or cold-water baths. Where practically possible, cold-water effects more rapid cooling in produce such as peaches, but

Table 14.1 Optimum storage temperature for fresh fruit and vegetables (reproduced from [2])

Optimum storage temperature	Commodity
Cold storage, 0–5°C (32–41°F)	Apples, apricots, artichokes, asparagus, beans (Lima), beets, broccoli, Brussel sprouts, cabbage, cantaloupes, carrots, cauliflower, celery, cherries, corn, dates, figs, grapes, Kiwi fruit, lettuce, mushrooms, nectarines, onions (green), oranges, peaches (ripe), pears* (ripe), peas, plums, radishes, rhubarb, spinach, strawberries, turnips
Cool storage, 5–10°C (41–50°F)	Avocados (ripe), beans (snap), blueberries, cranberries, cucumbers, eggplant (aubergines), melons (ripe), peppers, pineapple (ripe), squash (summer), tangerines
Slightly cool storage, 10–18°C (50–64°F)	Bananas, coconuts, grapefruit, limes, lemons, mangoes, melons (unripe), nuts, papayas, pears* (unripe), tomatoes, pumpkins, squash (winter), sweet potatoes
Room temperature storage, 18–25°C (64–77°F)	Avocados (unripe), nectarines (unripe), onions (dry), peaches (unripe), potatoes, watermelons

*The optimum storage temperature for pears is 3–7°C (37–45°F) (ripe) and 16–20°C (61–68°F) (unripe).

care needs to be taken to avoid cross-contamination with mould spores, etc.

The relative humidity during storage is also important, as the amount of moisture in the surrounding air will affect the water loss as well as possibly influencing decay development and uniformity of ripening, and causing wilting in leafy vegetables. Relative humidities of 85–95% are usually recommended for most fruits and 90–98% for most vegetables, although many are now stored at 100%.

Condensation of moisture on the surface (sweating) must be avoided as this enhances decay. For citrus fruits, a wax coating is sometimes applied to prevent moisture loss and plastic films or bags may be used for the same purpose. With films or bags, perforation is necessary to minimise decay due to fungi and moulds.

In the 1920s, it was found that storing apples in an atmosphere which contained more carbon dioxide and less oxygen increased their shelf-life. This led to the development of controlled and modified atmosphere (CA & MA) storage, which is now applied to storage rooms, containers or small-sized packages. Primarily it is used for apples and pears, and also for international and long-distance transportation of citrus, strawberries, bananas and cabbage [3], whilst MA packaging is increasingly being used for salad crops. The levels of oxygen, carbon dioxide, nitrogen, ethylene, etc. may be controlled. Reduction in oxygen lowers respiration and may delay ripening, but if it is reduced too far (<2%) off-flavours and odours can arise in many fruits due to anaerobic respiration, and uneven ripening may occur when fruits are removed from the atmosphere. High levels of carbon dioxide (10–20%) can have a similar effect. The required level of these two gases varies according to the product. In general, root crops such as carrots, beets and potatoes are best stored with no oxygen or carbon dioxide. Other vegetables such as tomatoes, peppers, cucumbers and lettuce need between 3 and 5% oxygen, but once again no carbon dioxide. Fruits almost all require both gases, but pears, lemons and apples have a lower recommended level of carbon dioxide (1–5%) compared with other fruits (5–15%). The actual optimum concentration for these gases will depend on the cultivar, maturity and temperature as well as the commodity itself.

14.3 Preparation

14.3.1 Cleaning

Impact between individual pieces of fruits or vegetables, dropping against a solid surface such as the floor, or pressure due to piling them too high, can all cause damage to the surface and internal bruising resulting in poor texture and shorter shelf-life. More severe injuries, such as cuts, penetrate the protective coating exposing new surfaces to outside contamination. In addition, enzymes may be released which will promote changes such as enzymic browning and chemical changes. Water loss and respiration also tend to increase, resulting in a short-term loss of sensory quality and nutrient value. For very fresh crops and a few tuber and root crops, some 'healing' can take place, but throughout all harvesting, transport, storage and cleaning care must be taken to cause the minimum amount of physical damage. Crops may be cleaned by shaking, or brushing, which by their nature involve physical impact thereby increasing the possibility of damage.

Some products require washing before packing. This is also necessary where previous processing such as peeling or cutting has taken place. Here the surface will be covered with enzymes and nutrients, which, if not removed, will result in premature decay. In this case, chlorinated water (hypochlorite) is frequently used to rinse or submerge the product. Following most wetting procedures, some drying process is required to reduce the risk of microbial spoilage. Methods include warm, dry-air jets, blotting with a porous material or centrifugation. This latter method can cause bruising and lead to more rapid deterioration of flavour and texture.

Where a product is being prepared for long term preservation, e.g. by freezing or canning, it is essential to remove all traces of soil, and 'foreign bodies' such as insects. Fertiliser and pesticide residues and the microbial load should be reduced as much as is reasonably possible. Root crops and low-growing vegetables in particular need very efficient cleaning. The most common cleansing chemical is hypochlorite solution at a level of 1–3 ppm for rinsing and about 50 ppm for sanitising. The higher concentration is used to sanitise the processing equipment, which must of course be maintained to a high standard of hygiene. High-temperature steam is another frequently used cleaning agent, as well as various detergent solutions. Other chemicals are sometimes used to aid peeling (see below).

14.3.2 Sorting

With the increasing consumer demand for high quality coupled with uniformity, grading machines are becoming more and more important. Traditionally, it was only necessary to remove foreign matter such as stones or wood and to grade according to size, by using processes such as vibrating tables and air suction, which sort according to density, to remove most waste material. This was followed by sieving, to give the correct size range, coupled with manual inspection to remove blemished or deformed product. Now colour and shape uniformity are becoming increasingly important and production rates are so high that manual inspection becomes inefficient. Fortunately, recent developments in the field of image analysis have enabled machines to be developed, capable of selecting product according to all these parameters and at the same time treating it very gently and thereby causing the minimum amount of damage. The product passes on a belt under a TV camera capable of distinguishing between several hundred thousand colours. Rotating the sample builds up a three-dimensional picture, which then enables individual items to be sorted according to colour, size, shape and the presence or otherwise of blemishes. The selection is carried out using pneumatic fingers which are able to channel each piece towards the appropriate subsequent process. Product such as peeled potatoes can be sorted at speeds of 7 tonnes per hour (15 700 lb/h), rejecting some and recirculating others for repeeling [4], with an efficiency of more than ten times that of two human sorters.

14.3.3 Peeling and cutting

As a result of the demands of automation, peeling is now normally carried out with steam or chemicals. The former involves exposing the fruit for a predetermined time to high-pressure–high-temperature steam and then washing away the skin by a pressurised water spray. This works due to the high temperature causing breakage of cells in the region of the surface, coupled with the heat on the tissue which results in loss of rigidity, biochemical changes, melting and breakdown of pectins and polysaccharides and disturbance to the cell structure [2].

Chemical peeling using caustic soda (NaOH, also known as 'lye peeling') is a commonly used alternative, which is economic and simple to use, although perhaps not so environmentally friendly as steam peeling. The hot solution dissolves and removes the epicuticular and cuticular waxes of the pericarp surface, penetrates the skin and diffuses into the fruit. Here it causes breakdown of the epidermal and hypodermal cells and by solubilising of the pectins separates the skin from the edible part of the fruit. The process relies heavily on using the correct sodium hydroxide concentration for the optimum temperature and time, so as not to destroy the fruit itself and yet get complete release of the skin. Abrasive peeling, where the product is tumbled against abrasive surfaces such as carborundum to loosen the skin, which is then washed away by water spray, is sometimes used in conjunction with a lye treatment to increase capacity and reduce weight loss. Peeling may also be effected by application of dry caustic combined with infrared radiation, by flaming, by freezing and by vacuum impregnation with hydrolytic enzymes.

Any cutting must, of course, destroy cells, release enzymes and leave the product in a condition where drying out can more easily occur. Cut surfaces have reduced barriers to gas diffusion and tolerate higher concentrations of carbon dioxide and lower concentrations of oxygen than intact fruits and vegetables. Important controls are to maintain sharp cutting edges to minimise cell damage and to ensure hygienic conditions so as to avoid the spreading of contaminants. Once the cells are broken, rapid deterioration of texture may take place. For certain crops, this can be offset by immediately following cutting by procedures such as calcification (calcium chloride treatment) or acidification. An important example is in the processing of diced fruits and vegetables such as tomatoes, carrots and apples, which can be firmed by treatment with calcium chloride solution (approximately 0.5%) under ambient conditions for several minutes. The firming effect is thought to be due to the calcium ions forming calcium pectates. This increases the rigidity of the lamella and cell wall and makes the cell tissue in general more resistant to attack and degradation. The effect is easily observable under the microscope using a stain to reveal the cell walls (see Figure 14.2).

Water knives (pressure of 3000 kPa, 405 psi) have been shown to be capable of high-capacity operation with fruits and vegetables and have the advantage that cell exudates are washed directly away from the cut surfaces.

14.3.4 Blanching

Blanching has been described as 'a pasteurisation technique that is applied to fresh fruit and vegetables prior to further preservation' [3]. It involves

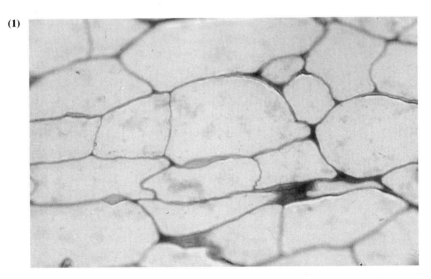

Figure 14.2 Light micrographs of carrot samples stained with toluidine blue. Plate 1. Raw carrot. Plate 2. Blanched and cooked carrot. Plate 3. Carrot blanched and cooked at 60°C (140°F) for 40 minutes at pH 7 with calcium. Reproduced from [5] with permission from Leatherhead Food RA.

exposing the product to hot water, 80–100°C (176–122°F), or steam for several minutes. Rapid cooling is necessary to avoid overcooking and prevent the growth of microorganisms which would take place under the warm conditions found with slow cooling. Satisfactorily blanched material will give a negative peroxidase test.

The blanching may be carried out for any of a large number of reasons.

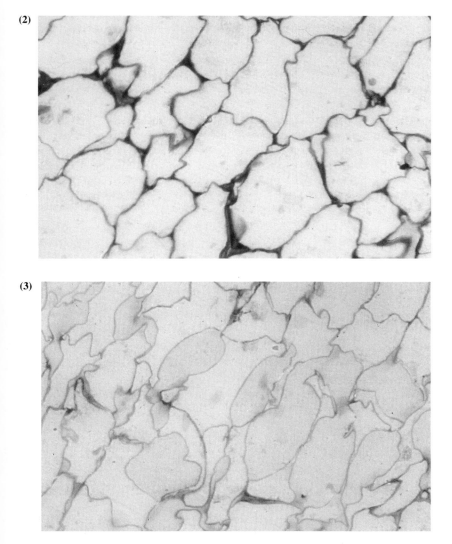

Figure 14.2 *Continued*

The main purpose, however, is to inactivate enzymes which might otherwise produce off-flavours. In addition, it prevents the polyphenoloxidase browning reaction responsible for causing discoloration in some fruits. Physically mobile starch, which might give rise to a cloudy appearance, is removed and the colour sometimes significantly improved. This will also aid the detection of defective product (section 14.3.2). The microbial quality is also improved and in certain cases it may prevent the development of bitter off-tastes during subsequent processing [2]. Certain negative effects, such as the loss of some vitamins, flavours, carbohydrates and other water soluble components, will, however, also occur during blanching.

14.4 Preserving

The purpose of preserving is to achieve a target shelf-life. Preservation processes cannot, however, preserve all of the desirable aspects of a fruit or vegetable, although preservation processes, such as those involving cooking (see below) can often have some desirable effects. The typical failure rate of a preserved product can be described by the 'bath-tub' curve shown in Figure 14.3 [5]. Early failure must, of course, be as close to zero as possible.

14.4.1 Minimal processing

The developing preference for minimal processing stems from the consumer's desire to have products that retain, as far as possible, the

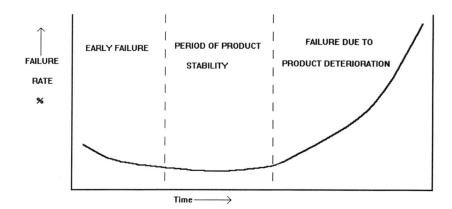

Figure 14.3 The 'bath-tub' curve describing failure rate with time.

nutritional properties and the organoleptic characteristics of fresh produce. Minimally processed fruits and vegetables have been defined as those prepared by one or more steps including the use of minimal heat, a preservative chemical or irradiation, in order to give sufficient shelf-life for distribution. Such products may be packed under gas and are always stored and transported refrigerated (typically at 4°C, 39°F). Minimally processed fruit and vegetable products are optimally prepared in custom-built facilities in order to apply strict hygiene and temperature control. The use of modified atmosphere packaging (see section 14.2) is being increasingly used to extend the shelf-life of minimally processed products.

14.4.2 Bottling and canning

The principles of bottling (whether in plastic or glass) and canning (now often employing semi-rigid and plastic containers) are the same, with the exception that the materials of construction of the containers will differ with the pH of the products, and that some fruits and vegetables in transparent containers will be susceptible to bleaching by light. In general, vegetables are of higher pH and lower sugar content than fruits. Furthermore, vegetables may require more cooking than fruit because of the presence in vegetables of more heat resistant micro-organisms and because they require more softening and flavour development. In order to preserve most flavour and texture, many fruits and especially vegetables are blanched as soon as they are prepared (section 14.3.4). For example, diced carrots may be blanched in live steam or water at 87°C (190°F) for 2–4 minutes. Blanching also serves the purposes of softening the product, so that more may readily be packed into the vessel, and of expelling gases to reduce the strain on can seams during retorting.

In order to expel gases, cans may be filled hot or heated immediately prior to sealing. Alternatively, the sealing operation may be accomplished in a vacuum, or the air may be displaced by live steam. Exhausting the can also gives the advantages of reducing oxidation and preserving vitamin C. Normally a vacuum of 0.4–0.5 atm. is desirable in a cooled can.

Cooking or retorting. Jams and similar preserves are, of course, cooked before filling into the jar or can (see chapter 15), and such products would not be retorted after sealing in their containers. All other conventionally canned and bottled products require retorting. Neutral or low acid products require higher temperatures for sterilisation. The higher the temperature, the shorter the time required. A more viscous product will require longer processing time in order to allow for slower heat transfer within the product. The size, shape, material and headspace of the container, as well as the physical properties of the foods themselves and the type of retorting used (static or agitated), all affect the rate of heat penetration. For low acid foods, processing conditions are carefully defined to effect adequate safety with respect to the growth and survival of spores of *Clostridium botulinum* and the process must also prevent spoilage by other heat-resistant non-pathogenic organisms. Thus the temperature and time of canning are calculated from the load of *Cl. botulinum* spores and the minimum process time at a reference temperature (usually 121.1°C (250°F), commonly called 'botulinum cook'). A minimum value of 3 minutes at 121°C at the slowest heating point in the container is usual. Heat may be applied by steam, water or direct flame. The unit of measurement used to compare the relative sterilising potential of differing heat processes, equal to 1 minute at 121.1°C (250°F), is designated F. As the temperature coefficient of destruction, z value, for the thermal death time (TDT) curve of *Cl. botulinum* is 10°C, also commonly observed for other bacterial spores, F10/121.1°C is more commonly written as Fo [7]. Cooling must be rapid to prevent overcooking.

In order to improve the retention of flavour of some tropical fruits, the cans are spun within the retort. The movement inside increases the rate of heat transfer and so reduces the processing time [8].

Cans and jars are now being replaced to an increasing degree by retortable pouches. These have the advantage of requiring shorter heat-processing time, thus saving energy and preserving nutritional value. Although not yet used commercially, in-pack pasteurisation with microwaves has been studied [8]. Another recent development is continuous heat processing followed by aseptic packaging. This high-temperature, short-time (HTST) processing takes advantage of the fact that high temperatures rapidly kill micro-organisms, giving commercially sterile products, but have less effect on the taste and texture than traditional lower-temperature, longer-time processes.

14.4.3 Freezing

Freezing of fruits and vegetables allows the retention of many of the attributes of fresh produce for periods often greater than a year. It was this technology that first enabled year-round availability of many crops, otherwise unavailable or available only heavily processed. The quality of unprocessed produce will, of course, be reflected in the quality of the products, but once varieties and producers of good freezing qualities have been identified, it is the freezing process itself that will determine the acceptability of the product. Crop maturity is of crucial importance for many products. For example, the sugar concentration in peas will reduce with age, as it is converted to starch, and peas at the sugary stage will give a sweeter and more tasty frozen product.

In general, the faster the rate of removal of heat, the better will be the quality of the product because there will be less opportunity for the growth of ice crystals. In addition, less time is available for deleterious enzymatic damage and for chemical reactions resulting from the disruption of cell membranes. Also, the temperature of the freezer for storage will vary with the product and the storage life required. The US Department of Agriculture has developed the concept of Time–Temperature Tolerances (TTT). This states that for each frozen product there is a relationship between the storage temperature and the time taken, at the storage temperature, for the product to undergo a certain quality change. It also states that quality changes during storage, over a range of temperatures, are cumulative and irreversible, and independent of the sequence of temperatures experienced by the product. In practice, deterioration is very slow below −20°C (−4°F). Although the manufacturer usually has no control over the rate of thawing, this will also have an influence on quality on the table, because ice crystal growth (recrystallisation) and chemical and biochemical action can occur during warming, unless the products are raised to cooking temperature instantaneously ([9]; see also section 14.6.1).

A typical freezing process may be divided into the phases 'pre-freezing', 'freezing' and 'reduction to storage temperature' (see Figure 14.4 and [10]). The freezing rate is the speed of movement of the ice front through the product. However, it is crucial to eventual product quality that the temperature continues to decline until it falls below the glass-transition temperature, where no further physical or chemical changes can occur (see chapter 2). Thus, individually frozen produce is of high quality only if the temperature is reduced sufficiently rapidly from the freezing point

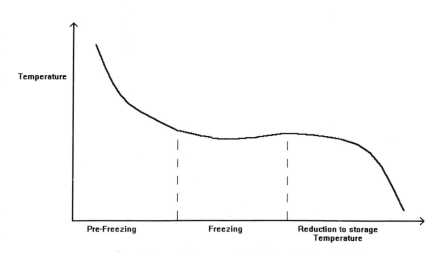

Figure 14.4 Schematic diagram of the phases of freezing [10].

to the storage temperature. Quality, however, has a price, and energy and capital costs are inevitably set against the improvements in quality that can be achieved. For example, cryogenic freezers, which employ direct contact between a liquid gas and the product, have relatively low capital costs but have high running costs compared with mechanically driven freezers. Thus they tend to find only specialist uses, but are more and more finding a place in combination with mechanical freezers because of their ability to achieve high heat-transfer rates [11]. Cryogenic freezing results in low dehydration losses compared with freezing tunnels.

14.4.4 Dehydration

Although dehydration may be the oldest preservation method known [12], the products made are frequently of limited quality. Difficulties in rehydration are being overcome, but flavour loss and texture modification remain. The techniques employed are often combinations of radiant and convection heat, although recent research has shown some potential for other energy sources such as microwave heating.

Heating causes browning (including the Maillard reaction, chapter 4), but as water is removed the rate of reaction reduces. The rate of browning has been found to be greatest at 15–20% moisture, whereas at 1–2% water the product will be stable even at quite high temperatures. Lipid oxidation will, however, continue unless air is expelled or the storage conditions maintain the food in the glassy state. Theoretically, lipid oxidation should cease if the water concentration is reduced to the point that the unfreezable water falls below its glass-transition temperature at the storage temperature.

Freeze-drying is worthy of special note here, since it is capable of producing products which retain their flavour and structure better than those made by other drying methods. If the product is frozen rapidly, minimum damage is done to the cells and if the unfrozen solute at the drying temperature is unfreezable water (i.e. below the glass-transition temperature), the solutes should be subject to no changes during water removal. The practical freeze-drying of produce is rarely straightforward, since the temperatures in the dryer will be a compromise between economy (rate of drying) and quality of the product. It is also practically impossible to achieve a uniform temperature throughout a mass from which water is subliming across a moving boundary and to which heat is being supplied from heated shelves. In accelerated freeze-drying, as much heat as possible is supplied in the early stages of drying by arranging the shelves to be at 150–200°C (300–400°F) and ensuring that the pressure is sufficiently low to maintain the product in the frozen state. Typically, freeze-dried products contain 2–3% water and are often hygroscopic and consequently, when the vacuum in the dryer is broken, dried air or nitrogen is allowed into the chamber. An understanding of the water activity and water adsorption isotherms (chapter 1) is also necessary in designing storage conditions. Although freeze-drying protects the fruits and vegetables from loss of flavour and does retain most of the structural features, the product is not protected from lipid oxidation; indeed the products are highly permeable to gases.

14.4.5 Candying of fruit and osmotic dehydration

One of the oldest methods of preserving fruit employs sugar or sugars as the preservative agent. Perhaps the most widely recorded candied fruits are cherries. Fresh cherries are heated to approximately 60°C (140°F) for 5–20 minutes to promote firming and calcium (at a level of about 0.04% of the finished product) may be added as a further means of retaining structure. For some fruits, a sulphur dioxide treatment is employed to assist in preservation and colour retention, whereas others may require the addition of pigments. The fruits are prepared by placing them in a boiling solution of sugar syrup. After simmering for a short period, the fruits are strained and may be coated with crystalline sugar. The increase in the internal sugar concentration brought about by this process prevents microbiological activity and preserves the structure of the fruits. However, they are relatively hygroscopic and so these products should be stored in cool conditions and at a relative humidity of <50%.

The process of candying effectively uses the osmotic pressure produced by the sugar solution to dehydrate the fruit and the increase in sugar concentration within the fruit to decrease the water activity. The concept of osmotic dehydration has been further developed in the osmovac process [13]. Here, fresh fruit is mixed with sugar or salt, and by this means 50% of the moisture can be removed from the tissues. Then the partially dehydrated fruit is dried further by other means, such as vacuum drying. It has been claimed that the presence of sugar in the tissues during the second dehydration stage prevents enzymatic browning and allows for significantly improved flavour retention. However, there are some problems with the process, such as the difficulty in finding a use for the excess diluted sugar syrup.

14.4.6 Irradiation

This technique uses gamma-rays or high-energy electron beams to treat the product and kill bacteria such as *Salmonella* and *Listeria* and to inhibit sprouting or ripening. Although originating in the 1930s and of well-proven effectiveness, its development has been slow owing to difficulty in persuading the public of its value and safety and because of the very high cost of production plant. Legislation now permits the limited use of irradiation in some countries, provided the operation is carried out under strict conditions and the product itself is appropriately labelled. Usually the dose is also limited. In the UK, no less than 98% of weight must be subjected to a maximum of 2 kiloGray (kGy) for fruit and 1 kGy for vegetables [14]; in the USA, a maximum of 1 kGy is allowed to inhibit growth and maturation of fresh fruits, vegetables and mushrooms as well as for insect disinfestation [2]. Spices and dehydrated vegetables are allowed much higher treatment doses in both countries.

Because legislation limits its use, there has been a lot of work carried out to try to develop methods for determining if foods have been treated by irradiation. This has proved very difficult as the changes caused are very similar to conventional processing. Electron spin resonance and thermoluminescence are, however, showing some promise for certain foods.

14.4.7 High pressure

Although the potential benefits of high-pressure treatments have been known for almost a hundred years, it was not until a development unit was set up at Kyoto University in 1989 that commercial use became a reality [15]. Thus, in Japan, high-pressure processes are used to make juices and jams having 'fresher flavours' than their conventionally prepared counterparts. A pressure of 300–1000 MPa (3000–10 000 atm.) applied for a few minutes at room temperature is able to destroy micro-organisms, but does not cause the loss of flavour and vitamins associated with higher temperature cooking. However, high-pressure processing is still at the experimental stage for all but the few products now available in Japan. Although the effects of high pressures on proteins and other food polymers are still the subject of research, there would seem to be much scope for product and process innovation. For example, gelatinised starch resulting from high-pressure processing retrogrades more slowly than heat gelatinised starch; also pressure-treated ovalbumin and soy proteins have enhanced foaming properties. However, the equipment to produce the high hydrostatic pressures for food processing is still in its infancy and widespread use is not imminent.

14.5 Cooking

14.5.1 Batch cooking

For pasteurisation/commercial sterilisation, where the purpose is to destroy pathogenic or spoilage organisms, yet retain as far as possible the original texture and flavour, heat treatment usually has to be at a high temperature and for as short a time as possible. This processing is normally carried out according to the pH of the product. Medium acid pH (>4.5) products are ideal for the growth of *Clostridium botulinum* and so must be treated to ≥121°C (≥250°F). Many fruits are more strongly acidic and only need a milder heat treatment at ≤100°C (≤212°F), so as to inactivate enzymes. If the pH is close to 4.5, as in the case of tomatoes, acid may be added to avoid the necessity of a high heat treatment.

Cooking, on the other hand, is normally carried out to change the taste, texture and colour (by breakage of cells, loss of volatiles and Maillard reaction, etc.) so as to make it more attractive to the consumer. In the fruit and vegetable industries, because each different product requires different processing conditions, batch processing is often preferred for both cooking and pasteurisation.

As was noted earlier, enzyme degradation starts to take place as soon as the cells are cut. This means that further processing should take place as soon as possible following the initial preparation, whether the product be a delicate one like tomatoes or a more robust one like French fries. Sometimes processes like marinating or heat sealing take place at this stage.

Steam-jacketed pans are often used as cookers. These may be heated from the outside, or steam may be introduced to provide direct heating. The latter has the advantage of reducing burn-on against the pan walls and also a reduced cycle time, as the latent heat from the steam rapidly gives large amounts of energy into the product. The amount of agitation used depends upon the ratio of the solid mass to the liquid present. Where this is high, undue agitation may destroy the solid particulates, but where it is low, as in the case of soups, more mixing is desirable to ensure a uniformity of cook. In general horizontal, single or multi-troughs containing ribbon mixers are used. These give a uniform distribution with limited agitation. Vertical mixers have to apply more shear to overcome sedimentation effects and are therefore more likely to break up the product.

Pressure cooking can be used to increase the temperature and reduce the cooking time. However, this often needs to be coupled with a rapid cooling system where the product is required for further processing, such as freezing. The use of vacuum cookers has been applied to pasteurisation because they operate at lower boiling points and give more rapid changes in temperature. They can, however, only be used for more robust produce as many fruits will explode under vacuum.

When carrying out cooking it is often desirable to determine the amount of processing which has taken place. This may be done by measuring the cook value (C_o). This is obtained by measuring the amount of a degradation ingredient, such as thiamin, which is present and comparing this with the amount which would remain after cooking for 1 minute in an open pan at 100°C (212°F). The calculation is similar to that for the sterilisation value (Fo) (see section 14.4.2) except that the reference temperature and z values are 100°C (212°F) and 33°C (92°F) (thiamin), respectively [16].

14.5.2 Continuous cooking

Where a large quantity of a single type of product is being manufactured, e.g. potatoes, a continuous system may be preferable. Here, however, the raw ingredients must be in a pumpable form, often a slurry. A knowledge of the rheology of this slurry is often required to ensure a uniform processing.

A very simple method is to extend the type of trough used for batch cooking and use the ribbon to transport the slurry. Here the action is gentle and little breakage occurs. Where the particle size is small, steam injection can once again be used to provide very rapid heating.

Where the product is in a semi-liquid state, for example purée or fruit juice, wide gap heat exchangers can be used. These are energy efficient, in that the heating medium can be regenerated, and are also able to carry out very fast cooling or heating. If lumps of material of approximate diameter 6–9 mm (0.25–0.4 in.) are present in a high proportion of carrier liquid, as for instance in soups or sauces, tubular heat exchangers are a possible method of cooking. Scraped-surface heat exchangers are a possibility for more robust fruits, such as pineapple or cherries, when preparing them for use in yoghurts and jams.

Steam infusion has been used for heating less-delicate particulate material for sauces. This technique involves pumping the product through an orifice or spinner into a steam atmosphere inside a chamber. As the product falls, the steam condenses and the product is heated. An extraction pump draws off the product from the bottom of the chamber.

The most energy-efficient method of cooking is ohmic heating. By directly passing the current through the material, losses are limited to about 3%, in addition particulate contents as high as 75% have been processed. In this case, the particulate size can be typically 25 mm (1 in.), or even larger where mechanical clearances allow. Where the product has uniform electrical properties, even heating is obtained. This contrasts with traditional cooking where the outside tends to be overcooked relative to the centre. However, the machines themselves have a very high capital cost and limited throughput. This tends to limit their use to high value-added, niche markets.

Microwave and radio-frequency (RF) heating also avoid the problems of surface heating. They are, however, less efficient than ohmic heating (ca. 65% cf. 97%) and have the major problem that many foods absorb the electromagnetic waves unevenly, resulting in hot spots (see also chapter 21). The higher the temperature range required, the worse is the problem. For this reason, these modes of heating are often used in combination with other forms of heating. The combined domestic oven used to given browning is very well known, and steam plus microwave is used to provide very rapid heating during sterilisation procedures. Microwave blanching of some vegetables has also been shown to be less satisfactory than traditional methods in that it has sometimes shown toughening as well as off-flavours and aromas, possibly due to enzyme activity [3].

Frying is also used to prepare fruits and vegetables. Here the high cooking temperature of the oil may shorten the processing time, giving a better texture when processing some types of products.

14.6 Reconstitution

14.6.1 Thawing

Fluctuations in temperature during frozen storage can cause severe damage and specially insulating packaging materials have been developed to minimise this effect. In addition, time–temperature indicators, based on melting point temperature, enzyme reaction polymerisation or liquid crystal properties, may be used to warn the consumer when unsatisfactory storage conditions have occurred.

Thawing of high-moisture products such as tomatoes and courgettes can result in a spongy texture, which may result in their rejection. This is due to the destruction of the ultrastructure of plant cells and a loss of moisture during thawing, brought about by ice crystal growth in the thawing zone. As a consequence, water and enzymes are released through the broken cell walls. Those enzymes which had become inactive owing to the low temperature will also start working again [9]. Recent work in France [17] has indicated that vacuum freezing can reduce these effects by partially dehydrating the product. This limits the moisture loss and also limits the activity of the enzymes involved in brown discoloration and peroxidation of unsaturated fatty acids. Freezing damage is not so serious for products containing gelatinised starch.

Care must be taken in the choice of thawing process used as food in general is a poor conductor of heat, and it is possible to have the outside boiling with ice remaining on the inside. Microwaves do not provide the solution, however, because ice does not absorb electromagnetic energy. Other materials will begin to get hot, and the hotter they become, the more energy they absorb. This produces a phenomenon known as 'thermal runaway', with extremely hot spots being created, which may burst the product. In certain cases, such as fruits, which are particularly prone to softening, it is sometimes possible to eat them in a semi-thawed state. The firmness of the remaining ice crystals will then help to compensate for the loss in texture due to cell damage.

14.6.2 Rehydration

Ideally rehydration involves restoring the product to its original moisture content and the soluble constituents of the cells are returned to their initial state. The process should also take place almost instantaneously as neither the consumer nor the food industry likes a long soaking period. This is obviously difficult as the air or other gas within the product must be displaced by water and the bonds between large molecular-weight molecules (especially polysacharides and proteins), that were formed during drying, must be broken. Partial replacement of the water, during drying, by small molecules can assist rehydration and special claims have been made for the properties of trehalose in this application [18]. Floating or sprinkling such pieces with water is usually faster than total immersion, as this enables the air to escape more easily and prevents it being trapped by surface swelling. Usually warm water should be used, and sufficient time must be allowed for complete hydration to take place, otherwise the quality is likely to be impaired.

Conventionally hot-air drying produces products which are difficult to rehydrate, particularly if they are case-hardened. Many products require >20 minutes in boiling water to fully rehydrate. Freeze-drying normally gives a much higher quality product which is much easier to hydrate. It is, however, much more expensive in terms of energy and capital costs.

14.7 Fruit juices

14.7.1 Manufacture

The market for fruit and vegetable juices has expanded remarkably during the past couple of decades, especially in Germany. Probably the most important product is orange juice, which is produced only in Mediterranean-type climates. Consequently, the juice must be transported great distances or made from freshly transported fruits. The cost of transport is reduced by concentration of the juice. The satisfactory operation of the concentration process, together with the quality of the fruit itself, is the key to the quality of most of the juice products sold.

The suitability of citrus fruit for juice production is usually defined by the degree Brix/acid ratio. °Brix is a scale relating the density of pure sucrose solutions to concentration. The measurement of °Brix is normally made by refractrometry with corrections being applied for fruit type, acid content and temperature [19]. Oranges are typically harvested at 12°Brix and the acidity (titratable acidity calculated as citric acid, w/w basis) will vary from region to region. Minimum °Brix:acid ratios are set for different regions; for example, 8 for California and 13 in Florida. The selected fruits are washed with water and the juice is pressed in one of several types of machine. All of these machines are designed to limit the amounts of pulp and oil that are carried forward with the juice. The pressed juice is 'finished' by removal of any pulpy matter and then may be pasteurised for local consumption or concentrated. The usual means of concentration is heat evaporation, but freeze concentration and ultrafiltration have been researched. Modern evaporators are likely to be of the type called the 'thermally accelerated short-time evaporator' (TASTE) which are multi-effect vacuum evaporators. Concentrated juice is made at 65°Brix and is stored at approximately −10°C (14°F) awaiting further processing. In order to give good flavour, the extraction pressure is limited and additional juice of inferior quality, called 'pulp wash', can be made by washing the pressed fruits. The concentrated pulp wash is used in the manufacture of citrus-flavoured drinks.

Freeze concentration is a method used to achieve high-quality concentrated juice through the removal of water without the loss of volatiles [20]. The process is, however, time-consuming and needs accurate control of temperature to achieve the desired concentration [19]. There is also a practical limit to the concentrations attainable of between 40 and 60°Brix. During separation of the ice crystals by centrifugation, loss of pulp may occur, altering the texture of the product.

Thawing of frozen juice must be done with great care in order to preserve the organoleptic quality. Slow thawing is best at 2°C (36°F) per day and the concentrate is often pumped around plant at −8°C (18°F). Air is excluded because it can cause deterioration, particularly the development of off-flavours through enzyme action. Water for reconstitution is usually purified through a charcoal filter and is chilled to 2°C (36°F) before blending with the concentrate. Citrus oil may be added to enhance the flavour.

Chemical changes, such as the production of furfurals, occur during the storage of citrus juices. These chemical changes and the flavour modifications that they produce occur more rapidly in the concentrate than in single-strength juice. Unconcentrated juices are usually flash-pasteurised and aseptically packed. Alternatively, they may be distributed chilled. Oxidation, especially of vitamin C, occurs if the products are packed in any container that is permeable to the air.

Many fruits, both tropical and temperate, are unsuitable for preparation of concentrate. In this case, a pulp is prepared that may be transported frozen or, in some cases, preserved with sulphur dioxide. However, other fruits such as apple are frequently converted to concentrates. In the case of apples, a clear product is frequently demanded and filtration or ultrafiltration is frequently employed immediately prior to concentration. In the manufacture of clear juice, enzyme preparations containing pectin methyl esterase and polygalacturonase may be used, but with good-quality fruit, fining alone is often sufficient to achieve sedimentation of the colloids. Fining of apple juice employs the positively charged protein gelatine. This protein reacts with the negatively charged colloids such as pectin and also complexes with tannin-like polyphenols. The choice of grade of gelatine is important because the presence of low-molecular-weight protein fragments can cause the slow development of haze during the storage of the clarified product. Storage haze may also be caused variously by arabinans or polyphenol polymers. Vacuum evaporation is the usual means of concentration, although freeze concentration and reverse osmosis have been researched.

Grape juice is extracted with the assistance of heat and enzymes. The broken grapes are heated to 60°C (140°F) and mixed with pectolytic enzymes and Kraft wood pulp, and the enzymes are allowed to act for approximately 30 minutes prior to pressing. The heat and enzymes ensure a high yield of juice and the wood pulp acts as a filter aid to give good colour. A further pectolytic step is required in the preparation of concentrated juice, where it is typically polished through diatomaceous earth prior to concentration.

14.7.2 Aseptic packaging

The development of hydrogen peroxide as a sterilant permitted the widespread introduction of aseptic packaging of juices. Aseptic processing has the advantages that the duration of the heat treatment is minimised and so the quality of the product is superior. In addition, the packaging materials are cheap relative to those used in hot fill operations.

Juice for aseptic packaging must be commercially sterile at the time of filling, the package must be free of micro-organisms and the sealing process must not permit any organisms or spores to gain entry to the product. For grape juice, which is a relatively acid product, the juice is typically held at 88°C (190°F) for 3 minutes prior to filling, or in high-temperature continuous processes, a temperature in the range 93–100°C (200–212°F) for 15–45 seconds might be selected. The sterile juice is dosed into laminated packs which have been sterilised on the filling line with a spray of hydrogen peroxide.

14.8 Conclusion

Changes in cell structure upon heating or freezing are most important in determining the conditions employed in fruit and vegetable processing.

Multiple steps are often required to achieve a product with the desired eating qualities, shelf-life and consumer safety.

Moreover, combination techniques are now increasingly being used. For example, dehydro-frozen vegetables are partly cooked, partly dehydrated and then frozen. These products are available to the industrial user of prepared vegetables and have the advantages of high quality, convenience and lower shipping costs. A further example of a recent development is that of the combination of microwave and vacuum drying. This so-called MIVAC process [21], when applied to grapes, retains the appearance and flavour of the grape and does not turn it into a raisin.

Thus, further developments can be expected, based upon increasing understanding of the physical chemistry and structural components of plant cells, and upon the emergence of technology such as the combination techniques.

References

1. Vaughan, J.G. (Ed.) *Food Microscopy*. Academic Press, London (1979).
2. Floros, J.D. The shelf life of fruits and vegetables. In *Shelf Life Studies of Foods and Beverages*, Ed. Charalambous G., Elsevier, Amsterdam (1993).
3. Klein, B.P. Fruits and vegetables. In *Food Theory and Applications*, 2nd edn. Ed. Bowers, J., Macmillan, USA (1992).
4. Loctronic International Limited, Danbury, Chelmsford, Essex, CM3 4NH, UK.
5. *Manipulation of fruit and vegetable cell wall structure and its implications for food processing*. Research Report No. 709, Leatherhead Food RA, Leatherhead, UK (1993).
6. Robertson, G.L. *Food Packaging. Principles and Practice*. Marcel Dekker, New York (1992).
7. *Guidelines for the Safe Production of Heat Preserved Foods*, Department of Health, HMSO, London (1994).
8. Selman, J.D. *Fruits and Vegetables in Food Processing*, Ed. Cottrell, R. Parthenon Publishing, Carnforth and New York (1989).
9. Edwards, M. and Hall, M. Freezing for quality. *Food Manufacture* (March) (1988), 41–45.
10. Persson, P.O. and Löndahl, G. *Freezing Technology in Frozen Food Technology*, Ed. Mallett, C.P. Blackie Academic and Professional, Glasgow (1993).
11. George, R.M. Freezing processes used in the food industry. *Trends in Food Science and Technology* (May) (1994), 134–138.
12. Meier, R.W. Vegetable dehydration. *Food Flavourings, Ingredients. Processing. Packaging* 7(10) (1985), 23, 25, 62.
13. Torreggiani, D. Osmotic dehydration in fruit and vegetable processing. *Food Research International* 26 (1993), 59–68.
14. Ranken, M.D. *et al.* Food preservation processes. In *Food Industries Manual*, 23rd Ed., Leatherhead Food RA, Leatherhead, UK (1993).
15. Johnston, D.E. High pressure – a new dimension in food processing. *Chemistry and Industry* (1994), 499–501.
16. Hersom, A.C. and Hulland, E.D. *Canned Foods, Thermal Processing and Microbiology*, 7th edn. Churchill Livingstone, Edinburgh (1980).
17. Varoquaux, P. Recent developments in the processing of fruit and vegetables in France. *European Food and Drink Review* (Autumn) (1993), 33–37.
18. Scher, M. Flavour savr tomatoes await FDA approval. *Food Processing* 54 (April) (1993), 165.

19. Chen, C.S. Physico-chemical Principles for the Concentration and Freezing of Fruit Juices. In *Fruit Juice Processing Technology* (eds Chen, C.S. and Show, R.G.), AG Science (1993).
20. Schwartzberg, H.G. Food freeze concentration. In *Biotechnology and Food Process Engineering*, Ed. Schwartzberg, H.G. and Rao, M.A. Marcel Dekker, New York (1990).
21. Anon. New drying technology makes dried fruits taste like fresh. *Food Engineering* (July) (1988), 81.

Further reading

Dalgleish, J. *Freeze Drying for Food Industries*. Chapman and Hall, London (1990).
Goodenough, P.W. and Atkin, R.K. (Eds) *Quality in Stored Processed Vegetables and Fruit. Proceedings of a Symposium held at Long Aston Research Station, University of Bristol, 1979*. Academic Press, London (1981).
Hicks, D. (Ed.) *Production and Packaging of Non-Carbonated Fruit Juices and Fruit Beverages*, Blackie Academic and Professional, Glasgow (1990).
Kimball, D.A. *Citrus Processing. Quality Control and Technology*. Van Nostrand Reinhold, New York (1991).
Luh, B.S. and Woodroof, J.G. (Eds) *Commercial Vegetable Processing*, 2nd edn. Van Nostrand Reinhold, New York (1988).
Woodroof, J.G. and Luh, B.S. *Commercial Fruit Processing*, 2nd edn., Avi Publishing, Westport, CT (1986).

15 Preserves and jellies

P. BOWLER, V.Y. LOH and R.A. MARSH

15.1 Introduction

Preserves have been consumed for many hundreds of years. Originally, they were produced domestically and made primarily to extend the life of fruit, beyond the few weeks following harvesting, by cooking in the presence of high concentrations of sugar, hence giving rise to the name preserves. Over the years the resultant product became a specialised food in its own right rather than a means of preservation, with the characteristics we expect of it today, namely that it should be a spreadable gelled or semi-gelled system. In many cases, fruit pieces are present and it is important that these should be softened and spreadable along with the gel.

The terms 'jams', 'jellies' and 'marmalades' refer to different classes of preserves. 'Jam' is the term which usually refers to non-citrus preserves, which have a total soluble solids content >60%. 'Marmalade' refers to the equivalent for citrus fruit based products. The term 'jelly' may be used for either of the above classes when the preserve is manufactured using only the juice of the named fruit. Jams with high fruit contents may be referred to as 'extra jams'. 'Reduced sugar jams' typically have a solids content of 30–55% and often incorporate preservatives to protect against yeast and mould spoilage. These different classes of preserve have differing legal minimum compositions, relating to both fruit and sugar content. For a more detailed description of the various classes of preserves and their composition from a legislative viewpoint, the reader is referred to a review by Broomfield [1].

Whatever the scale of the preserves manufacturing process, be it domestic or commercial, the primary objectives of the processing are essentially the same. Firstly, the fruit is heated to extract endogenous pectin, which will later take part in the gelling or setting of the product. Heat softens fruit, pasteurises the product and inactivates enzymes. Secondly, water must be evaporated off to increase the soluble solids content (to at least 60% total soluble solids) to cause the setting of the pectin on cooling. Reduced sugar jams are often soft set (i.e. the strength and number of bonds are decreased), or incorporate other gelling agents.

When a preserve is made in the domestic environment, the product will usually be boiled in an open pan until a small sample, which is taken and cooled, sets fairly rapidly. Preserves made in such a way will almost always produce some set when put into jars, though the set may be variable. Additionally, the fruit content of the product is likely to be high since this aids setting.

By contrast, the primary objectives of the commercial manufacturer are to make a consistent high-quality product, meeting legal requirements, as efficiently as possible. To this end, the processing received by the fruit during preserves manufacture is important, as is the condition of the fruit used. Pectin present in the fruit should be processed to yield optimum effectiveness as a gelling agent, whilst maintaining some integrity of any fruit pieces present, so that these are visible in the final product. If, as is normally the case, insufficient pectin is extracted from the fruit to bring about the desired set texture in the final product, then commercial pectin has to be added. The amount needed to be added will be dependent on the quality of the fruit used and on the effectiveness of the processing received. Since, on a cost per unit weight basis, commercial pectin is by far the most expensive raw material used in preserves manufacture, this processing along with control of quality of raw materials, will be major factors in determining the efficiency of the manufacturing process.

15.2 Pectin

The functionality of pectin derived from fruit will be dependent on a wide range of parameters, for example fruit type, age, time of harvest and conditions of storage, etc. Manufacturing conditions, such as time and temperature of processing, sugar concentration and pH are equally important. Some typical pectin contents of fruits are given in Table 15.1.

The general stucture of pectins is similar from all sources, namely a backbone composed primarily of galacturonic acid residues with occasional rhamnose units. A proportion of the galacturonic acid residues are methoxylated; this may be up to ~75% in the case of citrus fruits, but varies with fruit type. The importance of these differences will be discussed later. The rhamnose units have neutral sugar side-chains attached, fre-

Table 15.1 Pectin content of fruits

Fruit type	Pectin (%)
Apple	0.5–1.0
Orange	0.4–2.4
Lemon	0.8–3.0
Strawberry	0.3–0.6
Plum	0.6–1.0
Apricot	0.5–1.4

quently referred to as the 'hairy' regions of the pectin molecule. Once again, these can affect the characteristics of the pectin. Typically, the neutral sugars are xylose, arabinose and galactose. Several recent reviews of pectin structure are available (see for example [2–6]). Low levels of acetylation of the galacturonic acid residues may occur, though this is significant in only a few fruits of commercial interest such as strawberry where the galacturonic acid residues may be up to 15% acetylated. Acetylation is important because it reduces the gelling ability of the pectin in the final product.

The effectiveness of the pectin as a gelling agent in the preserve will be dependent on its molecular weight. The molecule must be large enough to allow the participation in several junction zones (see chapter 6). There must be an optimum size, however, because as the molecular weight becomes larger, the proportionate increase in gel strength might be expected to reduce. Equally, it is clear that as molecular weight becomes lower, there must be a lower limit below which it will not gel, i.e. when the molecule is not large enough to take part in two junction zones. As this point is approached, it would be expected that the pectin gelling efficiency would decrease dramatically. Our experience is that as the molecular weight of citrus pectin falls below around 100 000 Daltons (Da) then gelling efficiency is low. In practice, this would mean that the producer must add extra commercial pectin in order to maintain a constant product quality.

In most preserves, commercially available pectins are used to supplement the setting power of the naturally occurring pectin from the fruit used. These commercial pectins are extracted from either citrus fruits (lemon, lime and orange) or apples, usually from the byproducts of juice extraction. For preserves, the pectin will be high-methoxyl (HM) pectin, i.e. >50% of the galacturonic acid units substituted with methoxyl groups. These gel in the presence of high sugar concentrations and at low pH. For HM pectins to set, the soluble solids content of the system must be ⩾55% by weight, the approximate pH should be <3.5 and the temperature of the system must be below the gelling temperature of the pectin. Typical preserves have a sugar content of around 65% soluble solids by weight and a pH of 3.0–3.2.

The gelling temperature will be dependent on the degree of methoxylation (DM) of the pectin. This may vary from 85°C (185°F) for a pectin of 70–75% DM to 60°C (140°F) for a 55–60% DM pectin. Additionally, these differences in DM produce differing rates of set according to products in which they are used. Typically, a 55–60 DM pectin will be slow setting, whereas a 70–75 DM pectin will produce a rapid set. This setting rate–DM relationship is due to the charge on the pectin molecule. The charge itself comes from the dissociation of the non-methoxylated carboxylic acid groups of the anhydrogalacturonic acid residues. As the DM of the pectin is reduced, the molecule has a correspondingly greater charge at a given pH, and hence there is greater repulsion between molecules and thus a slower rate of association and resultant set. These effects may be modified to an extent by changes to the pH of the system. A lower pH will result in a faster set since this causes a reduction in ionisation of the carboxyl groups. This results in a reduction in charge on the molecule, which in turn allows a more rapid association of the molecules to produce a set. In practice, the usable pH range for all HM pectin types is approximately 2.9–3.5, since, both above and below this, the set strength becomes reduced or no set is formed. Such effects may be used to advantage in production. Some manufacturers prefer to add acid to adjust pH as late as possible in their processing, thus reducing risks of pre-set when using rapid set pectins, or indeed when the fruit pectin is high DM and hence has rapid setting properties. Pre-set is discussed in section 15.4.2.

The setting characteristics of the pectins present in preserves, either fruit derived or commercial pectins, will be affected by the formulation used. This is particularly true for changes in the type of sugar used. Traditionally, preserves contained sucrose, which was partially inverted during processing to glucose and fructose. Many modern formulations contain glucose syrups as partial replacers for sucrose in the formulation. Such changes alter the optimum pH range for pectin gelation: glucose syrups move this range downwards to lower pH relative to sucrose. Additionally, the absolute concentration of sugars present will affect the strength and rate of set of the product. Many of these, and other factors influencing pectin gelation, are discussed in references [6–13].

If the degree of methylation is <50%, the pectin is classed as a low-DM pectin. Here the gelling mechanism is due to cation bridging and is dependent on calcium concentration. Low-DM pectins are much less dependent than high-DM pectins on sugar concentration for gel strength. Low-DM pectins tend to be used for reduced sugar jams.

15.3 Manufacturing processes

Preserves and jellies are manufactured by the process of boiling to give an end-product having the required sugar content, flavour, colour, texture and shelf-life. Boiling may be carried out batchwise in open pans or vacuum pans. For higher throughputs, continuous vacuum systems may be used.

15.3.1 Open pan boiling process

A typical open boiling pan is shown in Figure 15.1. It is steam jacketed and may also have internal coils to provide additional heating. Vessels

PRESERVES AND JELLIES

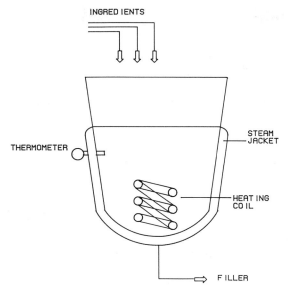

Figure 15.1 Jam boiling pan.

may vary in size, with the most common having a capacity of around 80 kg (176 lb) of product. Ingredients are either fed by gravity, air blowing or pumping into the pan, as selected on demand by the operator using automated feed valves. At the end of boiling, the finished product is discharged through a bottom outlet valve. A number of pans are often used together, in order to provide a continuous supply of product to the filler.

15.3.2 Batch vacuum pan boiling process

Preserves boiled under vacuum retain more flavour and colour due to the gentler heat treatment. Larger batches up to 2000 kg (2 tons) may be processed using vacuum boiling. An example of a vacuum boiling pan is shown in Figure 15.2.

A pre-mix is firstly prepared in a separate stirred heated vessel and then transferred by vacuum into the vacuum vessel. Once charged, the boiling is started and the steam produced by evaporation is drawn away by the vacuum pump to be condensed. Evaporation is maintained at a temperature of approximately 50°C (122°F) until the required solids content is reached, after which the vacuum is released gradually, so as to increase the temperature in order to pasteurise the product.

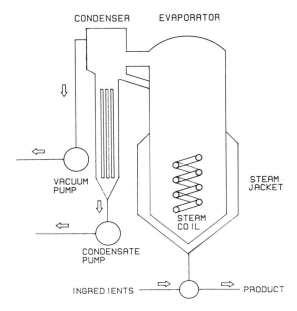

Figure 15.2 Vacuum jam boiling pan.

15.3.3 Continuous vacuum boiling process

Continuous vacuum boiling could be carried out using either plate evaporators or scraped-surface evaporators. Plate evaporators are only suitable for processing jelly or purée products, whilst scraped-surface evaporators can handle large fruit particulates.

The process for both systems is similar and a schematic is shown in Figure 15.3. Ingredients are charged into a pre-mix vessel, which supplies a continuous feed to the evaporator. As product passes through the heater, the temperature reaches boiling point very quickly. The mixture of steam and liquid is passed into a separator, where the steam is drawn off by vacuum, whilst the concentrated product is removed by pumping. After concentration, the product is transferred to a final heater for pasteurisation.

15.3.4 Processes for reduced-sugar products

As a general rule, reduced-sugar products do not require any water evaporation. The right level of solids is achieved by balancing the correct amounts of sugar, fruit and water used in the formulation. Therefore, the

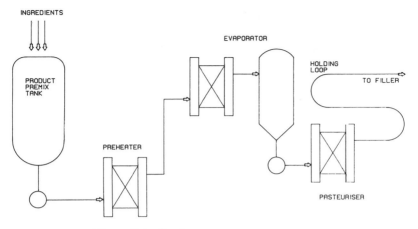

Figure 15.3 Continuous process for jam boiling.

boiling process discussed in the previous sections is not needed. However, heating is still necessary to soften the fruit, develop texture and pasteurise the final product.

Preparation of the pre-mix is similar to the standard preserves. After the product is ready, it may be pasteurised using either a plate heat exchanger or a scraped-surface heat exchanger, depending on whether the product contains particulates or not (see also chapter 14).

Due to low sugar levels, preservatives are recommended if hot filling is used. To eliminate the risks of product insterility, it is advantageous to fill using aseptic technology. However, capital costs for an aseptic plant are high. Alternatively, the product could be cold-filled and then pasteurised in a retort or water spray pasteuriser. This method tends to have a longer process time and consequently the product quality may suffer.

Low-methoxyl pectins are used for low-sugar products and calcium ions are usually added to set the pectin.

15.3.5 Selection criteria

When deciding on the type of boiling equipment or process to use, there are several criteria to consider. Usually the choice is dictated by capital costs, but product quality and consistency are becoming important factors in response to increasing consumer demands for premium quality preserves. Some of the features of the process discussed here have already been mentioned and Table 15.2 provides a summary of the advantages and disadvantages associated with each. These are only guidelines to help with equipment selection.

Table 15.2 A comparison of jam boiling processes

Atmospheric (open boiling pan)	Vacuum		
	Boiling pan	Plate evaporator	Scraped-surface evaporator
Applied to most products	Applied to most products	Applied to jelly, jams, purée, and sieved products	Applied to most products, but particularly those with large fruit pieces
	More suited for non-sulphited ingredients	More suited for non-sulphited ingredients	More suited for non-sulphited ingredients
Batch Low technology	Batch Advanced technology Automated control	Continuous Advanced technology Automated control	Continuous Advanced technology Automated control
Low capital costs Flexible	High capital costs Less flexible	High capital costs Limited to certain products	High capital costs
High-boiling temperature	Low-boiling temperature	Low-boiling temperature	Low-boiling temperature
More losses in flavours and colours	More flavour and colour retention	More flavour and colour retention	More flavour and colour retention
Product of average quality	Product of higher quality with more fruit pieces retained	Product of good quality	Premium quality product with whole fruit or large fruit pieces
Low throughput	Medium throughput	High throughput	High throughput
Less efficiency	Average efficiency	High efficiency	High efficiency

15.4 Set

One of the major objectives of preserves manufacturers is to provide appropriate rheology in the final product, i.e. to achieve the optimum set. The aim is to keep this set constant within and between batches. Hence some objective means of measurement is necessary.

15.4.1 Measurement of set

Several different instruments are used in the industry to make these measurements. Examples of these instruments are the Stevens LFRA Texture Analyser and the more recent Stable Microsystems TA-XT 2 Texture Analyser. These measure some aspect of the resistance to deformation of the gel system. The instruments drive a probe into the sample whilst measuring the sample resistance to the force. Typically, a curve of the form shown in Figure 15.4 is obtained. Sometimes the value for a single point on the curve is recorded and used as a quality-control measurement.

Two parameters may be taken from the data:

- The slope of the initial (straight) part of the line. This provides a value for the resistance to deformation, or the rigidity.
- Maximum force. This provides a measure of the break strength.

Either, or both, of these values may be used as a means of assessment of product set.

Our experience using trained sensory panels is that the rigidity value, rather than the break strength, provides the closer correlation with set. The optimum rigidity values, obtained from such tests, can be used to establish target rigidity values, and the acceptable range to be used for production.

15.4.2 The effects of process on set

There are a number of process variables which affect the setting characteristics. For example:

- Variation in any fruit pretreatment for pectin extraction.

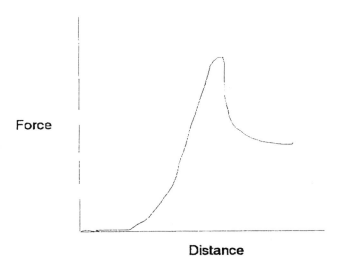

Figure 15.4 Measurement of set: general form of curve obtained from testing resistance to deformation.

- Pectin performance variation due to storage of the pre-prepared pectin solution.
- pH and citrate concentration.
- Sugar concentration.
- Extent of sucrose invertion.
- Time–temperature history during the evaporation process.
- Holding time prior to filling.
- Temperature of hold period.

All manufacturers have procedures to control many of these variables in their processes. Some factors may be manipulated to control the final set. This often involves the use of an accelerated test to predict the final set value and rate of set. Typically, added commercial pectin level is varied to control set value, and pH and citrate levels adjusted to affect rate of set. In circumstances where the fruit used has a high level of good-quality pectin, and no commercial pectin is added, it may be necessary to increase the pH in order to reduce the set to the desired value. Care must be taken, however, to ensure that this does not affect the flavour balance.

The final set quality is affected by heat degradation of pectin at elevated temperatures, and also by pre-set. 'Pre-set' is the term used to describe the partial gelling of the pectin network prior to the product being packaged into its container for sale. The inevitable result of pre-setting is the disruption of this set by the shear forces involved when the product is moved by pumping or other means during the subsequent processing. The product is then said to have a broken set. This broken set causes a reduction in final set strength, which is not recoverable and in extreme cases the product will not set, even remaining pourable.

Pre-set occurs when the product is held under conditions where the temperature is at or below the pectin gelling temperature for that system. In such cases the preset is obvious as, for example, a tank may be seen to be setting, and subsequent movement of the product through the process causes this to break, showing an uneven 'lumpy' surface on the product. Less obviously, pre-setting may occur whilst the product is being transferred through a system. In such cases there is no obvious evidence of any structure formation, since any structure which is formed is immediately broken by the shear forces caused by its movement through the system. This type of pre-setting is evident from the lower than predicted set strength in the final product.

15.4.3 Raw material effects on set

In addition to the factors which affect the rate and extent of gel formation in commercially produced preserves, fruit raw materials and their storage history can cause marked differences in control of set. For example, marmalade made from fresh Seville oranges, under typical conditions (65% sugar solids, pH 3.2) produces a strong resilient set with no addition of pectin.

If the same oranges are converted into orange peel and pulp (so-called 'orange dummy'), preserved by heat treatment or sulphur dioxide and kept for a commercial period of 6–18 months, in ambient warehouse conditions (5–30°C, 41–86°F), significant additional pectin is required to recreate the same gel strength. In order to understand the reasons, a procedure has been developed to reproducibly extract pectin from fruit.

Pectin molecular weight may be calibrated using equivalents of known molecular weight, e.g. pullulan. Our experience has shown that there is no significant variation in the equivalent molecular weight for pectin extraction with water in the 20–60°C (68–140°F) range. Extraction at 100°C (212°F) or with 0.01 M HCl reduces the mean molecular weight by ~15%. The variation in pectin molecular weight as a function of raw material batch is shown in Table 15.3.

The effects of extraction time and temperature on pectin molecular weight and methylester stability are shown in Figures 15.5 and 15.6 and the effect of sulphur dioxide is shown in Figure 15.7.

The analyses of all these data confirms that pectin is relatively stable even under adverse conditions. Of particular commercial importance are the various stages in the preparation and storage of orange dummy.

1. Initial washing. Oranges are washed at 60–80°C (140–176°F) with hot water or steam (≤3 minutes) for cleaning purposes at the start of processing.
2. Short time, high temperature. Aseptic or asepton (pasteurisation) processing of oranges typically involves a 10–30-minute heating–cooling cycle to 95°C (204°F).
3. Sulphur dioxide. Orange pulp is processed into dummy by heating to

Table 15.3 Comparison of pectin content, galacturonic acid content and molecular weight in commercial sulphited orange dummy (60°C (140°F) extraction with water)

Sample	Pectin extracted (w/w)	Galacturonic acid content of pectin (% w/w)	Molecular weight (kDa)
1	4.9	91.4	850
2	6.11	87.12	100
3	3.06	88.51	790
4	3.9	87.65	560
5	3.88	83.28	700
6	3.44	88.28	790
7	5.71	73.96	150
8	6.16	70.71	46
9	6.42	69.39	310
10	3.5	89.62	790

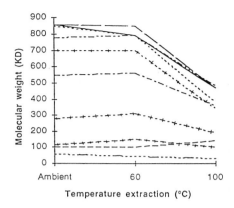

Figure 15.5 Molecular weight of extracted material as a function of dummy and temperature of extraction. (10 different samples)

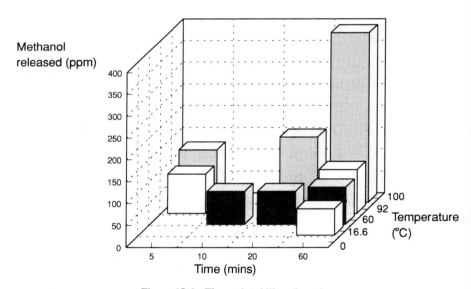

Figure 15.6 Thermal stability of pectin.

95°C (204°F) for 20–30 minutes, followed by sulphiting. Sulphur dioxide levels of 1500–3000 ppm are typical for sulphited dummy stored over a period of 1–2 years.
4. Long-time, moderate-temperature during storage. Storage of processed fruit in ambient temperatures in the country of origin (Spain, South Africa) exposes the raw material to higher (30°C, 86°F) temperatures than when stored in the UK.

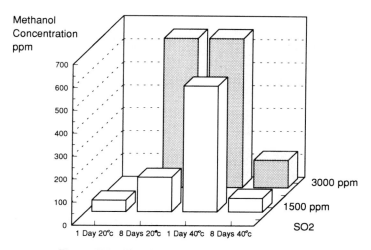

Figure 15.7 The effect of sulphur dioxide on pectin.

As the value of the pectin is more than the value of the fruit, but the cost of storage is cheaper in the country of origin, a clear understanding of the rate of break down of pectin as a function of temperature and storage conditions is crucial to the economics. Extensive work has shown, however, that provided the rate of cooling after processing from 95°C (204°F) to <60°C (<140°F) is rapid (i.e. <60 minutes), there is relatively little effect of time, temperature and sulphur dioxide concentration on pectin, molecular weight and DM resulting from long-term storage.

15.5 Effects of pectolytic enzymes

Two types of pectolytic enzyme may be present in fruit at high activity. Fruit may contain high levels of pectolytic and other enzymes when freshly harvested in a ripe condition. These are pectin methylesterase and polygalacturonase, which act sequentially.

Initially the pectin methylesterase demethylates a region of the polygalacturonic acid backbone of the pectin molecule; then polygalacturonase is able to hydrolyse the linkage between adjacent galacturonic acid residues. Additionally, it is possible that the naturally occurring microorganisms on the surface of the fruit may have pectin lyase present. This enzyme is able to hydrolyse the pectin backbone without prior demethoxylation.

It is vitally important that these enzymes are inactivated at the first stage of processing. Typically, this would be achieved by a rapid heating to above the temperature of inactivation of all the enzymes, such that no activity is present during further processing and storage. This is applicable to fruit which is to be stabilised by asepton processing, canning and sulphur dioxide treatment. An exception to this is when storage is by freezing, in which case great care is needed when the product is thawed to avoid enzymic degradation of pectin prior to, or during, the first heating stage. This must be rapid and sufficient to cause complete inactivation of the enzymes. The correct treatment of frozen fruit at the point of use is critical since the freezing process may have destroyed much of the cellular compartmentalisation of the fresh fruit, which kept the enzyme separate from its substrate, hence rapid degradation may occur on thawing, at ambient or even at chill-room temperatures.

An example where close control of enzyme activity is important is in stored Seville orange pulp for marmalade production. In laboratory studies, we have demonstrated that even very low levels of enzyme activity remaining after the fresh-fruit processing stage can degrade the endogenous pectin during storage to such an extent that it can no longer form a gel network. We have measured the molecular weights of the pectin molecules by size exclusion high-performance liquid chromatography and found size reductions from over 1×10^6 Da for pectin extracted from fresh fruit, to <20 000 Da after enzymic degradation.

Even when lower levels of degradation occur and most of the setting strength is retained, demethoxylation by pectin methylesterase may reduce the rate of set and setting temperature. Such changes may have adverse effects upon preserves production, such as allowing fruit to float or sink after filling, because it is not retained by the pectin set, and therefore causing product to be rejected.

15.6 Fruit flotation

Most types of fruits used contain >80% water and in some cases also have intercellular air spaces. As a result, when mixed with a 67% sugar solution which has a specific density of 1.3, the fruit will tend to float. The boiling process will help to minimise fruit flotation by expelling the air in the fruit and equalising its density with that of the surrounding sugar solution. Increasing the viscosity of the liquid by, for example, using a faster set pectin or high-pectin concentration, helps minimise fruit flotation.

For products containing whole fruit or large particulates, some form of pretreatment may be necessary to increase the sugar content of the fruit. This could be carried out using various methods but the basic principle is the extraction of water from fruit and transfer of sugar into the fruit as quickly as possible. However, the process is controlled by intraparticle diffusion and can be a lengthy one depending on the final solids required (Figure 15.8).

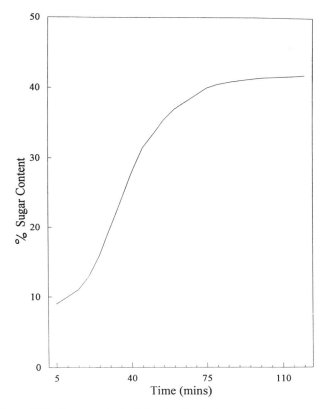

Figure 15.8 Sugar content of strawberries during pre-sugaring process.

15.7 High-pressure treatment

Recently, there has been considerable interest in the application of high-pressure processing as a food preservation technique (see also chapter 14). Research [14] has shown that micro-organisms can be destroyed when treated with pressure up to several kilobars. Unlike thermal processing, high-pressure processing minimises losses in natural flavours, colours and vitamins, and results in better quality food products.

Studies on the processing of jams using high pressure have been reported in the literature [15–17]. Mixtures of fresh fruit, sugar and pectin are pressurised at 4–6 kbar (580–870 k psi) for 10–30 minutes to inactivate the micro-organisms present. At the same time, the pressure also promotes gelation of the pectin, syrup penetration into the fruit pieces and softening of the fruit. This produces a jam containing soft-textured fruit, but with fresh colour and flavour which is considered to be organoleptically superior to a conventional heat-pasteurised jam.

In 1990, the Japanese company Meidi-Ya launched the first commercially produced high-pressure-treated jam [15]. However, enzymes present in the fruit are not completely inactivated by high pressure and the jam has to be kept chilled in order to achieve a two months' shelf-life.

Although the potential of high pressure as a food processing technology has been recognised by the food industry, there are as yet no high-pressure-treated jams produced outside Japan. This can be attributed to the fact that very little is known about the mechanisms in which micro-organisms are destroyed and whether high pressure has any secondary effects on the food itself. Furthermore, not all micro-organisms react to high pressure in the same way. Some are much more resistant to the treatment.

Several research groups in Japan and Europe are currently working to understand the process more fully.

15.8 Conclusion

Technological advances, coupled with changing market forces, will provide opportunities for significant developments in the jelly and preserve industries. A desire to preserve fruit in a way which is close to the natural texture, colour and flavour of fresh fruit, with all preservation and convenience characteristics of preserves, will lead to significant developments.

Combinations of technologies, which solve the following difficulties will be important:

- Dewatering of fruit with minimum damage to cell structure.
- Removal of viable micro-organisms with minimum damage to flavour and colouring components.
- Improvement to the flavour, colour and texture stability of fruit varieties and their tolerance to growing conditions and processing variation.
- On-line control of the rheological characteristics of the jellies and preserves through the critical processing stages and in the final product.

High pressure, electrical field and light sterilisation/pasteurisation, drying and osmotic processes, fruit variety control/breeding/genetic modification and packaging, filling and process advances will all have a role in the future industry.

References

1. Bloomfield, R.W. Preserves. In *Food Industries Manual*, 22nd edn. Ed. Rankin, R.D. (1988), Blackie and Sons, Glasgow, 335–355.
2. Pilink, W. and Voragen, A.G.J. Alginate and pectin. In *Gelling and Thickening Agents in Foods*. Eds Neukon, H. and Pilnik, W. (1980), Forster Publishing, Zurich.

3. Pilnik, W. Pectin – A many spendoured thing. In *Gums and Stabilisers in the Food Industry*, 5th edn. Phillips, G.O., Wedlock, D.J. and Williams, P.A. (1990), IRL Press, Oxford.
4. Christensen, S.H. Pectins. In *Food Hydrocolloids*. Volume 3. Ed. Glickman, M. (1986), CRC Press, Boca Raton, Florida.
5. Oakenfell, D.G. The chemistry of high-methoxyl pectins. In *The Chemistry and Technology of Pectin*. Ed. Walker, R.H. (1991), Academic Press, San Diego, California.
6. Robin, C. and Devries, J. Pectin. In *Food Gels*. Ed. Harries, P. (1990), Elsevier Applied Science Publishers, Amsterdam, 401–434.
7. May, C.D. and Strainsby, G. Factors affecting pectin gelation. In *Gums and Stabilisers in the Food Industry*. Volume 3. Eds Phillis, G.O., Wedlock D.J. and Williams, P.A. (1986), Elsevier Applied Science Publishers, Barking, Essex.
8. Crandall, P.G. and Wicker, L. Pectin internal gel strength: Theory, measurement and methodology. In *Chemistry and Function of Pectins*. Eds Fishman, M.L. and Jen, J.J. (1986), American Chemical Society, Washington DC.
9. May, C.D. Industrial pectins: sources, production and applications. *Carbohydrate Polymers* **12** (1990), 79–99.
10. Lohmann, R. Pektine. Zur Herstelling von Konfitüren, Marmeladen und Gelees. *Gordonian* **77**(10) (1977), 165–272.
11. Thiabault, J.F. and Petit, R. Pectin substances: general information and areas of application in the food industry [translated from French] *Industries Alimentaires et Agricoles* **96**(12) (1979), 1231–1240.
12. May, C.D. Pectins for the food industry. *Food Technology International, Europe* (1989), 269–271.
13. King, K. Pectin – an untapped natural resource. *Food Science and Technology Today* **7**(3) (1993), 147–152.
14. Ludwig, H., Bieler, C., Hallbauer, K. and Scigalla, W. Inactivation of micro-organisms by hydrostatic pressure. In *Proceedings of the First European Seminar on High Pressure and Biotechnology, September 1992, France*. Eds Balny, C., Hayashi, R., Heremans, K. and Masson, P., 25
15. Kimura, K. Development of a new fruit processing method by high hydrostatic pressure. In *Proceeding of the First European Seminar on High Pressure and Biotechnology September 1992, France*. Eds Bainy, C., Hayashi, R., Heremans, K. and Masson, P., 279.
16. Horie, Y., Kilura Iola, M., Yoshida, Y. and Okki K. Studies on pressure processing of jam. In *High Pressure Science for Food*. Ed Hayashi, R. (1991), 336.
17. Watanabe, M., Aria, E., Kumeno, K. and Honma K. A new method for producing a non-heated jam sample. *Agricultural and Biological Chemistry* **55**(8) (1991), 2175.

16 Sugar confectionery
M.F. EELES

16.1 Introduction

The characteristic sweet taste of sucrose, its ability to be processed in a number of physical forms (crystalline, amorphous glass and liquid syrup) and its easy combination with a variety of other food ingredients, enables a wide range of confectionery types to be produced. Better scientific understanding of the underlying processes allows consistent products of the desired taste, texture and shelf-life to be made and developed. Such understanding combines physical and chemical considerations from both a thermodynamic 'equilibrium' and a kinetic viewpoint. The importance of determining the rate at which a change or reaction takes place should not be forgotten and will be highlighted throughout this chapter.

16.2 Products and processes

The range of confectionery products available can be conveniently viewed as a progression from high-sucrose/low moisture to low-sucrose/high moisture goods (Figure 16.1). Additional ingredients such as milk products, hydrocolloids and flour change the organoleptic properties of simple sugar systems and may introduce new physical or chemical processes. For example high boilings (sugar/glucose/water) are characteristically clear, colourless (in the absence of added colouring) and require flavours to produce acceptable tasting sweets. On the other hand, toffees and caramels develop both colour and flavour due to Maillard reactions between the milk proteins and reducing sugars present in ingredients such as treacles, glucose syrup and invert sugar, etc. Hydrocolloids allow gel formation, which is stabilised by sucrose to form high moisture gums and jellies (15–20% water), whereas a mix of simple sugars alone, such as in fondants, results in a handleable sweet only of 10–12% maximum water content. Similar ingredient mixes can also be processed to yield differing physical properties. Fudges are deliberately allowed to grain, whereas in toffees and caramels this is avoided. In practice, this is accomplished by a change in the balance of the ingredients, i.e. higher sucrose levels are used in fudges, but excessive mechanical working during the processing of toffee can lead to grain developing during the storage life of the products.

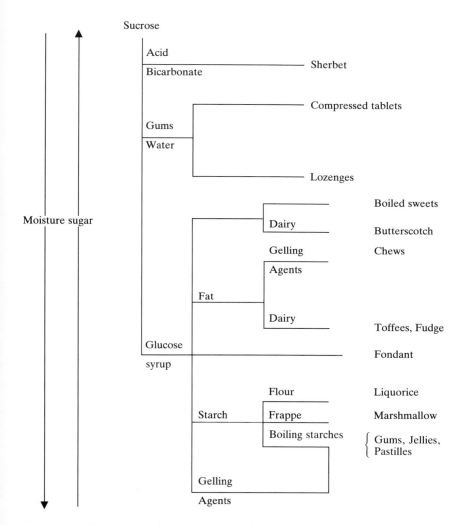

Figure 16.1 Typical range of confectionery products with respect to moisture and sucrose content.

The main processes used in confectionery manufacture are shown in Table 16.1. They can have important influences in the type of product produced. Cooking under vacuum to a given moisture content reduces the boil temperature required and thus the thermal-induced inversion and caramelisation of sucrose. This is required in the manufacture of high boiled sweets, but would produce pale and insipid toffees. Holding the sugar solution in hot jacketed open-topped vessels, in contrast, would be

Table 16.1 Main process operations in confectionery manufacture

Process	
Dissolving	Soluble ingredients.
Pre-mixing	Soluble and non-soluble ingredients. Emulsions will be formed with fats in the presence of emulsifiers or dairy products. Suspensions will be formed with starches.
Mixing	Liquid–liquid, liquid–solid and solid–solid.
Holding	Process requirement as buffer store for batch made operations. Product requirement for colour and flavour development in toffees.
Cooking	Atmospheric, vacuum batch or continuous.
Granulation	Pastes for compressed tablet manufacture.
Aeration	Batch or continuous.
Forming	Cutting and sheeting. Depositing. Plastic forming.
Mechanical working	Fondant manufacture.
Drying	Batch or continuous.
Wrapping	Individual sweets and final packs.

detrimental to cooked high boil mass but assist toffee colour and flavour development.

Sugar- and gum-containing pastes, on the other hand, can be granulated, dried and then compressed for tablet manufacture but rolled, cut and dried for lozenge manufacture. To some extent, therefore, the process defines the final product and vice versa. It is the challenge of confectionery technologists and engineers to configure the most appropriate process and, where changes are made, adapt as necessary in order to produce the required sweet property.

Consideration of the above examples leads to such questions as whether the chemical reactions observed during processing, such as inversion (the splitting of disaccharides such as sucrose into its monosaccharide components) and caramelisation, also occur during storage. In addition, some products behave in opposite ways. For example, boiled sweets absorb moisture and become sticky, but fudges dry out. The reasons behind these effects will be discussed in terms of the fundamental science described in the first ten chapters.

16.3 High-boiled sweets

The formation of high-boiled sweets (typically 2–3% moisture content) is principally a physical process of conversion of crystalline sucrose into an amorphous glass of random molecular structure. In practice, such a system would rapidly recrystallise during the manufacturing process and subsequent storage and so glucose syrup (acid or acid/enzyme partially hydrolysed starch) is added at around a 1:1 ratio of sucrose:glucose syrup by weight. This process is known as 'doctoring' and was originally carried out using cream of tartar (potassium hydrogen citrate). The 'doctoring' materials act by maintaining the sugar in solution at higher concentrations than would be possible if only sucrose were present, where it would crystallise out of solution as the water evaporates. Both rely on the fact that a mixture of sucrose and other sugars is able to remain in solution at much higher concentrations. In the case of glucose syrup, this is carried out via direct replacement of sucrose, whereas cream of tartar works by hydrolysing (inverting) some of the sucrose to dextrose and fructose.

In the production of good-quality boilings, especially colourless mints or lightly coloured fruit sweets, Maillard browning must be avoided. Reducing sugars from glucose syrup (i.e. dextrose and its oligomers and invert produced by thermal hydrolysis of sucrose) can react with trace proteins present in the glucose syrup or sugar to produce a yellow discoloration in the final sweet.

Simple saccharides such as fructose can also degrade and polymerise to produce discoloration. In the past, this has been minimised by the use of sulphur dioxide (SO_2) added to the glucose syrup by the manufacturers permitted under UK law up to a maximum of 400 ppm (Specified Sugar Products Regulations, 1976). With the development of deionised syrups, lower levels of SO_2 in the region 100–200 ppm can be used to the same effect. The SO_2 probably acts by forming colourless addition products with reaction intermediates thus preventing further degradation and colour formulation, although most is evaporated off during the boiling process. Currently, the European Union Directive on Additives, other than Colours and Sweeteners, will permit a residual SO_2 content of 50 ppm in glucose syrup-based confectionery; this level should be sufficient to cover normal practice. In the USA, 40 ppm SO_2 in glucose syrup (corn syrup) is permitted.

After cooking, the hot supersaturated syrup is either deposited as a liquid into moulds or cooled sufficiently to a point where it becomes mouldable in plastic-forming machines, such as a Uniplast (Figure 16.2). In both processes, a centre can be introduced, although this is usually made with the plastic-forming process. Subsequent cooling for both processes then locks in the random molecular structure as a highly viscous glass. The sweet is now essentially stable provided it is protected from atmospheric moisture uptake.

Chapter 1 described the effects of water activity or equilibrium relative humidity (ERH) on keeping properties. The ERH of high boiled sweets lies around 30% relative humidity; storage atmospheres above this there-

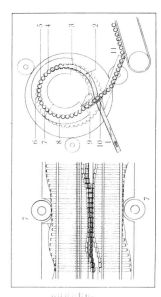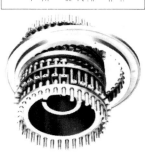

Figure 16.2 Uniplast die-forming system for high-boiled sweets. 1 = Rope feed; 2 = rope infeed; 3 = pre-forming; 4 = sectioning; 5 = insertion; 6 = pre-embossing; 7 = embossing by roller or cam (160/A); 8 = tension equalisation; 9 = discharge; 10 = sliding section; 11 = feed chute to distributor belt.

fore result in a net moisture gain. Use of the Money and Born equation [1] shows the effects of moisture content on ERH and thus the final moisture content possible from a given storage humidity (Table 16.2). Under high humidity conditions, the sweet can gain sufficient moisture to form a sticky surface syrup layer. For this reason, these products must be protected from normal ambient conditions, which usually lie in the range 40–70%, in order to avoid moisture uptake. Traditionally, such sweets were packed in jars or rolls and now, increasingly, in bags. Further protection is afforded either by individual wraps, i.e. twistwraps in jars, bunchwraps in rolls or by coating the sweets with sugar. The so-called 'crystallised boilings' are produced by rolling wetted boiled sweets (steam or sugar syrup) in caster or granulated sugar in a coating drum. Frosted sweets are made by panning (tumbling the sweets together in an air-conditioned drum or pan) in a sugar syrup and drying to produce a fine sugar crystal coating. Both methods produce sweets that physically separate easily and tend not to stick and clump together. Since the coating sucrose is predominantly crystalline in structure and only the sugar in solution will contribute to the water activity, the ERH of the crystallised sugar coat is relatively high, around 60%. This means that moisture uptake by this coating is less than for uncoated sweets.

As well as the possibility of sweets becoming sticky during storage, graining, i.e. the crystallisation of sucrose, can occur, leading to a loss of gloss and a softening of the surface which can easily be penetrated by pressure by the finger nail. Breaking the sweet in half shows that this graining begins on the outside of the sweet and gradually moves inwards with time. It has long been recognised that moisture absorbed onto the sweet surface initiates this process by partially dissolving the sugar glass and allowing crystallisation to occur in the resulting less viscous syrup phase. With the recent expansion of interest in the glass-transition point and rubbery state (chapter 2) this phenomenon is now more clearly understood.

Table 16.2 Calculated ERH values of a 50:50 sucrose: glucose syrup using the Money and Born equation. No crystalline material present

Moisture (%)	ERH
2	22
4	37
5	42
7.5	53
10	61
12.5	66
15	71

The glass-transition temperature (T_g) has two important implications. It represents in one sense a thermodynamic limit below which molecular mobility is frozen and in which the material structure remains as an amorphous solid, so supersaturated glasses do not crystallise and remain stable. In another sense it represents a boundary of kinetic change due to rapidly falling viscosity above T_g. Graining during storage above T_g is thus thermodynamically favoured but kinetically dependent on the difference between the holding temperature and T_g. Since viscosity and hence molecular mobility follows a fivefold order of magnitude change for a 20°C (36°F) range above the glass transition (chapter 2), crystallisation or graining rates will change dramatically over quite small temperature differences above this. For high-boiled sweets at 2-3% moisture, the T_g is sufficiently high (40-50°C, 104-122°F) to ensure that graining does not occur at ambient temperatures. Moisture uptake, however, can reduce the glass-transition temperature to ambient or below and so allow crystallisation to proceed. Prevention of graining can therefore be achieved in the following ways:

- Control of initial cooked moisture content.
- Cooling of sweets (below T_g) during processing before packing.
- Protection from atmospheric moisture uptake by packaging.

The rapid rise of viscosity near T_g will also retard chemical reactions due to decreasing molecular mobilities and so inversion rates dramatically fall as the mass becomes plastic and will cease during normal storage of these products.

16.4 Toffee and fudge

In compositional terms toffees, caramels and fudges can be classified together, but separated from boilings in that they contain dairy ingredients and vegetable fats in characterising amounts. However, within this classification, fudges are physically distinct due to their grained structure. There is no agreed nomenclature to distinguish toffees and caramels and for the purposes of this chapter, they will not be discussed separately. They can cover a relatively wide range of moisture levels, ranging from around 6-8% at the 'hard' end and sold as individual pieces, up to 12% or even higher when used as fillings in, for example, chocolate bars. Although the principles of processing are similar to those used for boilings, these products require the cooking process to generate flavour. Virtually all other sugar confectionery is dependent on added flavourings to provide the organoleptic properties, and so the importance of the Maillard reaction cannot be overstressed.

Toffees and fudges are composed of emulsified fat as an oil in water emulsion, the sugar syrup being the continuous phase and which can under appropriate conditions crystallise (see below). Milk proteins, notably casein which can form small aggregates called 'micelles', act to stabilise oil droplets, effectively surrounding each fat globule and maintaining the emulsion. Figure 16.3 shows an electron photomicrograph of absorbed casein protein at a fat-droplet surface. It is important at the outset to ensure that the milk used, whether bought in as condensed or reconstituted on site from milk powder, is well emulsified and that the pre-mix is well mixed before cooking commences (see also below). During cooking, some coalescence of oil droplets occurs and the flow behaviour of the final sweet is probably related to the final globule size distribution, which is itself dependent on the original emulsification level of the milk used. A large number of small droplets is likely to result in high viscosities,

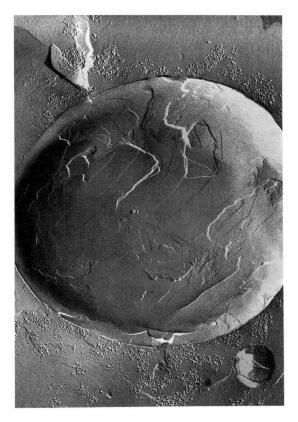

Figure 16.3 Adsorbed casein protein at a fat-droplet surface (courtesy of Dr Ashley Wilson, CCTR, University of York, UK).

giving rise to flow difficulties in subsequent processing. Sometimes additional emulsifiers such as soya lecithin or glyceryl monostearate are added. These appear to improve the handling characteristics of the cooked mass and may assist release performance from cutters.

Manufacture of toffees and fudges is initially similar to boilings; all ingredients (other than flavours, nuts or fruit) are pre-mixed. Milk is used, as mentioned above, in liquid form as sweetened condensed milk (traditionally this is the material originally used when toffees were first developed in the 19th century) either purchased in bulk or in steel drums or reconstituted locally from skimmed milk powder. Whichever is used, a complete solubilisation of the milk solids is required in order to avoid roughness in the final product and good emulsification with the other ingredients for reasons described above. In addition, all of the sucrose should be completely dissolved, as any remaining sugar crystals can also cause roughness in the product and can act as seeds for subsequent graining. To make matters worse, undissolved sugar or milk can caramelise or burn during the cooking operation. (Such caramelisation of sugar is not due necessarily to the Maillard reaction as such, since sucrose can undergo decomposition reactions without the involvement of proteins or amino compounds.) Pre-mixes may be made to relatively high solids (80–85%) before cooking commences.

The cooking processing conditions determine the final colour and flavour developed as a result of the Maillard and associated reactions (chapter 4) between the reducing sugars present (invert sugar in treacles and brown sugars, dextrose in glucose syrup and lactose from milk) and milk proteins. Conventional toffees used brown sugars and treacles; however, white sugar may be used, provided a source of reducing sugars is available. In the absence of reducing sugars, therefore, sugar substitutes such as sorbitol and other polyols cannot react with the milk proteins and browning reactions will not occur. (There will, however, be some browning due to reaction between lactose and proteins, although this is a slower reaction than between the reducing sugars dextrose and fructose, found in conventional sugar-containing formulations.) Fats are also known to undergo Maillard reactions (probably due to carbonyl-containing trace-degradation compounds) and the quality and flavour of butterfat used does affect the final cooked toffee flavour, although reactions other than 'Maillard type' are undoubtedly also involved. Any change of ingredient formulation therefore can be expected to affect the development of the final flavour.

Apart from starting ingredients, the other important factors for the caramelisation reactions are time and temperature. Since moisture removal is the prime physical reason for cooking toffees and the final moisture content is dependent on the boiling temperature, the rate at which this is done will determine the extent of the chemical reactions involved. Cooking can be done either batchwise or continuously with or without vacuum, but process conditions must be specified in order to achieve the desired product.

In the traditional batch process, individual boils are cooked to the desired temperature or 'crack' corresponding to the specified moisture content. Crack testing was used before reliable thermometers were available and is a measure of how much a scoop-shaped piece of toffee snaps or cracks when cooled and bent between the fingers under water, skilled operators being expected to cook to grades of toffee hardness 2°F (~1°C) apart. As soon as the specified boil height is reached, the steam supply into the cooker jacket is turned off and the boil dropped into trolleys or slabs. It is now ready for cutting and wrapping or for press-cut work or to be pumped to a depositor for moulded toffees. These operations are similar in principle to that described for high boilings.

Continuous cookers speed up the boiling process and moisture removal by reducing the bulk mass-to-surface ratio, for example by the use of scraped-surface heat exchangers. Typical continuous toffee cooking is illustrated in Figure 16.4. Although the final moisture content is quickly attained, the kinetics of the Maillard and associated reactions are slower and so a post-cooking caramelisation stage is used which holds the cooked mass until the correct colour and flavour is achieved. It follows that for both batch and continuous cooking, the boiling time, temperature, moisture, resulting colour and flavour are related for a given recipe and a change to the process conditions will affect the final product.

It is, of course, possible to cook a toffee-type recipe and minimise the caramelisation reactions either by quicker heating as in continuous cooking or by a reduction in the cooking temperature by the use of vacuum. Understanding the underlying principles therefore enables advantage to be taken of the possible interacting factors that affect final sweet properties. Fudges can be considered as 'grained toffees' but in order to make a short, rather than a longer chewy texture that would result from graining a toffee a higher level of sucrose is used in order to increase its supersaturation and thus promote quicker graining. In addition, a fondant seed is added at the end of the boil in order to act as a nucleus for subsequent graining.

Sugar-to-glucose syrup ratios of around 2:1 or higher can be used with an addition of fondant at the rate of about 20% of the boil weight. Care must be taken to ensure that the fondant seed is not dissolved and so cooling is required after the boil height has been reached. This is done by turning off the steam at completion of cooling and either allowing it to cool naturally with stirring, or by the use of a cooling water jacket. The fondant and egg frappe (sometimes used to lighten the textural qualities of the finished product) can then be added in the range 90–95°C (194–203°F).

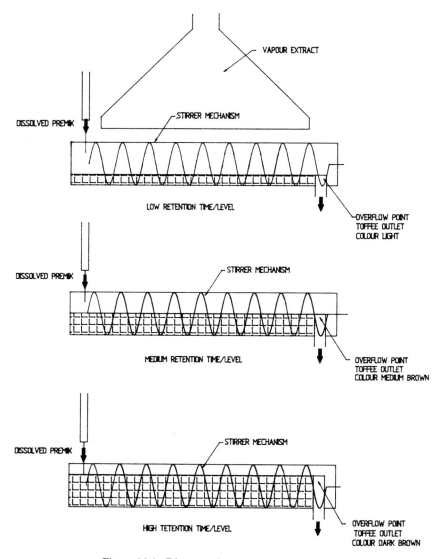

Figure 16.4 Diagram of continuous toffee cooker.

Since fudges are deliberately grained (by the addition of fondant), similar considerations to that for making satisfactory fondant also apply. Crystallisation proceeds quicker at higher temperatures (subject to solubility considerations) than lower temperatures, due to increased molecular mobility, which also results in larger coarser crystals. Since large crystals can detract from the eating quality, the graining process is carried out at lower temperatures and for this reason, once seeded, fudge boils should be quickly deposited or slabbed out and cooled to room temperature. The slower crystal growth that results, however, requires a maturation time to stabilise the crystal structure and a holding storage area is required. In addition, if individual pieces are subsequently cut before wrapping, the disruption to the crystalline matrix exposes a film of syrup at the cut surface and this requires a further storage period in order to recrystallise and 'dry out' before twistwrapping can be successful.

Wrapping offers protection from atmospheric conditions as described for high boilings. Since moisture contents are higher and glass-transition temperatures lower than for boiled sweets, the shelf-life of toffees and fudges is shorter. Fudges tend to dry out and harden since, although overall moisture contents may be similar to that of some toffees, the presence of crystalline sucrose effectively releases more water to the syrup phase and so increases water activity. Thus fudges tend to have ERH values in the region 60–70% whilst toffees are generally <50%. As moisture is lost more sucrose will crystallise and T_g will rise until a balance is reached between the level of sucrose supersaturation and viscosity (which is related to T_g), the rising syrup phase viscosity thus causing a hardening of the sweet texture. This textural change can be minimised by sealed wrapping, for example, in a flow pack format, which will preserve the characteristic short fudge texture longer than in a twistwrapped sweet.

Toffees with their lower ERH than fudges grain by the same mechanism as described for boilings, i.e. by moisture uptake which lowers the T_g and increases molecular mobility. Since moisture is gained, the syrup-phase viscosity falls, sucrose crystallisation 'opens' the syrup-phase structure and effectively increases the moisture content of the remaining syrup phase, so overall firmness decreases. Toffees have traditionally been wrapped in two layers, an underfoil which covers most of the sweet other than the ends and covering this a conventional twistwrap. Examination of these sweets as they age will show a greater development and depth of grain in areas not covered by the foil, due to moisture uptake from the atmosphere through the twistwrap, whose ends are not sealed and are therefore permeable to moisture vapour transfer.

Mechanical working during processing also provides energy to allow crystallisation to proceed; this can sometimes be seen in depositing heads for processing boilings or toffees. Handling of cooked toffees must seek to minimise such mechanical working; the higher sugar-ratio compositions particularly should be worked as little as possible and are better suited to depositing or slab-cutting rather than cut and wrapping (similar in principle to that used for plastic boilings described above) which are more suited to lower-sucrose content sweets. The technologist will therefore need to match shelf-life requirements with recipe and processing route.

16.5 Fondant

Fondants are made from supersaturated syrups of high sucrose content, typically at ratios of around 6:1 sucrose:glucose syrup. Rapid cooling and beating ensures an even dispersion of crystals throughout the mass. Similar considerations of crystal size growth apply, as described for fudge. Fondants mature as this crystal size distribution equilibrates, the larger crystals growing at the expense of the smaller ones, under the storage parameters of time and temperature, and a crust forms as surface moisture is lost. In manufacture, a pre-mix solution is cooked to around 12% moisture, fed through a fondant beater (a cooled jacketed intense shear mixer) which produces a white creamy texture of fresh fondant. It is then deposited, usually with the aid of cooked but ungrained syrup or 'bob' to aid flowability either into blocks or final sweet-shaped moulds. These blocks are then used as intermediate raw materials in the manufacture of grained confectionery such as fudges.

The ERH of fondant is high, around 80% and therefore moisture is lost and the mass hardens with time. For this reason, the fondant must be protected from the atmosphere and is coated to reduce drying out and textural change. This may be done either by crystallising a layer of sucrose crystals from a supersaturated sugar syrup onto the sweet surface or by enrobing with chocolate. In the former case, moisture changes will be reduced in much the same way as that described for boilings (although opposite in direction); for the latter, by providing a moisture impermeable barrier. Chocolate-coated fondants are sometimes softened by the use of invertase, an enzyme which slowly hydrolyses the sucrose to dextrose and fructose (the same reaction which is so carefully avoided for other types of confectionery; see, for example, discussion on high boiled sweets, section 16.3) and raises soluble solids to produce a softer flowing creamy texture. Such a process is clearly kinetically controlled and will be optimised in the manufacturing plant, but will continue during subsequent storage until enzyme activity becomes inhibited by the raised amounts of soluble sugars. Factors such as temperature and pH affect enzymic activity which have been more fully described under the particular case of fermentation reactions in chapter 9.

16.6 Hydrocolloid-based sweets

Hydrocolloids such as carbohydrate-based starches and gums or protein-based gelatine enable the formation of relatively high-moisture sweets due to their water-binding properties. However, their resulting ERH values of 60–70% result in drying out on prolonged storage, a property which is usefully employed during manufacture. These are frequently produced by a process called 'starch moulding', in which the liquid hydrocolloid mixture is poured into imprints made into a tray of starch, which are the size and shape of the final sweet. Low-moisture mixtures are very thick and cannot be easily poured into the imprints, or take up their shape. By using high-moisture syrups of ≤25% moisture, the viscosity is reduced sufficiently for this process to be carried out. This moisture must then be removed for the sweet to attain its proper texture. By placing the starch trays in temperature- and humidity-controlled rooms or stoves, this moisture leaves the sweets naturally. After a predetermined time, the sweets reach their required hardness and are sieved out of the starch ready for wrapping.

Hydrocolloid sweets have a wide range of hardness, from the relatively soft high-moisture wine gum, through the intermediate moisture and textured pastille, to the harder drier fruit gum. From these examples, it can be seen that moisture plays a role in the final texture. This is also true for the ingredients, as varying the proportion of, say, starch and gum, will have a very significant effect. This may in part be due to the different water-binding properties of these two ingredients.

16.7 Tableting

Tableting is used to make sweets which are composed almost entirely of sugar and which are often flavoured with peppermint. This involves placing finely milled sugar into a mould and applying a high pressure. If only sugar were present then the sweet would not properly form. For this reason, a small amount of binder must be added to the sugar. The mixing process and the rheology produced play an important role in the texture of the final sweet. The rheology and surface wetting of sugars, in particular lactose, has been the subject of considerable research in the pharmaceutical industry, where tableting is of major importance. Figure 16.5 illustrates different stages in a wetting process (adapted from [2]). The addition of binder to dry powder results in the formation of liquid bridges, which increase in number and strength as the mixing continues. As more and more surface becomes wetted, the mixture becomes much thicker, until if there is sufficient binder to form droplets (final diagram), it becomes very thin again.

16.8 Conclusion

The above considerations show how detailed examinations of physical properties, such as glass-transition temperatures and ERH, and chemical processes, such as the Maillard reaction, can both determine the type of

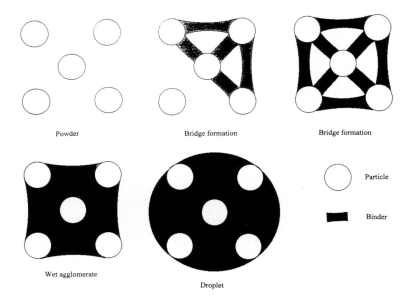

Figure 16.5 The stages of wetting of powder by a binder.

sweet produced and set limits on the way a particular product should be manufactured. In addition, it has been shown how certain principles can be applied to all types of sweets. Although such physical laws do predict the direction of change, for example water loss or water gain, the rate at which this occurs is critical in assessing their importance in relation to the manufacturing process and shelf-life stability. Further advances in this area will equip the confectioner and food manufacturer to maximise product texture and quality against consumer expectations.

References

1. Money, R.W. and Born, R. *J. Sci. Food Agric.* (1951), 180.
2. Hancock, B.C., York, P. and Rowe, R.C. An assessment of substrate–binder interactions in model wet masses. *Int. J. Pharmaceut.* **102** (1994), 167–176.

Further reading

Jackson, E.B. (Ed.) *Sugar Confectionery Manufacture*, 2nd edn. Blackie Academic and Professional, Glasgow, UK (1995).

17 Chocolate confectionery
S.T. BECKETT

17.1 Introduction

Chocolate has been separated from sugar confectionery in this book, as many of the processes and problems are very different. It is also a relatively large industry, with over 2 million tonnes being consumed annually in Western Europe and over 1 million tonnes in North America alone [1]. Currently, many companies are changing from high-labour/craft-oriented product to highly automated systems, which require a greater scientific and technical knowledge of the processes involved, in order to maintain a consistent, high-quality product from constantly changing ingredients.

There are three distinct areas with the making of chocolate confectionery: the growing and processing of the cocoa beans themselves, the manufacture of the chocolate itself and then its use to make bars or to coat other materials. The processes themselves are summarised in Figure 17.1 and this chapter will concentrate on those where physico-chemical effects play a major role.

17.2 Cocoa beans

17.2.1 Growing [2]

Cocoa beans are the seed of a small tree (6 m; 20 ft) known botanically as *Theobroma cacao*, which grows mostly within 20° latitude of the equator. Good varieties bear cocoa pods after 2–3 years, giving a full crop a few years later and continue to do so for a further 10–20 years. Pods, which are produced on the trunk or main branches of the trees, are normally 150–300 mm (6–12 in.) long containing 30–40 seeds embedded in a sweet mucilaginous pulp.

The seed has an outer shell, germ and two cotyledons, known as 'nib', which are over 50% by weight fat. This is contained within the cells within the nib and may be in an oil-in-water or a water-in-oil emulsion (see Figure 17.2). This is critical for the chocolate maker, and in general the higher the level of nib/fat within the bean the greater its value. This means that larger beans are often preferred to smaller ones, which have a

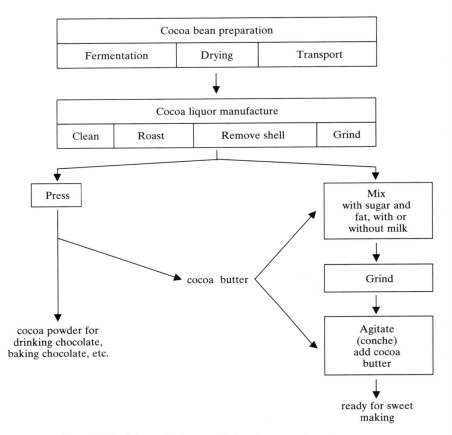

Figure 17.1 Schematic diagram of chocolate manufacturing process.

much higher shell-to-nib ratio. The fat, known as 'cocoa butter', is almost solid under temperate ambient conditions and yet liquid at body temperature. This gives chocolate its pleasant eating sensation, when it melts easily in the mouth. Not all cocoa butter is the same however. Trees grown nearer the Equator produce a fat which melts over a higher range of temperatures (chapter 7) than that from trees grown at higher latitudes. Consequently, chocolate produced from beans originating in Malaysia is very much harder than one made from Brazilian beans. Those from West Africa (Nigeria, Ivory Coast and Ghana) lie between these two extremes.

The taste of the chocolate will also vary according to the type of bean grown, the country of origin, season and processing, in particular the fermentation and roasting.

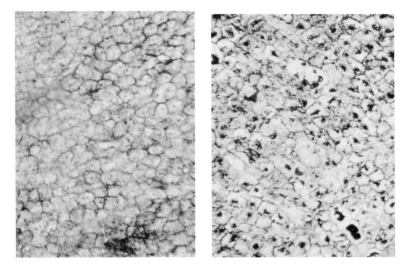

Figure 17.2 Microscopical sections through two cocoa beans (the fat is stained black). (a) Fat in large droplets; (b) fat in multiple droplets.

17.2.2 Fermentation and drying

When ripe, the pods are collected and then opened with a machete or wooden club, with care being taken not to cut the beans themselves. The beans must then be fermented; failure to do so will result in a slaty grey-coloured bean with a flavour totally different from normal chocolate. There are many methods of fermentation, but all involve heaping together the fresh beans together with some of the pulp and leaving them long enough at elevated temperatures for the bean to die and for the required chemical changes to take place. In West Africa, smallholders carry out the process in heaps under banana leaves for about 5 days, whereas estates with larger quantities of material may use boxes and ferment for 6–7 days. The flavour of the final chocolate has also been shown to depend upon the storage period of the pods before fermenting [3] and upon the depulping process [4] used. Overfermentation can also result in the development of undesirable flavours.

The process is a very unusual in that it is not the product itself which ferments, but the pulp around the beans. Flavour change is brought about by the fermentation products diffusing into the beans. In common with many other fermentations, however (chapter 9), the reaction begins with yeast converting the sugars in the pulp into ethanol. Bacteria then convert this initially into acetic acid and then into carbon dioxide and water, as well as producing much heat. As the temperature rises to about 45°C (113°F) and the remains of the pulp drain away, bacterial activity increases in the more aerobic conditions producing lactic and acetic acids. These acids, and the other chemical changes taking place, do not themselves produce the flavour of the final chocolate, but are necessary precursors for further reactions which take place in the subsequent roasting and conching processes.

The moisture of the beans before fermentation is about 65% and this must be reduced to about 7% before transport and subsequent processing. Failure to do so, or later wetting of the beans, will result in the development of moulds, which impart an objectionable flavour to the chocolate (probably due to monocarbonyls), and render the beans unsuitable for this use. Drying may be carried out in the sun or artificially. Care must be taken with the latter to avoid smoke contamination, which like mould imparts flavours to the beans which cannot be removed by further processing.

The characteristic colour of chocolate develops during the drying stage. With the death of the bean and the break up of the cell structure, enzymes are released, which are able to carry out oxidation reactions leading to brown colour formation.

17.2.3 Roasting

Before roasting, it is essential that the beans are cleaned thoroughly. Stones and other hard materials, transported in the sacks containing the beans, will damage subsequent milling procedures, whereas organic matter may produce volatiles during the roasting, which will result in off-flavours in the chocolate. The shell surrounding the bean is also extremely hard and will produce less-desirable flavours, in addition to which it contains the majority of the microbiological contamination. It must not be used in chocolate, in other than very small amounts, for legal reasons. Consequently, shell removal and roasting are closely linked.

There are three different processes currently used to carry out the roasting: whole bean, nib (shelled beans) and liquor (finely ground nib in liquid form) roasting. Whole-bean roasting is the traditional method and involves heating the bean to an air temperature of 125–200°C (257–392°F) [5], although the actual bean temperature may be considerably lower and this will be chosen together with the roasting time to give the required end flavour. This heating kills any potentially harmful microbiological contamination, causes chemical reactions which lead to chocolate flavour, lowers the moisture to about 2% and loosens the shell ready for subsequent removal.

The beans are then broken so as to produce pieces of shell and nib. The shell is in the form of low-fat platelets and consequently behaves aerodynamically very different from the more spherical, higher fat, pieces of nib. This enables the shell to be removed by sieving and suction in a process similar to grain/chaff separation and consequently called 'winnowing'. For nib and liquor roasting, this is carried out before the heat treatment. As the shell is more firmly attached to the nib, some form of surface heat treatment, sometimes called 'micronising', is applied to increase the efficiency of the separation.

Each of the three roasting processes has its own advantages and disadvantages [6]. Probably the most important difference lies in the fact that in bean roasting, some cocoa butter melts and migrates into the shell, where it is subsequently thrown away. By pre-winnowing, this is saved within the product, together with the energy wasted in heating the shell itself. In addition, because of the different sizes of beans, not all are uniformly roasted, whereas because of its liquid state, all liquor is able to receive the same treatment. Nib roasting lies between these two extremes.

The reactions taking place during roasting are extremely complicated [7], and the flavour changes normally, becoming much less acidic, probably due to the loss of volatile acids. One of the most important mechanisms for the development of chocolate flavour is the non-enzymatic Maillard reaction (see chapter 4). Analysis has shown at least 400 compounds to be present; no one group is responsible for the unique flavour of chocolate. Roasting is, however, critical in obtaining the final flavour, and although grinding and conching bring about some changes, these are relatively small in comparison, because of their lower operating temperatures. Off-flavours can also be produced owing to incorrect roasting or to the inclusion of non-cocoa material or mouldy beans. In the higher moisture and temperature regimes of the roaster, these can cause lipid breakdown and the generation of potent flavour compounds.

17.2.4 Cocoa liquor and powder production

Cocoa liquor is used in the production of cocoa powder and chocolate. It is produced by grinding the cocoa nib so as to destroy the cell material and release the fat. It is often stored and used at temperatures about 40°C (110°F) when it is a thin liquid, owing to its high content of cocoa butter (about 55%). Once released from the cells (Figure 17.2), this fat can contribute to the chocolate's texture and flow properties. Liquor is, in fact, composed of cocoa cell material in a continuous fat phase. During milling, some cell walls are broken and new surfaces are created, which must be coated in fat before those particles can move easily within that phase. Initially, however, the amount of fat released from within the cell is greater than that required to coat the surface, so the liquor becomes thinner. Eventually, most of the fat has been released and subsequent grinding makes the liquor thicker (see Figure 17.3). Most commercial mills, however, will not reach this extreme fineness.

The Dutch produced the first cocoa powder by pressing cocoa liquor and also developed the idea of alkalising the nib to make the powder disperse more easily in water, a process called 'Dutching'. An alkali solution, normally potassium carbonate, is added to the nib or less frequently to the liquor or compressed solids. This results in a milder taste owing to the neutralising effect upon the acids formed during fermentation and also darkens the colour due to more complex reactions with the cocoa tannins. Because the composition of the beans themselves is continually varying, it is a skilled operation to obtain a powder with a consistent colour and flavour. In addition, the amount of potassium carbonate used is limited by law, and also by the fact that it can react with the cocoa butter, by saponification or inter-esterification, which gives rise to off-flavours in the product.

The cocoa powder is made by pressing hot cocoa liquor (90–100°C, 194–212°F) against a filter. The cocoa butter passes through and is used later in the manufacture of chocolate or for other products such as cosmetics. The liquor used must not be ground too finely, as although this releases the fat, it also produces very small solid particles which would block the sieves. The cocoa liquor is pressed until 10–20% of fat remains, when the hard block of material, known as 'cocoa press cake', is removed from the filter. This is then cooled and milled to produce cocoa powder, which is used to make drinks, chocolate-flavoured coatings, etc.

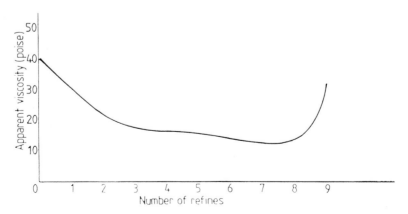

Figure 17.3 The change in apparent viscosity ($24 s^{-1}$) of cocoa liquor with number of passes through a five-roll refiner.

17.3 Chocolate manufacture

17.3.1 Grinding

Dark chocolate is almost entirely manufactured from sugar and cocoa liquor and butter. Granulated sugar is normally used and this must be reduced in size to <30 μm (1×10^{-3} in.) so that it does not feel gritty in the mouth. During breaking, many very small particles are created, and in a chocolate where 10% of the mass of particles is above this size, in number terms most of the particles are smaller than 5 μm (0.2×10^{-3} in.). These very small particles must be coated in fat for the chocolate to flow correctly in the liquid state and be made into bars, etc. The more very fine particles present in a chocolate, the more fat is required to coat them, which in turn increases the cost of the product. The grinding should therefore be carried out so as to minimise the production of the smallest particles, whilst ensuring the correct 'top end' fineness is achieved.

When milling sugar very finely much of it may change from the normal crystalline into an amorphous, glassy state (see chapter 2). This is unstable and will absorb moisture to become crystalline again. Once this has happened, the moisture is expelled and taken up by neighbouring particles. Where finely milled sugar is stored in a container, this may result in all the crystals joining together to form a solid lump. Amorphous sugar not only takes in water, but it very readily absorbs flavours or odours present. Care must be taken to ensure that the mill and grinding elements are free from such contaminants. Dark chocolate is able to be produced by milling the sugar to its final particle size and then mixing it with finely ground cocoa liquor in a conche or mixer. More usually, however, cocoa liquor and butter are added to granulated sugar in a pre-mixer. This is then ground using a roll refiner.

In order to achieve the fineness of finished chocolate, two milling stages are usually required. This may consist of two five-roll refiners (Figure 17.4), or more usually, a two-roll refiner followed by a five-roll one. By grinding in the two stages, the manufacturer is able to reduce the number of fine particles produced compared with a more traditional method, which pre-milled the sugar and then used a single five-roll refiner. The rheology of the mixture being processed is critical for the roll refiner. If the fat does not coat enough of the sugar, the mixture is dry, and will not grind properly; it may also be thrown off the rolls. Too fatty mixtures may result in the sugar sedimenting in the refiner hopper and uneven milling.

The roll refiner crushes by a mixture of shear and pressure. Each roll revolves faster than the one below it, and as the sugar–liquor–fat mixture adheres preferentially to the faster-moving surface, it moves to the top of the refiner where it is removed by a scraper bar. The film of material is

354   PHYSICO-CHEMICAL ASPECTS OF FOOD PROCESSING

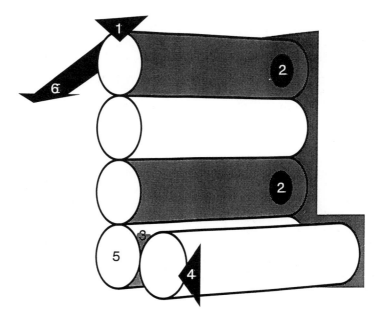

Figure 17.4 Diagram of a five-roll refiner. 1 = Roll stack pressure; 2 = chocolate film; 3 = chocolate feed; 4 = feed roll pressure; 5 = fixed roll; 6 = chocolate from scraper.

continuous, so it becomes thinner in proportion to the ratio of the speeds of the rolls. This factor, coupled with the width of the gap between the first two rolls (from the feed hopper) controls the final particle size of the chocolate. The rolls are also temperature-regulated, and this has an effect upon the rheology of the film and the roll diameter, thus altering the amount of crushing taking place. Because of the high energies involved in breaking the sugar, it is likely that some amorphous sugar will be formed. At this stage, it is in contact with cocoa particles and so is able to absorb their flavour. For this reason, dark chocolate produced by grinding the components together will normally taste quite different from one produced where they are ground individually. The rheology of the final chocolate will also be different, as when milling the ingredients separately it is possible to optimise the procedure for each component. Milling a mixture usually gives an inferior particle size distribution.

Milk chocolate is made by adding milk powder (either full cream or skim milk and butter oil) to the pre-mixer, before the refining stage. This is normally more difficult to break than sugar and refining conditions must be adjusted accordingly. Some manufacturers, especially in the UK and USA, make their milk chocolate from chocolate crumb. This is a mixture of condensed milk and cocoa liquor, which is dried to a moisture content of <2%. It was developed to provide a long shelf-life ingredient for chocolate making, at a time when milk supplies were very seasonal and milk powders had poor keeping properties. The method of crumb manufacture imparts unique flavours into the final chocolate, and this makes chocolates such as Cadbury's 'Dairy Milk' or 'Yorkie' taste very different from many of the French and Belgian powder-based products. White chocolate is produced from sugar, milk powder and cocoa butter, omitting any cocoa liquor or powder.

17.3.2 Conching

The roll refiner creates a lot of freshly broken surfaces in the sugar. This means that the material being removed by the scraper is not flowable, but in a crumbly or pasty form. One of the purposes of the conche is to coat these surfaces in fat so that the chocolate becomes a pourable liquid at temperatures, above the melting point of cocoa butter. This is equally true where the sugar and milk components have been milled separately.

Historically, the conche contained slow-moving granite rolls which smeared the chocolate over the base of the conche, thereby coating the particles and producing a large surface area from which moisture and other volatiles could escape. Processing took several days for dark chocolates and this type of machine has been replaced by large tanks containing paddle wheels which may be intermeshing, as shown in Figure 17.5. The ends of the paddles are often wedge-shaped, so that when the material is pasty it is able to use the sharp end to cut into it and smear the product against the sides. The flat side gives a high shear for processing the chocolate when it becomes liquid. These conches contain up to

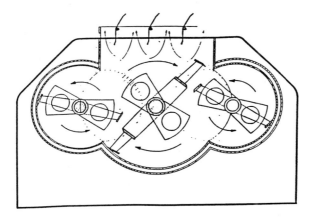

Figure 17.5 Schematic diagram of a Frisse conche [8].

10 tonnes of chocolate and owing to the high energy input can process the contents in about 12 hours.

Although it is a continuous process, many authors divide conching into three phases [8]:

Dry phase. Many of the surfaces are still uncoated with fat so moisture and other volatiles are able to escape.

Pasty phase. The frictional heat put in by the conche stirrers and from any jacket promotes some flavour development. There is a further loss of volatiles due to the higher temperature.

Liquid phase. This is largely homogenisation in order to produce the required chocolate viscosity. Final additions of ingredients is made during this stage.

As was noted previously, fermentation and roasting produce all the precursors to chocolate flavour, but subsequent treatment, normally in a conche, is required to remove some of the undesirable acidic notes and develop the more chocolatey ones. The actual chemical changes which take place are far from being fully understood. The air surrounding a conche usually smells of acetic acid, and this was generally thought to be a major cause of flavour changes during conching. Although the concentration of volatile fatty acids does decrease during conching, the fact that their lowest boiling point is 118°C (224°F), which is well above the conching temperature, means that other reactions are likely to be equally, if not more, important. Phenols, for instance, have been shown to decrease markedly during the conching of some types of cocoa [7]. It is also known that polyphenols form complexes with amino acids, peptides and proteins due to oxidation and enzymatic mechanisms. As phenols are associated with astringent flavours, this is in agreement with the fact that chocolate has a more mellow flavour following conching.

One of the ways in which the individual manufacturers are able to obtain distinctive 'house' flavours for their chocolate is by their choice of conching temperature. The literature gives many temperatures between 45°C (115°F) at the lower end and 85°C (185°F) for hot conching dark chocolate [5]. Maillard-type reactions can take place, giving cooked flavours at some of the higher temperatures and this must be avoided if a more milky flavour is required. At higher temperatures, the milk proteins may cause the chocolate to become very viscous, although this is sometimes reversible on cooling if the chocolate has not been too hot.

The loss of moisture, during the initial stages of conching, is very important in order to obtain good flow properties in the final chocolate, but must be carefully controlled. The moisture is initially present in the cocoa liquor or milk powder. As the conching proceeds, the temperature rises and the moisture tries to evaporate from the chocolate mass. If the texture is crumbly with many surfaces uncoated by fat, it is relatively easy for this to happen. Later in the liquid phase it is far more difficult and so

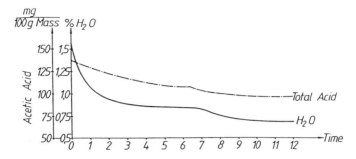

Figure 17.6 Changes in moisture and acidity of the air in a conche with time [8].

most moisture loss occurs at the early stages of conching (Figure 17.6). If, however, the conche is inadequately ventilated, the moisture cannot escape and becomes entrapped within the chocolate mass. Similarly, if the temperature is raised too quickly, there is insufficient time for the moisture vapour to be transported through the mass and it condenses inside it. In both cases the water is attracted to the sugar particles where it binds them together to form agglomerates. This means that although the particles have been adequately milled on entering the conche, the chocolate will now have a very gritty texture and will need to be reground.

If extra fat is present during the early stages of conching, it tends to make moisture loss more difficult. In addition, by making the chocolate thinner, it is more difficult for the conche to put work into the mass. Consequently, fat which is added towards the end of the conching cycle produces a thinner chocolate than if the same amount were added at the beginning of this process. There are, of course, limits, as there must be sufficient fat present for the mass to pass up the roll refiner, and the conche must have a big enough motor to be able to process the lower fat material.

17.3.3 Factors affecting chocolate viscosity

The point of addition of the fat is not the only factor which affects the flow properties of the final chocolate. Some of these can be manipulated by changes in recipe or processing, so as to tailor the chocolate's viscosity according to its subsequent use. The rheology of chocolate is complex in that it is a non-Newtonian fluid (see chapter 5). The International Organisation of Cocoa, Chocolate and Confectionery (IOCCC) have a recommended method to measure its viscosity [9] in which a viscometer measures the shear stress at several shear rates within the range $5-60\,\mathrm{s}^{-1}$. The data are then used to calculate the Casson yield value (YV) and plastic viscosity (PV). The former relates to the amount of energy required to start the chocolate moving, and is important when it is flowing slowly. Where markings are being put on the product, a high YV helps the chocolate retain its position as it sets. The PV corresponds with the amount of energy required to keep the chocolate moving, and is therefore important when carrying out calculations with regard to pumping chocolate. The two are, however, somewhat interrelated with respect to their use within a production department, as a low PV can partly offset a high YV and vice versa. The factors affecting either or both of these viscosity parameters will now be discussed.

Fat content obviously affects the flow properties of the chocolate. When the temperature is above its melting point, the more fat that there is present, the easier it will flow. In addition, as the temperature rises, the thinner the fat becomes and the lower the chocolate PV. The YV, however, is likely to increase, especially for milk chocolates and dark ones which do not contain an emulsifier (section 17.4). Chocolate is, however, used in a tempered state (see section 17.4.1), i.e. with a few percent of the fat crystallised, so this restricts the temperature range which can be used. The fat content has a major effect on the PV, but only a minor one on the YV [10].

Particle size distribution is also a very important factor in determining the flow properties of chocolate, but unlike fat, it has a great effect on the YV and very much less on the PV. As was described earlier, a large proportion of very fine particles will need more fat to coat them and cause a high YV. The actual particle size distribution can be measured by a laser light-scattering instrument. The chocolate is dispersed in a liquid such as trichloroethane or a low-melting point oil, like sunflower oil. The particles diffract the light, and from the measured spectrum the instrument calculates the particle size spectrum. Although not necessarily exact, because of the assumptions made in the calculations, the results are very useful to the chocolate maker. Of particular interest are the 90th percentile and the specific surface area. The former relates to the largest particles present and hence to how rough the chocolate will feel in the mouth. The specific surface area, normally expressed in square metres per centimetre cubed $[(m^2/cm)^3]$, is an indication of the total surface area of solid material to be coated with fat within a given volume of chocolate. This tends to be proportional to the YV. Sometimes the PV can in fact decrease with increasing fineness, which may be due to further fat being freed from cocoa liquor within the mass.

Moisture has a very detrimental effect on both flow parameters. It might be expected that as water is a liquid, it would add to the liquid phase of the chocolate and hence aid its flow properties. The exact opposite is in fact true, and, very approximately, for every 0.3% of water added to chocolate, 1% of fat must be added to compensate for it in flow terms. This is probably because the water attaches itself to the sugar and increases the friction between the particles. This theory is supported by the fact that where the sugar is coated with a surface active agent, like lecithin, far more water can be tolerated, although a very significant increase in both YV and PV still takes place. The great effect of moisture on the flow properties of chocolate underlines the importance of removing as much as possible during the early stages of conching. Most commercial chocolate contains between 0.5% and 1.5% of moisture, which makes its water activity very low, and consequently there is a negligible possibility of mould growth unless water condenses on the surface.

Emulsifier (surface-active agents) are used in most chocolates and may be used to affect the YV and PV independently. Lecithin is the most widely employed, since its introduction in the 1930s. It is thought to form a monomolecular layer on the sugar, thus helping the particles slide past one another. At a concentration of 0.5% lecithin, it has been found that about 85% of the sugar surface is coated. Up to this concentration, both YV and PV decrease with increasing lecithin concentration. The effect is particularly dramatic in low-fat chocolates. Over about 0.3%–0.5% (depending upon composition and particle size distribution) the PV may continue to fall slightly, but the YV will start to increase again (see Figure 17.7). This has led to the development of synthetic lecithins such as YN

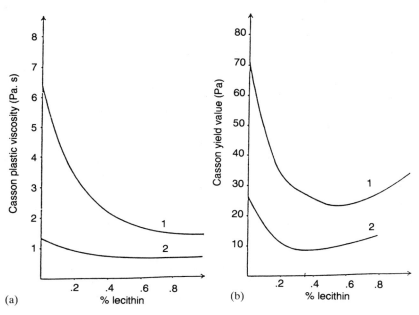

Figure 17.7 Influence of soya lecithin on Casson yield value and plastic viscosity [10]. 1 = Dark chocolate (33.5% fat); 2 = dark chocolate (39.5% fat).

where the increase in YV does not occur until much higher levels. It is, however, not permitted in some countries. Also of interest is polyglyceryl polyricinoleate (PGPR) also known as Admul-WOL. This almost only affects the YV (the PV may indeed sometimes increase) and at high levels (0.7%) can remove the YV altogether. This means that by choice of emulsifier, it is possible to control the two parameters independently over quite a large viscosity range. Admul-WOL, however, like YN, has legal restrictions on its use in some countries.

As in the case of fat, the point of addition of lecithin, or other emulsifier, is critical. Once again, lecithin added towards the end of conching has a much greater effect than that added at the beginning. This may be due to its strong water-binding properties, which will limit moisture removal, or as is claimed by some authors due to some being absorbed into the cocoa particles. Lecithin is known to have little effect upon the flow properties of cocoa liquor. High temperatures are also said to have a negative effect upon the performance of lecithin.

Finally, *work input* depends upon conching time and the power of the conche itself and relates to the conche's ability to coat the solid particles with fat, so that they will flow past one another. Traditionally, the conche was able to put large amounts of energy into the chocolate mass during the dry phase, only for the power input to fall dramatically as the mass became more liquid. This meant that long conching times were used and the particles were never fully coated. More recently, engineering developments in process control have enabled the conches to speed up automatically as the viscosity of the chocolate decreased. This has produced thinner chocolate in a very much shorter processing time.

17.4 Chocolate usage

17.4.1 Tempering

If liquid chocolate is placed in a container in a room, it will be many hours before it solidifies. This is of course impractical for commercial chocolate production, where it is desirable for it to set as fast as possible. In addition, it is not possible to apply severe cooling as cocoa butter can set in six different crystalline forms (see chapter 7 and Table 17.1). The lower melting-point ones are relatively unstable and will not give the gloss and breakability (snap) associated with chocolate products. If a chocolate product is not set in the correct form, or if it is placed in the sun where it melts and resets, it will become covered with a white sheen. This looks like mould, but is in fact quite harmless, being made up of a lot of fat crystals. This phenomenon is known as 'fat bloom'.

In order to ensure that the fat is in the correct crystalline form,

Table 17.1 Polymorphic forms of cocoa butter [12]

Wille and Lutton	Larsson	Melting point	
		°C	°F
Form I	β'_2	16–18	61–67
Form II	α	21–22	70–72
Form III	Mixed	25.5	78
Form IV	β'_1	27–29	81–84
Form V	β_2	34–35	93–95
Form VI	β_1	36	97

chocolate is subjected to a process known as 'tempering' before being used to make the final product. Following conching, chocolate is normally stored in liquid form at a temperature of about 40°C (104°F) when the fat will be in a fully melted state. If it is stored in solid form, it must be thoroughly melted before tempering.

One way of batch tempering is to cool the chocolate to a temperature where the desirable type of crystal can form, and then grate solid chocolate into it. This should be carried out under stirring so as to ensure that the soild chocolate seeds the rest of the mass.

For mass production, however, continuous temperers are needed. These cool the chocolate to a temperature where stable and unstable crystals are formed. The unstable ones are removed by raising the temperature to above their melting point, thus leaving the correct crystalline form as seeds. Research by Ziegleder [11] has shown that the rate of crystallisation depends very strongly upon the shear rate being applied to the chocolate during cooling. Temperers therefore consist of heat exchangers containing blades or discs which rotate at high speed, and are able to temper the chocolate in a matter of minutes. The rotating elements are limited in their speed due to the fact that the shearing action develops heat, which may in turn melt the crystals. A balance must therefore be achieved between this and the rate of cooling through the heat exchanger surfaces (see also chapter 5). The actual temperatures used will depend on the type of fat present in the recipe. Because of fat eutectics (chapter 7), milk chocolate must be cooled to a lower temperature than plain chocolate to strike seed (i.e. create small stable and unstable crystals).

In section 17.3.3 it was noted that it was desirable from the point of view of having a minimum viscosity to use the chocolate at as high a temperature as possible, whilst retaining the required number of seed crystals present. In order to do this, the seed crystals must be very stable, and this is not necessarily the case from all tempering machines, especially those with very small residence times or low shear. In these cases, it is possible to place the chocolate in a stirred tank and slowly raise the temperature as more crystallisation takes place. Failure to raise the temperature would result in very thick chocolate.

The degree of temper present in a chocolate is usually determined from a cooling curve. A thermometer is placed in a small volume of chocolate, which is then cooled rapidly, either electrically or placing the other end of the container in iced water. The rate of cooling is then monitored. Initially a uniform temperature drop occurs, but if seed crystals are present this rate will suddenly slow down, or even stop, or the temperature increase again. This is due to the release of latent heat from the fat as it solidifies. The change in cooling rate and the temperature at which it first occurs are noted. In general, the higher the temperature, the more stable the temper. Where the chocolate heats up again is normally called 'under temper', and a period of uniform temperature is referred to as 'correct temper'. In practice, plants may operate better on any of these conditions, depending upon the product and the subsequent processing. If no inflection point occurs, this indicates that the chocolate is untempered.

17.4.2 Moulding and enrobing

The two most common methods of processing tempered chocolate are 'moulding' and 'enrobing'. As its name implies, moulding involves depositing a known amount of chocolate into a mould. The surface of the mould should be at a similar temperature to that of the chocolate, as if it is too hot it may detemper the chocolate or if it is too cold cause some of it to set giving a poor finish. Normally, however, the smooth surface of the mould gives the chocolate a much shinier finish than enrobed produce. A tempered fat will not only give a good gloss, but will also contract better than a poorly tempered one and so come out of the mould much more easily.

Once in the mould it is shaken to remove air bubbles, which happens most easily if the YV is relatively low. For hollow-moulded products such as Easter eggs, the mould is filled and partially allowed to set. It is then inverted and the chocolate, which is still liquid, allowed to run out. This needs careful control of chocolate viscosity, in order to obtain a shell with uniform thickness.

Enrobing is the process of pouring tempered chocolate over the solid centre of the product. A schematic diagram of an enrober is shown in Figure 17.8. The chocolate from the temperer enters the large stirred tank at the base, from where it is pumped into the upper trough. A slit in the bottom of this trough lets the chocolate pour over the centres, which are moved through it on a grid conveyor. The excess chocolate falls through the grid back into the tank below. The product passes through a shallow tray of chocolate so that it is coated underneath, and also shaken or blown so that the coating comes to the correct weight. The blower or a

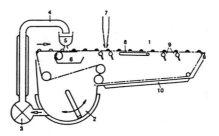

Figure 17.8 Components of a chocolate enrober. 1 = Wire grid conveyor; 2 = reservoir tank; 3 = pump; 4 = 'riser pipe'; 5 = smaller tank; 6 = bottoming trough; 7 = air nozzle; 8 = shaker; 9 = licking rolls; 10 = extension trough.

mechanical marker may also be used to put a pattern on the top of the product.

17.4.3 Cooling

Although tempered chocolate sets a lot more rapidly than untempered chocolate, under normal ambient conditions the average moulded or enrobed procuct would take several hours to become sufficiently hard to package. This is far too long for large-scale confectionery production, so cooling tunnels are employed. These may take the form of long, single-tier coolers, or multitier coolers, which are shorter and higher. In both cases, the chocolate passes through zones set to different temperatures.

The time spent in a cooler may vary from <10 minutes to about 30 minutes. This will depend upon the amount of heat that it is necessary to remove, its rate of removal and the quality required for the final product. This means that the time will depend upon the amount of chocolate in the product and its crystalline state (temper). For enrobed sweets, the sides and top set much more quickly than the base. If the latter is not regarded as being of importance, the sweet can be removed sooner than when a shiny base is desired.

Cooling can take place by conduction, convection and radiation. As the moulds or plaques/cooler belts (for enrobed sweets) are normally made of poor heat conducting plastic materials, and because radiation cooling is only really rapid with large temperature differences, convection is the most effective. In order to obtain rapid processing, the air may be blown at high speeds over the product to give an extra 'chill factor' during part of the cooling stage.

The initial cooling stage, however, must not be too fierce, so slower moving, not too cold, air is normally used. Failure to do this may result in fat being drawn to the surface, where it sets, producing a white bloomed appearance. Subsequently, the temperature can be lowered and the convection rate increased. This, however, is also limited, but this time by the humidity of the air. If the air temperature falls below its dew point, condensation will occur either on the walls of the cooler or on the product itself. If the latter happens, or water drips from the cooler onto the sweets, it reacts with the sugar causing a white sheen on the surface, which looks very like fat bloom. It is, however, composed of sugar, and feels slightly rough to the touch – unlike fat bloom which melts. This is known as 'sugar bloom'.

Condensation can be a problem as the sweets enter the packing room. This is normally at a higher temperature than the cooling tunnel and where the humidity is high, the moisture from the air may condense on the sweet surface. For this reason, the final sections of the coolers are normally operated at a higher temperature, so that the product is more approaching packing room temperature when it leaves. There are, therefore, three stages: gentle initial cooling, maximum convection with minimum temperature and then a slight rewarming. For some coatings, which contain fats other than cocoa butter, very different cooling regimes may be required.

17.5 Multiple component products

17.5.1 Modified chocolates

Because of the conditions under which chocolate is used or stored, it is sometimes necessary to alter its ingredients or processing. Although its property of melting rapidly in the mouth is one of the most distinguishing and desirable qualities of chocolate, this can also give rise to product deformation and fat staining of the packaging in hot climates.

There are, however, numerous patents for producing the so-called 'heat-resistant chocolate'. Many are based on the use of water. As was noted earlier, water has an extreme thickening effect upon chocolate, so much so that if it is added at about the 1% level, it becomes impossible to carry out the normal moulding and enrobing processes. Some patents overcome this by using water–fat emulsions. Another allows the chocolate to absorb water once the product has been manufactured by wrapping it in moisture-permeable material and then storing it under humid conditions.

For chocolates used in baking, etc. where it is desirable for the pieces to retain their shape even at very high temperatures, other techniques are often used, such as substituting the cocoa butter, perhaps by a non-tempering higher melting point fat. It is often useful to increase the yield value, by techniques mentioned in section 17.3.3, so that once it has melted, the chocolate is much less likely to flow into the surrounding material.

At the other end of the temperature scale, chocolate-coated ice creams present different problems. Here the centre is very cold so the chocolate will set very quickly making it difficult to produce a thin even coating. Also the low temperature makes it brittle and very likely to crack and fall from the product when it is eaten. Once again these problems can often be reduced by replacing the cocoa butter, but this time by a softer lower melting point fat. Recently however, real chocolate ice creams containing cocoa butter have become very popular. These tend to use much higher fat levels than normal chocolates. Because there is more liquid fat, there is more heat to take out, so the chocolate remains flowable longer enabling a thin uniform coating to be produced.

17.5.2 Fat and moisture migration

Many chocolate-coated products have fat-containing centres, for example pralines or nuts, where the fat has a high liquid content under temperate storage conditions. Because it is so soft, the fat is able to migrate relatively easily into the chocolate (via liquid-phase exchange) taking with it many of the centre's flavour components, thus removing the separate flavour sensation perceived when eating fresh product. Once in the chocolate fat eutectics (chapter 7) causes softening and may also result in bloom on the surface. At the same time some of the fats in the chocolate may migrate into the centre, once again producing texture changes. It is possible to put a layer of higher melting-point fat, or a sugar-based coating between the two components, so as to reduce migration, but this itself may result in undesirable eating characteristics. Where the centre is manufactured to have a continuous fat phase, it is possible to use the so-called 'structuring fats'. These set up a sponge-like system within the centre, which greatly restricts fat migration, by holding the liquid phase more *in situ*.

Although chocolate itself is a good barrier to water migration, it does have a relatively low equilibrium relative humidity (ERH; see chapter 1) and so is able to absorb moisture from humid air or from a higher ERH centre. Also once the chocolate is cracked, it is very easy for moisture to be absorbed by a hygroscopic centre, such as a wafer, resulting in a very undesirable change in texture. Moist, high ERH centres such as marzipan, can be equally adversely affected by the cracking of the chocolate coating, but this time due to drying out. Care must also be taken, when developing multicomponent centres for chocolate products, to ensure that the ERHs of each section are as similar as possible. It is very common to include toffee, with a high moisture, together with cereal or wafer products, which are very dry. Due to the differing ERHs, moisture will transfer to the cereals, making them soggy and leaving a much harder toffee. Processing and recipe optimisation must be used to minimise the difference in ERH so as to extend the shelf life of the product.

17.6 Conclusion

The majority of the topics covered in the first ten chapters of this book are of great relevance to the production of chocolate confectionery. Vapour pressure and water activity (chapter 1) are particularly important in the operation of cooling tunnels and in the design of multicomponent chocolate products. Glass transitions (chapter 2) on the other hand, may play a part in the development of the final flavour of the chocolate, particularly due to the way amorphous sugar is formed during some grinding processes.

The role of emulsifiers (chapter 3), especially lecithin, lies in their ability to modify the flow properties of the chocolate. They are normally added during the conching stage, where the Maillard reaction (chapter 4) may occur. Its effect on the final taste of the chocolate is also present, however, due to it taking place during roasting.

Current high-speed production plants rely on the chocolate having the correct rheology (chapter 5) and crystallisation (chapter 7) in order to achieve consistent quality and weight control.

Good chocolate requires high-quality, well-processed beans. This means that the fermentation (chapter 9), drying, roasting (chapter 10) and grinding must all be carried out correctly.

Only chapter 6 (foams, gels and gelling) has not been mentioned and even this is of importance in the manufacture of many aerated product centres, such as nougat, aerated chocolate itself and for scum treatment in wastewater systems. Thus all these different branches of science are now finding a major role in aiding the chocolate confectionery industry.

References

1. Nuttall, C. and Hart, W.A. Chocolate marketing and other aspects of the confectionery industry world-wide. In *Industrial Chocolate Manufacture and Use*, 2nd edn. Ed. Beckett, S.T., Blackie Academic and Professional, Glasgow (1994).
2. Hancock, B.L. and Fowler, M.S. Cocoa bean production and transport. In *Industrial Chocolate Manufacture and Use*, 2nd edn. Ed. Beckett, S.T., Blackie Academic and Professional, Glasgow (1994).
3. Duncan, R.J.E., Godfrey, G., Yap, T.N., Pettipher, G.L. and Tharumarajah, T. Improvement of Malaysian cocoa bean flavour by modification of harvesting, fermentation and drying methods – the Sime–Cadbury process. *The Planter* **65**, (1990), 157–173.
4. Nestlé-Golden Hope Improved Cocoa Fermentation Process. European Patent Application No. 91101882.8 (1991).
5. Jackson, K. Recipes. In *Industrial Chocolate Manufacture and Use*, 2nd edn. Ed. Beckett, S.T., Blackie Academic and Professional, Glasgow (1994).
6. Kleinert, J. Cleaning, roasting and winnowing. In *Industrial Chocolate Manufacture and Use*, 2nd edn. Ed. Beckett, S.T., Blackie Academic and Professional, Glasgow (1994).
7. Hoskin, J.C. and Dimick, P.S. Chemistry of flavour development in chocolate. In *Industrial Chocolate Manufacture and Use*, 2nd edn. Ed. Beckett, S.T. Blackie Academic and Professional, Glasgow (1994).

8. Ley, D. Conching. In *Industrial Chocolate Manufacture and Use*, 2nd edn. Ed. Beckett, S.T. Blackie Academic and Professional, Glasgow (1994).
9. Office International du Cacao et du Chocolat. *Analytical Methods* (E/1973) **10** *Rev. Int. Choc.* (1973) (September), 216–218.
10. Chevalley, J. Chocolate flow properties. In *Industrial Chocolate Manufacture and Use*, 2nd edn. Ed. Beckett, S.T. Blackie Academic and Professional, Glasgow (1994).
11. Ziegleder, G. Verbesserte Kristallisation von Kakaobutter unter dem Einfluss eines Schergefälles *Int. Z. Lebensm. Techn. Verfahrenst.* **36** (1985), 412–418.
12. Talbot, G. Vegetable fats. In *Industrial Chocolate Manufacture and Use*, 2nd edn. Ed. Beckett, S.T. Blackie Academic and Professional, Glasgow (1994).

18 Breakfast cereals and snackfoods
R.C.E. GUY

18.1 Introduction

Breakfast cereals and snackfoods [1,2] have many features in common, both in their manufacturing processes and in their product characteristics. Many of these products are made from similar types of raw materials and tend to be processed as doughs at much higher temperatures than those used for conventional baked products [3]. Although the final products in the breakfast cereal and snackfood markets are distinctly different in their appearance, they all tend to have low residual moisture contents and similar brittle crispy textures. Most of the products are based on starch-rich formulations of raw materials, in which starch biopolymers may provide from 50 to 80% of the dry solids. The most common methods of manufacture involve the addition of a limited amount of water to a dry mix, which serves to hydrate the starch and protein polymers and to act as a plasticiser, followed by a thermal process. The physical manipulation of these biopolymers occurs at high temperatures, usually >100°C (>212°F). Under such conditions, the base raw materials are transformed into viscoelastic melts (chapter 5), from which flakes, shreds or foams may be formed by further physical processing. These new structures are dried to low moisture levels to form crisp glassy textures, which provide the characteristic mouthfeel to the products. Snackfoods tend to be eaten in this dry form, so that they feel brittle and crunchy in the mouth, but breakfast cereals are usually consumed after immersion in milk, where they may absorb water and soften. The complex physico-chemical interactions of the biopolymers with water, during the development of the raw materials into doughs or fluids, and in their subsequent expansion and stabilisation processes are difficult to envisage. However, the understanding of these phenomena is essential for the control of the individual process and product variables. In recent years, research in the field of food polymer science [4] has begun to reveal some of the relationships which exist between the polymers, small sugars and the plasticisers. At present, the limited number of measurements which have been collected only provide sufficient detail to make rough predictions for changing the processing conditions or formulation, and each product type requires further extensive studies. In this chapter some of the most important physico-chemical reactions are reviewed for a range of products from the breakfast cereal and snackfood markets.

18.2 Physico-chemical changes taking place during the technological processes used for breakfast cereals and snackfoods

18.2.1 A general view of the initial changes in mixing and dough formation

Among the manufacturing processes employed for breakfast cereals and snackfoods, there are several different types of process which have been developed empirically over the last century. These processes may be grouped into two broad categories. The processes in the first category employ whole grains, or large pieces of grain endosperm, known as 'grits', as their basic raw materials. They generally are operated with steam cookers, pressure vessels or boiling water vessels as their main processing units. Such processes are used for the manufacture of well-known products such as corn flakes, wheat biscuits, shredded wheat and tortilla snacks. In each case, water is added to the raw materials and

Table 18.1 Breakfast cereal and snackfood products

Product type	Main raw material	Moisture level for processing (% w/w)	Type of process	Structural form
Cornflakes	maize grit	30–35	steam cooker	rolled flake
Shredded wheat	wheat grain	35–45	steam cooker	mat of rolled shreds
Wheat biscuit	wheat grain	30–35	steam cooker	layered rolled flakes
Puffed wheat	wheat grain	18–20	puffing gun/chamber	expanded foam in form of grain
Ready-to-eat hoops, etc.	cereal mix	16–20	extrusion cooker	shaped foams
Tortilla	maize grain	35–45	hot water/steam vessel	baked as flat biscuit
Corn chips	maize grain	35–45	hot water/steam vessel	fried as flat biscuit
Popcorn	maize grain	15–16	hot air microwaves	fine foam shaped by grain
Corn puffs	maize grits	16–18	extrusion cookers	shaped foams
Pellets snacks	cereal or potato mix	25–28	extrusion cookers and air dryer	dense pellet/expands to foam when heated in air or oil
Traditional prawn crackers	cereal mix	40–50	hot-water boiler and sun drying	dense pellet expands in hot oil
Tubes and hoops	potato mix	30–35	forming extrusion and frying process	slightly foamed shape

allowed to soak into the particles to achieve a fairly high moisture level throughout the particles, before thermal processing is completed. The water levels are generally in the range 30–40% of the total weight of the mixture. However, popcorn and some puffed cereals may be processed at the much lower moisture contents of 16–20% w/w. Popcorn is closely related to the puffed wheat product and it will be included in discussions of the puffing process.

The second category of processes is based on flours, smaller forms of grits, or even precooked flours. In their native form, these materials are mixed with water and processed at moisture levels ranging from 14 to 18% w/w. The native raw materials may be processed as powders directly in the barrel of an extrusion cooker, where after a brief mixing process, they undergo a thermal transformation into a fluid dough. Precooked flours must be processed at higher moisture levels, 35–40% w/w. They may be formed into a moist dough, either in a dough mixer or an extrusion cooker. Following dough formation, the intermediates for product manufacture can be shaped by sheeting and cutting or by forming extrusion. Alternatively, the doughs can be expanded into foams by extrusion at high temperatures so that water is vaporised within the extrudate.

In the initial stages of processing, the most important factors are the physical state of the starch polymers within the system and the water levels used. Mixtures or doughs, which contain predominantly native starch, behave very differently from those which are based on the precooked flours, containing pregelatinised starch granules. Initially they must be treated as two distinct groups in terms of polymer science. However, later in the processes it will become apparent that their initial differences have disappeared.

18.2.2 *Processes involving native grains or grits for breakfast cereals*

The native form of starch, in cereals or tubers, is a mixture of the natural polymers, amylose and amylopectin laid down in the endosperm of the seed or tuber as large aggregates, known commonly as the starch granule. The native granules are partly crystalline, with 20–40% of the polymers being involved in crystalline junctions zones and the remainder, present in an amorphous form, linking the junctions to form rigid structures. Water can penetrate the granules and hydrate or plasticise the polymers in the amorphous regions. The presence of the crystalline junctions limits the amount of water which can enter the granules to ca. 30% of the weight of the starch. Thus, the initial behaviour of the starch in the amorphous regions is restrained to this limit, even if the starch is placed in a large excess of water, as in the case of tortilla production.

The polymers within the amorphous regions of the granules have a glass-transition value, T_g (see also chapter 2), which is related to the water levels used and this value will vary as the moisture is increased, as is shown in Figure 18.1. At the 30% level, no more water can enter the granules and dilute the polymers any further. Whole-grain processes with high water additions are employed for the manufacture of products such as flaked or shredded breakfast cereals and tortilla-type Mexican snackfoods. In these processes whole cereal grains are heated in water at levels of 35–40% by weight, to achieve a physical change in the endosperm of the grain. At these high water levels, the grains may be cooked in boiling water vessels. For flaked breakfast cereal products, the addition of water is usually controlled to keep the operating moisture level <35% w/w, to ensure that the process is efficient and does not entail excessive drying costs. In some processes, the grains may be placed in an excess of water and allowed to absorb water freely during the heating process. In commercial practice, with the lower moisture levels, the grains are heated in sealed vessels, at pressures of ca. 200 kPa (30 psi) overpressure, to achieve a softening effect throughout the grain. After this cooking process, the grains are dried to moisture levels of 20–22% w/w, and allowed to cool to a temperature of 20–25°C (68–77°F), before being rolled under high pressure into the thin flakes, or shredded on special grooved rolls. The basic transformations, which occur within the grains during processing, are related to the hydration of the cereal endosperm and its major components, the starch granules and the endosperm proteins.

Water is added to grains which are relatively dry, having only 14–15%

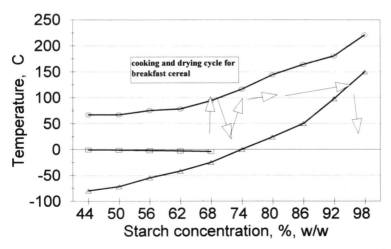

Figure 18.1 Diagram of the changes in the starch relative to T_g and T_m during the processing of flaked or shredded cereals.

moisture w/w during storage and transit to the factory. Therefore, the added water has to diffuse into the grains and equilibrate within the endosperm components in the initial stages of processing. This diffusion process is hindered in certain grains or parts of grains by the nature of the endosperm. The endosperms of cereal grains vary from hard dense vitreous materials to more loosely-bonded, porous chalky textures [5]. Maize grains display both types of endosperm in a single grain for Dent and Flint maizes, whereas wheats may be either hard and dense, or soft and porous. The water will hydrate parts of the proteins and the amorphous regions of the starch granules as it diffuses into the endosperm. In cereals with hard vitreous endosperms, such as flint maize used in cornflake processing, the diffusion of water may be slow and a 2-hour cooking period in the steam cooker is required. However, for soft wheats, with their more porous endosperm, the cooking time is much less, at ca. 30 min.

At the start of the cooking process, the amorphous regions of the starch granules will be able absorb water up to a maximum of ca. 30% moisture w/w and the remainder of the moisture will be available to hydrate and plasticise the proteins. Cereal proteins are varied in their nature, but contain a large proportion of high-molecular-weight glutenins. These molecules have a polymer form which permits them to swell in water to form hydrogels. These gels can hold water in their interstices, but it is still free to diffuse to other components if there is sufficient driving force such as osmosis or the hydration of new hydrophilic centres formed within the polymer system as the crystalline regions melt. At moisture levels of 30–35% w/w, the physical form of the proteins and starch polymers may be assessed from the published data on the measurements for the glass-transition temperatures [4]. The proteins of the endosperm will be present in a glassy state at temperatures of ca. 0–20°C (32–68°F) [4]. Thus, as they hydrate from their initial level of 14–16% w/w, they will pass through the glass transition and become a viscoelastic material, similar to the gluten proteins in cereal doughs. The starch polymers within the granules will be plasticised by the water in the amorphous regions. For a 30% moisture level in the starch polymers, the glass transition is at about 35–40°C (95–104°F). Therefore, the starch polymer chains in the amorphous regions will be in the fluid state above such temperatures. The starch granules will retain their rigid structure until their crystalline junction zones are broken down. This phenomenon occurs at temperatures above the melt transition, T_m, when the crystalline regions melt and the junction points controlling the structure of the granule are removed. The melting temperature of the polymer junctions in starch at 30–35% w/w, moisture levels is shown in the Figure 18.1, as being in the range 100–110°C (212–230°F). Such temperatures can only be achieved in the heated vessel by containing the water vapour under pressure. It was for this reason that the earlier breakfast cereal technologists used pressure vessels, rather than the simple boiling water processes employed in some forms of food manufacture. At the highest temperatures attained in the cooking vessels, the starch granules lose all their crystalline junction zones and these polymer regions hydrate and cause water to diffuse into the granular structures. As the cooked grains are dried to 22% moisture w/w, and cooled to 20–25°C (68–77°F) for their conditioning and rolling processes, the polymers will change in their physical form. The changes to the starch polymers can be charted on the polymers map of their glass-transition temperatures against moisture levels, as in Figure 18.1. The critical path for the starch polymers is marked on the graph from the cooking temperature through a drying process to 22% moisture and a final temperature of 20–30°C (68–86°F). It can be seen that the starch polymer mass always remains above the glass transition during processing, but that in the final stages, where the grain has been prepared for flaking, its temperature is much closer to the glass transition than immediately after cooking. It has been shown, in the more fundamental reports on the glass transition, that the fluid viscosity of the material above the glass transition increases rapidly as it approaches the T_g values. This means that the materials being flaked will be highly viscous, but will still flow as a plastic fluid under the compression of the flaking rolls. There may be a significant holding time at ambient temperatures in the process before flaking or shredding occurs. During this time, the starch polymers are able to recrystallise to some extent in the phenomenon known as 'retrogradation'. The introduction of crystalline regions within the starch fluid will stiffen the material and increase its bulk viscosity.

The final stages of processing for the flaked or shredded products is a drying process, to reduce their moisture to <4% w/w. This reduction in moisture content is achieved at high temperatures, so that the moisture is removed while the starch polymers are well above their T_g. In the toasting process used for corn flakes, the air temperature is very high at 230–315°C (446–600°F). This causes the temperature of the flakes to rise very quickly with the result that the water in it may be superheated until it vaporises within the flake and forms a bubble or blister. Cornflakes and similar products have a critical thickness for this phenomenon to occur. If they are too thin, the blisters do not form readily and the flake may be smooth and hard. If the flakes are too thick the blisters may be too large and there may be areas which are too hard in the flake.

Shredded wheat is cooked at a moisture level of 40–45% w/w, and on cooling to ambient temperatures prior to shredding on serrated rolls, it may undergo retrogradation. The shredded product is dried from a higher moisture range of up to 40% w/w in a dense mat of shreds. This process causes the mat to lift in the traditional biscuit shape. The final product becomes crisp and brittle on cooling.

18.2.3 Processes involving maize grains for tortillas and tacos

In the production of the tortilla and taco products, a pre-gelatinised flour is produced from whole grains [6]. Traditionally, the process is carried out by heating whole grains of maize in water to soften the grains, which are then milled into a dough with a stone mill (Figure 18.2). This masa dough is then sheeted and cut into the required shapes, or formed in a ram extruder into strips and cut to make a similar product. The dough pieces are baked at high temperatures for tortillas or fried at 170–180°C (338–356°F) to produce corn chips. The grains are placed in water, with a little added lime to help soften the husks, at levels of >40% w/w and heated to temperatures of 80–90°C (176–194°F). An excess of water is used because of the primitive conditions in which the process originated and the simplicity of heating the maize in a vat. The physical phenomena occurring within the process are similar to those observed in the breakfast cereal process, except that the moisture content is allowed to reach

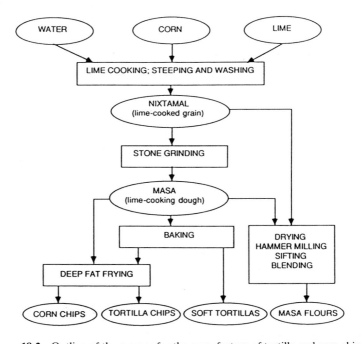

Figure 18.2 Outline of the process for the manufacture of tortilla and corn chips.

higher levels, by using excess water around the grains. Thus the melting temperature, T_m, of the starch within the grains is lowered to the temperature range being using in the heating vessel (Figure 18.1) and the starch granules begin to lose their crystallinity and absorb water. The process is continued for a period of several hours to allow the diffusion of the water from outside the grains to the starch granules within the endosperm. After steeping overnight, the grain is washed and allowed to cool before milling. A large number of granules lose their crystallinity and become soft and plastic at the end of the cooking period. Their starch polymers will remain within the granular structure, because the water is still only available in a limited amount due to the competition with other granules and the restricted access through the grains' outer coatings and endosperm. On cooling, the starch polymers form an increasingly viscous and leathery material as they approach the region close to the glass transition, which lies at around $-5°C$ (23°F). The stiffening of the starch granules in the endosperm allows the mill to break down the grains into a coarse-textured dough. This dough may be packed into a ram extruder and formed into strips, or reduced with gauging rolls to finally form into a thin sheet, which may be cut into tortilla chips or the larger tacos. The dough pieces remain soft and pliable until they are baked in a very hot oven or fried in hot oil to remove their moisture to levels of ca. 2–3%. As the water is removed from the starch polymers their T_g rises as in Figure 18.3, and eventually the starch polymers reach the glassy state at temperatures above 50°C (122°F). Therefore, in all the products which are consumed at ambient temperatures, the polymers are in a glassy state.

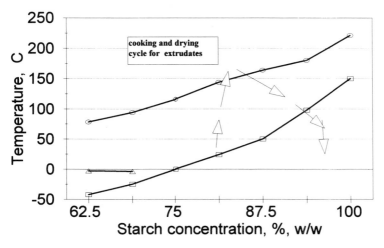

Figure 18.3 Diagram of the changes in the starch relative to T_g and T_m during the processing of directly extruded snacks at 16–18% moisture w/w.

However, the microstructure within tortilla chips still consists of dispersed starch granular aggregates, with only very weak bonding by protein bodies and starch extrudates between the aggregates. Thus, when the products are broken by a normal biting action with the teeth, there are many points of weakness or fracture lines in the product and it breaks up easily into large fragments.

The texture of the baked tortilla chips is usually denser than the fried chips due to a greater number of bubbles which are formed in the fried products. In the high temperatures of the tortilla oven the loss of moisture from the chips is not accompanied by the retention of water vapour in gas cells. This is probably due to the speed of the drying process and the lack of any true bubble structure in the coarse doughs. However, the frying process is known to induce the formation of a foam structure in snack doughs, when their dimensions are thicker than about 0.5 mm (0.02 in.). A crust layer is formed at the oil–dough interface which allows some water to be superheated, provided that the chip is thick enough to retain moisture in the central region. If the chip is too thin it becomes dehydrated before any expansion can occur. Thus, the blistering of fried chips and the formation of a foam structure in similar products, such as reformed chips, potato hoops and tubes, is possible provided the dough mass has suitable dimensions and is in a fluid state.

18.2.4 Processes from grain involving puffing guns and chambers

The use of whole-grain processes in the breakfast cereals and snackfood industries includes a range of products which have very fine and well-expanded foam structures. These products are said to be formed by an explosive puffing process from the natural grains. They include the well-known products, puffed wheat, puffed rice and popcorn. During processing, the grains are equilibrated at moisture levels of ca. 14% w/w during storage and may be processed at this moisture, or at higher levels up to 20% w/w in the more sophisticated equipment.

The basic principles of puffing popcorn, which have been studied in detail [7], appear to be similar to those involved in breakfast cereal puffing. It was shown for popcorn that the outer layer of the grain serves to act as the wall of a pressure cooker. An intact pericarp retains the water within the endosperm of the grain as it is heated. Very high temperatures can be measured within a grain, before the resistance of the pericarp to the pressure is overcome and it ruptures, dramatically releasing the pressure. This explosive pressure release occurs at temperatures in the region of 170°C (338°F). Water, which has been superheated within the endosperm, is released as vapour to expand the system. Surprisingly, it forms a fine open foam structure with thousands of cells per ml. The bubbles appear to expand instantaneously to their full extent and rupture leaving a finely textured foam with a specific volume in the range 20–30 ml/g (0.5 ft^3/lb). It was proposed [7] that popcorn bubbles are formed out of existing holes at the centre of the maize starch granules. At the high temperatures inside the endosperm of the grains, the starch granules are heated above their T_m values for the given moisture level (14–16% w/w). Their polymers all then become amorphous, existing in a fluid state before the rupture of the husk occurs. Therefore, when the water is vaporised within the endosperm, the water vapour expands within the existing bubbles or escapes through cracks and fissures. In the vitreous endosperm there are few cracks and most of the vapour is released into the bubbles at the centre of each granule. The individual granules can expand to the limit of the starch polymer available in the granule itself. This source of material provides the cell-wall material for each bubble, but at the limit of the polymer resource within each granule, the expansion stops and the bubble ruptures. However, there also appears to be a form of collective behaviour between the bubbles, in that only the bubbles within the vitreous phase of the maize endosperm are able to expand into a foam structure. Those granules which are present in the floury endosperm do not expand and form bubbles as the rest of the material is puffed. They were found to have retained their crystallinity and to be spread over the surface of the expanded vitreous endosperm [6]. It was thought that they had either lost their moisture prematurely, or had failed to attain their T_m values for some other reason.

In the processing of puffed wheat, the traditional methods employing specially designed puffing guns or chambers are still used today. The grains of wheat are pretreated to weaken the husks by a milling process known as 'pearling' or by soaking in 2–4% w/w salt solutions [8]. They are loaded into the puffing gun with a small amount of extra water making an overall level in the range 18–20% w/w before being sealed into the chamber and heated rapidly with steam or direct fired systems. The temperature within the grains is rapidly raised to high levels, in the range 150–170°C (300–340°F), so that the starch granules in the grains exceed their T_m values (Figure 18.1). Thus, in a short heating time the physical state of the starch granules is changed to a soft viscoelastic form. Under the pressure of the water vapour in the chamber, the grains are compressed slightly and the bulk of the water within the grains is held in a liquid form. On firing the gun, by opening the vessel to the atmosphere, the water vaporises and causes a fine foam structure to form in each of the grains. The close similarity in the physical nature of puffed wheat and popcorn, would suggest that the mechanisms of expansion are similar in these two processes, but to date no detailed studies have been published on the microstructure of puffed wheat. The advantage of the puffing gun or chamber, over the popping of natural grain (where the grain is heated in oil at a high temperature until it puffs), is that they allow the puffing to be controlled and to be performed with a wide variety of grains. Popcorn is very dependent on the quality of the pericarp of the grain, the moisture content within the grain and the amount of the vitreous endosperm, but the puffing gun replaces the outer layers of the grain with a steel chamber. The most modern forms of puffing equipment permit a wider range of moisture contents to be employed and enables the operators to puff cereals with endosperms which are not dense and vitreous, such as soft wheats.

18.2.5 Low-moisture extruded products made from flours or grits – directly expanded breakfast cereals and snackfoods

The manufacture of products such as flakes, masa doughs or expanded foam structures from cereals using extrusion cookers was found to be possible at much lower moisture levels than those used in the older steam cooker and boiling water technologies.

Moisture levels of 25–27% w/w may be used for breakfast cereal flakes and shreds and moisture levels as low as 14–18% w/w are used for directly expanded breakfast cereals and snackfoods. Such processes reduce the need to use expensive drying equipment and have other benefits in terms of efficiency. The processes involve three stages, a transformation of the native raw materials into new physical forms, the shaping or expansion of a foam structure from the fluid extrudate and the secondary processing such as drying.

At low-moisture levels, the starch granules within cereal mixes are only partially hydrated and appear to be unchanged from their native state, in terms of size and rigidity, until the temperature is raised to exceed their melting points, T_m. In the extrusion cooker, where the high pressures are contained within the barrel, the mass of starch and water can be heated to temperatures of over 180°C (356°F). At these temperatures, the pressures within the extruder must exceed 2 MPa (300 psi). The T_m values for wheat starch, shown in Figure 18.3, may be exceeded at ca. 140°C (284°F) for the moisture contents mentioned above. It should be noted that the T_m values quoted in the literature are normally measured by techniques such as differential scanning calorimetry (DSC) under static conditions in a sealed cell. In an extruder, the starch granules are under considerable mechanical strain as well as the thermal effect. It is clear from experimental observations on the extruder [9] with wheat flour and starch, that the melting point of the crystalline structures in the granules lies 5–10°C (9–18°F) below the DSC value. At the melting temperature the starch granules lose their crystallinity and the polymer mass becomes amorphous. Since it is above its glass transition at these temperatures, it behaves as a viscoelastic fluid. The starch granules become soft and deformable and are unstable when subjected to strong shearing forces. They do not swell until the water level in the system becomes ca. >30% w/w. Therefore, at moistures <30% w/w, the granules occupy the same volume as the native form, but may be deformed into flatter pancake shapes [9]. These granular aggregates begin to lose their integrity under continual shearing action within the extruder and eventually a continuous starch polymer phase develops. Some molecular degradation of the amylopectin and amylose also occurs [10], but any change in the T_m and T_g values for the starch polymers is thought to be small. The amount of degradation occurring in a normal process does not reduce the polymer size sufficiently to reach the critical levels which affect the T_g. The amount of residual granular aggregates which remain in the continuous phase, depends on both the severity and time of application of the shearing forces within the extruder. At the exit of the extruder, the pressure created by the fluid as it is forced through the die is related to its viscosity, flow rate and the resistance of the die passage. The presence of the starch granules in the fluid has a large effect on the die pressure because they behave as giant molecular structures. As they are removed by the mixing/shearing action of the extruder, the die pressure falls rapidly until they are all dispersed (Figure 18.4). The high pressure acting on the fluid at the die entrance and on the screws, is released at the die exit. This causes a foam to be formed in the fluid mass, if the temperature is greater than the boiling point of the water, by the release of water vapour from the superheated liquid water. At temperatures <100°C (<212°F), a dense rope of extrudate is formed which may be cut, shaped or rolled to make products such as flakes, half-products, etc.

At temperatures >100–102°C (212–216°F), bubbles of water vapour are formed in the extrudate by spontaneous nucleation. These bubbles

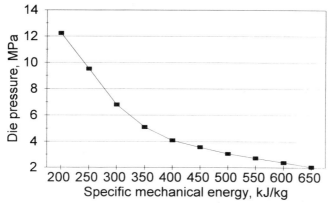

Figure 18.4 An illustration of the reduction in die pressure as increasing specific mechanical energy inputs degrade the starch on an APV MPF 50D twin-screw extruder.

may either expand until they rupture, thereby releasing their vapour to the atmosphere, or expand until the gas pressure is balanced by a combination of the viscous resistance to flow and the energy stored in elastic rheological forces by the extension of the cell walls (chapter 5). In normal extrudates manufactured from cereals and potato recipes, extensive rupture occurs [9] and the excess pressure is released from the vast majority of the cells. This can be confirmed by penetration studies [10] and the examination of the loss of moisture from extrudate. For example, in an extrusion of maize grits at 18% moisture w/w, 40–60% of the water is lost, depending on the actual expansion of the cell-wall materials. A general hypothesis for normal extrudate expansion [9] has been proposed, which states that the controlling for expansion in normal extrudates is the extensibility of the cell walls, rather than the gas pressure within the cells or other factors such as surface tension. It was shown that under comparable circumstances with circular dies, normal die pressures >2 MPa (300 psi) and moisture levels of 14–20% w/w, that the extrudates which had the highest concentrations of dispersed starch polymers gave the greatest extension of their cell walls and the greatest overall levels of expansion. The greatest moisture loss, producing the driest extrudates, also occurred at the highest level of expansion. It was also shown that the addition of substances which reduced the extensibility of the starch polymer films in the cell walls, such as added fibres or bran platelets in wholemeal wheat flour, reduced the expansion of extrudates (Figure 18.5). It was thought that the extensibility of the cell wall is reduced by the presence of the bran, which introduces points of weakness and causes premature rupture in the cell walls [11].

The failure to release the starch polymers from their granular aggregates also reduced the expansion of the extrudates. This is illustrated in

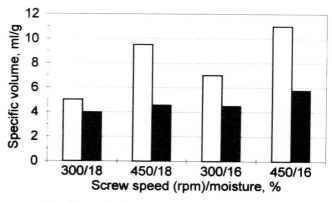

Figure 18.5 Reduction in the expansion of flour due to the presence of bran.

Figure 18.6, where the expansion of a wheat flour extrudate may be seen to increase as the specific mechanical energy (energy input to starch expressed as kJ/kg or kW h/lb) applied to the starch is increased.

During the expansion and stabilisation of extrudates, some shrinkage of the foam structure occurs after the point of maximum expansion. It was recognised [11] that the shrinkage of cereal extrudates was small (<10%) for extrusion moisture levels <20% w/w, but increased markedly as the moisture level increased. For example, the shrinkage was ca. 40–50% and 50–60% at moisture levels of 25% and 35% w/w, respectively. Two mechanisms have been proposed to explain this shrinkage phenomenon, the elastic recoil of the starch polymer phase [14] and the closed-cell hypothesis [14,15]. In the first mechanism, it proposed that the elastic energy stored in the starch polymer during the expansion of the gases is released when the cells rupture. Thereafter, the cell walls shrink until their increasing viscosity freezes the elastic memory. Once cooled to below the T_g value, no further change in dimensions can take place. However, if extrudates are exposed to water vapour and allowed to take up moisture to exceed their T_g by a significant amount, they begin to shrink once more and lose much of their expansion. This phenomenon confirms the presence of an elastic characteristic in extrudates.

In the second hypothesis, proposed for extrudate expansion [15], the cells are considered to be closed. The extrudates expand until they are restrained by the rheological forces. The cell walls are thought to stabilise by the increase in viscosity and eventually to be fixed by the high viscosity close to the glass transition. This hypothesis does not explain the loss of moisture from the extrudate or any of the phenomena caused by the physical changes of the starch or the presence of particulate material

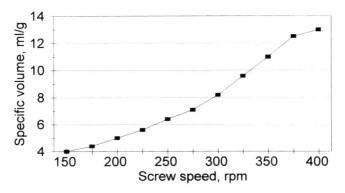

Figure 18.6 An illustration of the expansion of starch extrudates (under similar extrusion conditions to those shown in Figure 18.4) on an APV MPF 50D twin-screw extruder, with increasing screw speed.

such as bran. However, this phenomenon has been observed in certain extrudates, such as starch and sugar mixture 15–25% w/w sugar contents. In this case the extrudates, which did not rupture, collapsed completely after extrusion during a period of 1 minute. On cooling, the structures were not viscous enough, when the water vapour condensed at 100–103°C (212–217°F), to support their structures and they collapse from specific volumes of 20–30 ml/g to 0.8–1.0 ml/g (0.3–0.5 ft^3/lb to 0.01–0.02 ft^3/lb).

In normal snackfood and breakfast cereal extrudates, at the point of rupture, the expanded structure of hot fluid polymer melt may contract under the influence of two physical phenomena. The contractions may be either due to the elastic recovery of the polymer melt, which was stretched during the expansion of the bubbles [14], or due to a small number of bubbles within the structure which had not ruptured and which collapse as the water vapour cools. In either case, the contraction is slowed and then inhibited by the increase in viscosity within the cooling cell-wall material. The extrudate may stabilise and remain in the rubbery state above the glass transition for a brief period during which time it is still deformable under high pressures. Once it has cooled to temperatures below the glass transition, it becomes hard and brittle, the characteristic form of snackfoods and breakfast cereals.

The finished products will have moisture contents of <5% w/w, which means that their water activity is low in the range 0.2–0.4 expressed as a ratio of water vapour pressure to pure water (see chapter 1). Therefore the products will absorb water from the atmosphere unless they are packed in moisture-impermeable packaging. Any failure to protect the products from the atmosphere will cause changes in their texture as the water plasticises the polymers and increases the volume and potential freedom of movement within the glassy state. At moisture levels of 7–10% w/w, the glass transition will be exceeded and the polymer systems will become leathery or rubbery and the brittle nature of the product's crumb structure will be lost. This results in a loss of those sensory qualities in snackfoods which consumers find to be desirable. The snackfoods [13] tend to have a very open porous structure with very thin cell wall and therefore all parts of the product are rapidly transformed on exposure to moisture.

Breakfast cereals are also manufactured as crisp products which must be kept in that form for the maximum enjoyment of the consumers. However, they are different in their structural form due the manner in which they are consumed. In the traditional method, the breakfast cereals are mixed with milk and are consumed over a period of a few minutes. The diffusion of water into the structure causes a softening of the crisp texture, but if this occurs too quickly the breakfast cereal becomes soft and loses its appeal. The ideal structure for the products is one which has cell walls which are thick enough, and of sufficiently low porosity, to resist hydration for a short period of a few minutes. Thus, the glassy state can be maintained within parts of the structure to give a crisp bite, in contrast to other areas which are softening. This combination of contrasting taste sensations provides the character of good-quality breakfast cereals.

18.2.6 Intermediate-moisture extruded products – half-products for snackfoods

One of the oldest types of snacks was made by simple traditional methods in the Far East and India from a cooked flour dough. After cooking in a high water level to gelatinise the starch, the dough was sliced and dried in the sun to form a vitreous half-product which could be stored until required. On heating in hot oil or sand, the half-product expanded into a fine foam structure which was known as Java cracker, Malaysian kerapok and other local names. This traditional process has been transferred to the extrusion cooking, to mass produce this form of snack more efficiently for the modern markets.

In the simplest process, a flour dough at 25–30% w/w, moisture content is mixed and heated in a low shear extruder until the temperature of the dough is 120–130°C (248–266°F). At this stage, the starch granules lose their crystallinity and become soft and deformable. They are compressed together into a dense mass during the processing by the extruder screw and at the same time the compressed fluid dough is cooled to 90–95°C (194–203°F). On extrusion from the die, the extrudate remains unexpanded and may be cut into thin slices. The moist slices have a moisture content of ca. 18–20% w/w and remain plastic at temperatures of 40–50°C (104–122°F). They are dried to 10–12% w/w moisture in a slow process resembling pasta drying to obtained a stable half-product. This has a dense structure formed by the starch granules and some dispersed starch phase. All the polymers are in the glassy state, forming a glass throughout the structure.

The expansion of the half-products is carried out by the rapid application of heat, either by frying in hot oil at ca. 180°C (356°F) or heating in hot air. It is thought that the mechanism of expansion is similar to that of the popcorn, utilising gas bubbles within the vitreous glass to act as the nuclei for the bubbles which expand in the foam. The polymer glass, which is present throughout the half-products, retains the moisture within its structure so that it becomes superheated before the viscosity of the starch polymer melt falls sufficiently to flow under the gas pressure. Once the cell walls within the half-product become fluid and flow, the bubbles expand as in the popcorn and create a foam structure. It was shown that the residual structure of the starch granules is very important in providing the air-bubble nuclei within the glass [16]. If the starch granules are completely dispersed in the extrusion process, the number of gas cells appearing in the half-products is very small and the expansion is poor.

18.3 Colour and flavour formation

In the field of breakfast cereals and snackfoods, the colours and flavours developed by processing are still an empirical art, which has been practised for many years without any fundamental understanding. It is well known that reactions such as caramelisation, the Maillard reaction and oxidative decomposition contribute to the colour and the flavour in fried and baked snackfoods and in most toasted breakfast cereals. Colour formation in snackfoods is very important in the fried products where the colour appears at the surface of products such as crisps, potato sticks, reformed chips, tubes, hoops, etc. The reactions occur at high temperatures in the crust, which forms at the edge of the crisps, when the moisture levels have been reduced to very low levels and the temperature rises towards the temperatures of the oven or frying oil. There are similarities with the colour and flavour reactions found in baking processes and the same effects can be observed for precursors such as sugars, amino acids, peptides and fats (chapter 12).

In extruded products, which are directly expanded or expanded from pellets in hot melts, the conditions for colour and flavour formation may be different to those found at the surface of baked or fried products. The reactions take place at constant moisture levels of 14–28% w/w and temperatures of 120–170°C (248–338°F). If a flour is processed under such conditions on its own, the colours and flavours will be fairly weak, but with the addition of precursors such as the components of milk solids, intense colours and flavours can be produced in a short residence time of 45–60 s. There is an increasing reaction rate for all the main reactions as the temperature rises and it is noticeable that lipid degradations, relating to shelf-life instability, also increase with higher temperature extrusion [17].

The products which are made by the various breakfast cereal and snackfood technologies may be listed in terms of the temperature and moisture processing histories as in Table 18.2. This brief summary shows that colour is only formed in the products which are subjected to both high temperatures and low moistures. In the processes with steam cookers, such as corn flakes, the main colour reactions occur in the toasting ovens rather than in the steam cookers, where the moisture is at 35% w/w. However, colours may be developed in the higher moisture products by adding precursors to the cereal or potato mixes. For example, if commonly available reducing sugars, such as glucose or lactose, are

Table 18.2 Colour reactions in breakfast cereals and snack foods

Process	Temperature		Moisture (% w/w)	Location of colour	Residence time at high temperature (s)
	°C	°F			
Corn flakes	130–140	266–284	2–4	throughout flake	10–15
Expanded cereal	140–160	284–320	16–18	little	30–60
Puffed wheat	150–170	302–338	18–20	little	30
Extruded snacks	140–160	284–320	14–18	little	30–60
Fried snacks	140–160	284–320	2–4	throughout with more at surface	15–20
Tortilla chips	160–180	320–356	2–4	throughout with more at surface	30
Half-products	140–150	284–302	10–12	little until fried	10–15

added with some peptides and amino acids as found in whey and milk solids, red-brown colours are formed. The intensity of the colours increases with the temperature within the process and to a lesser extent with the residence time.

At the temperatures of the frying oil, 180–190°C (356–374°F), the reactants within the raw materials will be reducing sugars such as glucose, maltose and fructose and amino acids and peptides. In baked products such as tortilla chips, the same phenomenon occurs at the surface due to the high oven temperature and the loss of moisture from the surface layer. This closely resembles the toasting process used to dry corn and other types of flakes in the breakfast cereal processes. In this case, the formulations for the flakes are varied with the additions of syrups and milk solids to provide precursors for the Maillard reaction. Milk solids contain the reducing sugar, lactose, and the peptides and amino acids, which are the reactants of the colour and flavour reactions.

18.4 Conclusions

As with the making of bread and cakes, a knowledge of the glass-transition temperature, T_g, greatly aids the food technologist in understanding and controlling breakfast cereal and snackfood production. Texture is of critical importance for most of these products, and this is greatly affected not only by the type of processing used, but also the moisture present at each manufacturing stage and the melt temperature of the ingredients present.

References

1. Fast, R.B. and Caldwell, E.F. *Breakfast Cereals; and how they are made* (1990), American Association of Cereal Chemists, St. Paul, Minnesota.
2. Matz, S.A. *Snack Food Technology*, 3rd edn. (1993), Avi Publishing Co., Westport, Connecticut.
3. Guy, R.C.E. Extrusion cooking versus conventional baking. In *Chemistry and Physics of Baking*, chapter 12. Eds Blanshard, J.M.V., Frazier, P.J. and Galliard, T. (1986), The Royal Society of Chemistry, special publication No. 56.
4. Slade, L. and Levine, H. Structure–function relationships of cookie and cracker ingredients. In *The Science of Cookie and Cracker Production*. Ed. Faridi, H. (1994), Chapman and Hall, London, 23–142.
5. Guy, R.C.E. Raw materials for extrusion cooking processes. In *The Technology of Extrusion Cooking*. Ed. Frame, N.D. (1994), Blackie Academic and Professional, Glasgow, 52–60.
6. Serna-Saldivar, S.O., Gomez, M.H. and Rooney, L.W. *Advances in Cereal Science and Technology*, Volume X (1992), American Association of Cereal Chemists, St. Paul, Minnesota, 243–302.
7. Hoseney, R.C., Zeleznak, K. and Abadelrahaman, A. Mechanism of popcorn popping. *J. Cereal Sci.* $1(1)$ (1983), 43–52.
8. Fast, R.B. Manufacturing technology of ready-to-eat cereals. In *Breakfast Cereals; and how they are made* (1990), American Association of Cereal Chemists, St. Paul, Minnesota, 29–34.
9. Guy, R.C.E. Extrusion cooking of wheat flour. *Flour Milling and Baking R.A., Research report No. 154*, (1993), 17.
10. Colonna, P., Tayeb, J. and Mercier, C. Extrusion cooking of starch and starchy products. In *Extrusion Cooking*. Eds Mercier, C. Linko, P. and Harper, J.M. (1990), 247–319.
11. Guy, R.C.E. and Horne, A.W. Extrusion and co-extrusion of cereals. In *Food Structure – its creation and evaluation*. Eds Blanshard, J.M.V. and Mitchel, J.R. (1987), Butterworths, London.
12. Guy, R.C.E. Wheat bran in extrusion cooking. *Extrusion Communique*, April–June, $5(2)$, supplement $1(2)$ (1992), 10–11.
13. Barrett, A.H. and Ross, E.W. Correlation of extrudate infudibility with bulk properties using image analysis. *J. Food Sci.* $55(5)$ (1990), 1378–1382.
14. Tharrault, T-F. Contribution à l'étude du Comportment des Farines de Ble Tendre en Cuisson-Extrusion bi-vis. PhD thesis (1993), Université de Paris XI, Centre D'Orsay.
15. Mitchel, J.R., Fan, J. and Blanshard, J.M.V. The shrinkage domain. *Extrusion Communique* (1994), (March), 10–12.
16. Lai, C.S., Guetzlay, J. and Hoseney, R.C. Effects of baking powder on extrudates. *Cereal Chem.* **66** (1989), 69–73.
17. Maga, J.A. Flavour formation and retention during extrusion. In *Extrusion Cooking*. Eds Mercier, C., Colonna, P. and Linko, P. (1990), American Association of Cereal Chemists, St. Paul, Minnesota, 387–396.

19 Sauces, pickles and condiments
C.J.B. BRIMELOW

19.1 Introduction

This chapter is concerned with acid preserves, their essential characteristics in physico-chemical terms and how these characteristics are achieved through processing. A wide range of products can be categorized as acid preserves, including thin and thick sauces, mayonnaises and salad dressings, clear pickles, sweet pickles, relishes, fruit chutneys and condiments. It would be impossible therefore to cover all the individual features of each of these product groups in one chapter. The important characteristics, however, relate to such aspects as distinctive vegetable or fruit texture, sauce consistency and rheology, product colour and appearance, emulsion stability and mouthfeel of oil-based sauces, and product preservation through the contribution of interactive factors such as acidity and water activity.

The science and technology of acid preserves has not been extensively covered in the literature, but the reader is referred to early books and reviews by Cruess [1], Desrosier [2], Binstead et al. [3], and Bhasin and Bhatia [4], and more recent reviews by Campbell-Platt and Anderson [5] and Anderson [6].

19.2 History

The preservation of vegetables and fruits in the form of sauces, pickles, relishes, condiments and chutneys began as a kitchen art, the origins of which are lost in antiquity. The earliest writings from many parts of the world indicate that vinegar, produced from fermentation of apple, grape or other fruits, was already well known many centuries BC and that it was widely used to preserve fruits, vegetables, fish and other foods. It is probable that addition of herbs, spices and aromatic fruits and vegetables to such preserves, in order to provide distinctive flavour and aromas, was also common early practice. Writings of the ancient Greeks and Romans, for example, explain the making in the kitchen of many seasonings, relishes and sauces. Mint vinegar was a favourite seasoning of the Romans.

A wider appreciation of the culinary art of pickles and sauces came with the opening up of the spice routes. Many recipes date from the times of the Crusades, when soldiers returning from the Middle East brought back with them exotic spices and herbs and a knowledge of their use. Interest was stimulated in growing and using indigenous spices and herbs. The 16th century saw the start of world sea trade. Spices in bulk from the East became regularly available in all European countries and pickle and sauce making in the kitchen and in cottage industries became well established. The origins of many of the products familiar in the modern marketplace date from these times. The early settlers in North America and Australia took with them from the Old World their recipes for pickles and sauces. More recently, a pickle and sauce industry, based often on traditional formulations, has grown up in many parts of the world.

19.3 Definitions

The common property of pickles, sauces, relishes, chutneys and condiments, is that they all contain acid as part of the preservation system. In most cases, the principal acid is acetic derived from vinegar, but there are a few exceptions such as lactic acid derived from lactic fermentation (olives, sauerkraut), citric acid from citrus fruits or malic acid from apples. Acidity may be the main contributor to preservation but a number of other factors may also be operating, such as the salt or sugar content, other dissolved solids, the water activity, the presence of preservatives (either natural or artificial), the thermal process given to the product and the storage temperature.

It used to be that all pickles and sauce products were self-preserving, in other words, the bottle or jar containing such a product could be opened and closed by the user without compromising the product safety. More recently consumer preferences for less acidity and less saltiness have led to products which are no longer self-preserving. Often such products have to be refrigerated after opening.

In the pickle and sauce industry, products are defined in terms of such parameters as the size of the vegetable or fruit piece, the appearance and viscosity of the surrounding liquor or sauce, and the relative amounts of key ingredients such as sugar, spices or acid.

Thick sauces. These consist of a mixture of comminuted vegetables, fruits and spices suspended in a thick medium of sugar, salt and vinegar. Flours, starches or vegetable gums may be added to give additional viscosity and mouthfeel to the product. A typical example is tomato ketchup.

Thin sauces. These are essentially vinegar extracts of a variety of flavouring and spice materials. Starches or other thickening agents are usually absent. A popular example of a thin sauce is Worcestershire sauce.

Mayonnaise. This is an emulsion of vegetable oil, egg yolk and vinegar with spices.

Salad dressing. This is a derivative of mayonnaise consisting essentially of mayonnaise mixed with a cooked, starch or gum thickened, spicy paste.

Clear pickles. These are whole or sliced vegetables or fruits, often fully fermented, and preserved in a clear vinegar, sometimes containing spices and sugar. Examples are pickled onions, olives, beetroot, cauliflower, capers (the flower buds of *Capparis spinosa* L.) and cucumbers.

Sweet pickles. These consist of large pieces of vegetables, often diced or sliced, suspended in a thick, sweetened, spicy sauce. Examples are Branston^{TM} pickle and piccalilli.

Relishes. These are similar to sweet pickles, but the vegetables or fruit pieces are smaller and may even be finely chopped. The pieces are often suspended in a clear, thickened, sweet vinegar or in a mayonnaise-style sauce. Examples are cucumber and red-pepper relish, and corn relish.

Chutneys. These are highly spiced sweet pickles, containing large fruit pieces more usually than vegetable pieces. The fruit pieces are suspended in a sweet sauce or syrup. Examples are mango chutney and green tomato chutney.

Condiment. This is a thick medium of finely comminuted vegetable or aromatic seasoning in vinegar; it can be utilized to provide aromatic flavour to meals by addition either at the cooking stage or on the plate. Examples are mustard condiment and black olive condiment.

19.4 Acid preservation and preservative indices

It has been mentioned that much of the contribution to the preservation of pickles and sauces comes from their organic acid content. An important feature of these acids, which influences their effectiveness in acting as antimicrobial agents, is the property of dissociation. Dissociation is the separation of a chemical compound into component parts capable of existing in equilibrium with the parent. For acetic acid, the dissociation equation can be represented as:

$$CH_3COOH \leftrightarrow CH_3COO^- + H^+ \qquad (19.1)$$

undissociated acetic acid · · · acetate ion · · · proton

The ratio of the concentrations of the undissociated form to the dissociated form of an organic acid is described by the dissociation constant, K_a. The pK_a values (where pK_a is the negative logarithm of K_a) of a number of organic acids commonly used to preserve sauces and pickles are given in Table 19.1.

Dissociation is affected slightly by temperature, but is profoundly affected by pH. As the pH decreases, the amount of undissociated acid increases. Thus at pH 3.0, >98% of acetic acid present is undissociated, at pH 4.76 (i.e. numerically equal to the pK_a) 50% is undissociated and at pH 7.0 only 0.6% is undissociated. This is an important factor because it has been shown that it is predominantly the undissociated form of the acid that provides the antimicrobial effect, possibly due to the greater ease with which an uncharged molecule can penetrate microbial cell membranes.

A number of empirical formulae have been evolved over the years to calculate the amount of acid required in a pickle or sauce formulation, in order to provide preservation. These formulae are known as preservation indices. Binstead et al. [3] report an index advanced by Macara and Morpeth:

$$\text{Acetic acid concentration} \geq 3.6V/100$$

where V = percentage of volatile constituents obtained by drying. This relation indicates that it is not the content of acid in the whole formulation which provides a reliable indication of stability, but rather that which is acting in the aqueous phase.

Another preservation index, due originally to Tuynenberg-Muys, has been proposed by the Comité des Industries des Mayonnaises et Sauces Condimentaires de la Communité Economique Européenne (CIMSCEE, [7]) for determining the microbiological stability of mayonnaises and similar products:

P.I. (spoilage):
15.75[(proportion of undissociated acetic acid)(% total acetic acid)] + 3.08(% salt) + (% hexose) + 0.5(% disaccharide) $\geq$ 63

Table 19.1 pK_a values of various organic acids used as part of the preservation system in sauces and pickles

Acid	pK_a at 25°C (77°F)
Acetic	4.76
Lactic	3.86
Citric	3.09
Malic	3.40
Tartaric	2.98
Benzoic	4.18
Sorbic	4.76
Propionic	4.85

P.I. (safety):

15.75[(proportion of undissociated acetic acid)(% total acetic acid]
+ 3.08(% salt) + (% hexose) + 0.5(% disaccharide)
+ 40(4.0 − pH) ⩾ 63

This index, in its two forms for spoilage and pathogenic micro-organisms, is of interest in that it attempts to incorporate the combined effects of undissociated acetic acid content, pH, sugar and salt contents.

It has been noted by Campbell-Platt and Anderson [5] that some caution should be exercised when applying these preservation indices. Certain micro-organisms, such as the acid tolerant *Monilliella acetoabutans* can survive at very low pH values outside the limiting values provided by the indices. Brocklehurst *et al.* [8] have noted some specific limitations of the CIMSCEE index relating to its pH range of usefulness and to situations where products contain low sugar and salt contents.

Before leaving the subject of prediction of microbial stability of pickles and sauces, mention must be made of recent work on stability modelling. Some of the models recently evolved can be applied to a wide range of heterogeneous foods including those, such as pickles and sauces, which rely on interactive preservative effects of different factors including pH, a_w (see below) and salt content. One such system of models has been developed in software form by a UK consortium including MAFF, Campden FDRA and Leatherhead Food RA [9].

19.5 Vegetable/fruit treatments for sauces and pickles

Manufacture of acid preserves may sometimes involve the use of fresh fruits or vegetables delivered directly to the manufacturing site, but it is more common to utilize raw material that has been pre-processed into brined, sugared, fermented or concentrated forms. In each case, one of the essential requirements of the pre-process is to preserve or enhance some of the original unique properties of the fruit or vegetable, such as colour, texture, structure and flavour.

19.5.1 Fermentation brining

Brining is the preservation in sodium chloride solution of raw vegetables prior to their use in pickle or sauce formulations. Brining is normally carried out at one or other of two salt concentration ranges (6–15% or >18%). If salt solutions giving between 6% and 15% equilibrium salt content overall are used, then this permits fermentation to occur by certain desirable natural bacteria during the brining process. Homofermentative bacteria convert the sugars present in the vegetable or fruit almost entirely to lactic acid; heterofermentative bacteria produce lactic acid, carbon dioxide, alcohol and acetic acid. In either case, the acid produced provides extra preservation over and above that given by the salt.

Fleming [10] has reviewed in some detail the different types of fermentative bacteria that can be present during brining fermentation of vegetables. Four stages of fermentation are described: initiation, primary fermentation, secondary fermentation and post-fermentation. During the initiation and primary fermentation stages, the predominant species are the heterofermentative types, such as *Leucanostoc mesenteroides* and *Lactobacillus brevis*, and the homofermentative types *Pediococcus cerevisiae* and *Lactobacillus plantarum*. In the secondary fermentation stage, the acid-tolerant fermentative yeasts, which become established during the primary stage, take over and grow until the fermentatable carbohydrates are exhausted. During post-fermentation, microbial growth is restricted to those oxidative yeasts, moulds and spoilage bacteria which can grow at brine surfaces exposed to air. It is normal to seal brining tanks and casks, or achieve anaerobicity in some other way after the second-stage fermentation, to prevent this surface growth, thereby avoiding spoilage and off-flavour problems.

Fleming [10] indicated that a number of chemical and physical factors influence the rate and extent of growth of these various micro-organisms, including salt concentration, acidity, temperature, natural inhibitory compounds present in the vegetables, chemical additives, presence of air and sunlight, fermentatable carbohydrate in the vegetable and availability of nutrients in the brine.

Anderson [6,11] has described the essential requirements for a traditional fermentation brining: sound and unblemished raw material, suitable vessels in which the vegetables can go through a 'shrinking' stage and then the final fermentation and storage stage, a salt solution which will result in a final concentration of 6–15%, prompt brining when the vegetables arrive at the brining station, careful control of weights of vegetables and brine, placing part of the brine ahead of the vegetables in the vessels to prevent clumps of vegetable not being penetrated by brine, suitable intermittent mixing to prevent salt stratification, careful analytical and visual quality checking during fermentation and storage, topping up with brine when necessary and airtight storage after completion of fermentation. The sequence of process steps is shown in Figure 19.1 (note that in this and others depicting typical manufacturing processes for pickles and sauces, optional steps are represented by boxes with dashed outlines). Traditionally, brined fermented vegetables have been stored in 40-gallon (145-litre) wooden casks or plastic drums, but more recently bulk storage in large fibreglass or plastic composite containers, fitted with air-locks, has been practised. These containers may hold

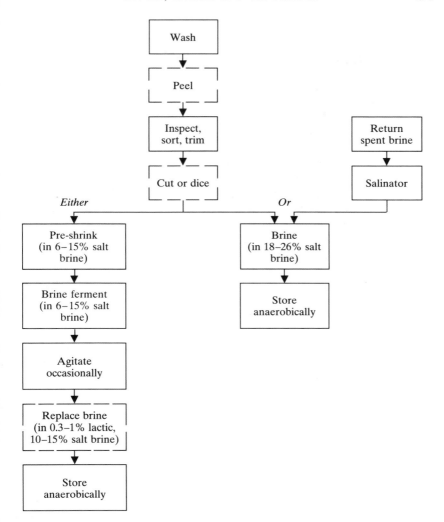

Figure 19.1 Sequence of process operations for fermentation brining, or brining without fermentation, of vegetables and fruits.

10 000–50 000 gallons (45 000–225 000 litres) of brine and vegetables. A typical installation is shown in Figure 19.2.

Recent progress in fermentation brining of vegetables has involved control of the fermentation process in such a way as to favour growth of an added culture rather than that of the naturally occurring microorganisms. One method, proposed by Etchells *et al.* [12,13] originally involves washing the vegetables, adding an acidified brine at a pH below

Figure 19.2 Bulk storage containers for brined vegetables.

3.0, culturing, purging the carbon dioxide gas with nitrogen, adding further salt and finally raising the pH. In other methods, calcium salts are added to the brines to improve vegetable firmness [14,15].

Sodium chloride has one other function in fermentation brining of vegetables, in addition to preservation, and this is to produce desirable sensory quality, particularly flavour and texture. Texture changes during brining are caused by plasmolysis (contraction of the protoplasm) and rearrangement of cell-wall materials, such as pectins and hemicelluloses (chapter 10 and [16]). Water is expressed during the brining process as the internal turgor pressure collapses and internal pressures come to equilibrium with the vapour pressure of the brine solution. This expression accounts for the dramatic shrinkage of vegetables in the early stages of brining (up to 30–40%); it is therefore normal practice in batch brining systems to transfer the vegetables after two or three days from shrinkage vessels to final storage vessels, in order not to waste valuable storage volume.

The sequence of events at the cellular level, and the resulting impact on texture, has been reviewed by Edgell and Kilcast [17]. At commencement of brining, plasmolysis starts to occur and the texture gradually changes from the brittle state characteristic of fresh vegetables. After three or four days, texture is recovered somewhat due to stabilization of turgor pressure. At five or six days, the protoplast is ruptured and turgor pressure is lost. This results in decreased hardness. The final texture of a pickled vegetable, such as cauliflower for example, has been shown by Saxton et al. [18] to be provided entirely by a network of modified and ruptured cell walls, with no contribution coming from turgor pressure.

Flavour is another characteristic which is influenced during fermentation brining. The actions of the dominant micro-organisms impart typical desirable flavour to the vegetable host. The concentration of volatile acid is important to the flavour of sauerkraut [19], whilst ethanol and various aldehydes may influence the flavour of cucumbers [20]. When the fermentation process is improperly managed, microorganisms can also give rise to problems. Certain moulds, naturally present on the vegetable or fruit, or remaining on vessel surfaces due to poor hygiene, can produce pectolytic enzymes, such as polygalacturonase. If these moulds are permitted to grow during the fermentation, then the enzymes attack the vegetable tissue and cause softening. Certain gas-forming bacteria can grow under the skin of vegetables if the salt concentration is too low, causing white spots in olives [21] or bloating of cucumbers [22,23]. Other quality problems include pink/brown discolorations in vegetables, caused by pink yeasts, bleaching caused by exposure to sunlight and off-flavours caused by growth of various undesirable micro-organisms.

Scholey [24] has described various parameters desired in pickles by consumers. The feel of the outer surface, the crispness and elasticity between the teeth and the crunch during chewing are all important. The flavour released during chewing is also important. Most of these characteristics are influenced by the physico-chemical changes taking place during fermentation and brining.

19.5.2 Brining without fermentation

The other salt concentration used to preserve vegetables is 18% or above. At this concentration growth of most of the naturally occurring micro-organisms is inhibited, fermentation does not occur and there is no conversion of the natural sugars to acid and alcohol. The inhibition is largely due to the reduced water activity of the highly concentrated brine solution and the influence of this reduced water activity on microbial growth. Water activity (a_w) is defined as the ratio of the vapour pressure (P) above a solution or a material containing a water phase, to that of pure water (P_0), at a given temperature (see also chapter 1):

$$a_w = P/P_0. \tag{19.2}$$

This value is related to the amount of water available for microbiological growth and for transport mechanisms in various biological reactions. The lower the value compared to unity (the water activity of pure water), the lower is the amount of available water. Different micro-organisms exhibit different abilities to survive at reduced water activities. Most bacterial species do not grow below a_w 0.91, yeasts below a_w 0.88 and moulds below a_w 0.80. Some osmophilic yeasts and halophilic bacteria can grow down to much lower water activities, but these strains are rare. The general effect of increased solute concentration in water is to reduce its water activity. Thus a brine solution containing 18% salt has a water activity of approximately 0.88 at 25°C (77°F), whilst a saturated salt solution (26.4% w/w) has a water activity of approximately 0.75. It follows that above 18% salt concentration the potential for growth of most species of bacteria, yeasts and moulds is minimal, whether this growth would be desirable or undesirable.

Brining of vegetables or fruits to provide a salt content >18% can be achieved by the batchwise in-cask or in-tank method described earlier for preparation of fermented materials. Brimelow and Brittain [25,26], however, describe a more convenient continuous plant for vegetable brining, using a tower system fed by fresh vegetable, and brine from an automatic saturated brine-making installation. Brined vegetable at 18% salt content or above is continuously transferred to bulk storage tanks equipped with float devices to exclude oxygen. The process steps are illustrated in Figure 19.1.

19.5.3 Sugaring/syruping

Sugaring or syruping is sometimes used as a technique for preparation of fruits or vegetables for inclusion in sweet pickles, chutneys or sweet clear pickles. Syruping, like brining, is a method of reducing water activity and thus improving microbiological stability. Syruping is one method under the general heading of osmotic preservation (i.e. the transfer of moisture across a membrane by placing a more concentrated solution on the outside whereby moisture will naturally migrate from the lower to the higher concentration). It has been stated [27] that osmotic dehydration offers some advantages compared to the other methods of reducing water content of fruits and vegetables: heat damage, resulting in colour and flavour changes, is greatly minimized; enzymatic oxidative browning effects are reduced; harsh acidic flavour notes are removed during the syruping process.

Pre-syruping of fruits and vegetables before addition to a sweet sauce prevents floating of the pieces in the sauce and also prevents diffusion of water from the pieces, which would cause localized dilution of the sauce [6]. Syruping is usually carried out using multi-batchwise systems in which sugar content is gradually increased. However, the continuous plant described by Brimelow and Brittain [25,26] for brining can also be utilized for syruping. Examples of syruped fruits or vegetables used in the production of acid preserves include mangoes, citrus fruits, tamarinds (the fruit of the tropical tree *Tamarindus indica*) and apples.

19.5.4 Concentration

Concentration is a method which can be utilized for pre-processing a number of fruits and vegetables, but principally tomato, prior to their use in various sauces. As well as tomato, concentration is sometimes used for citrus fruits, apple, mango and squash.

Tomato paste and purée are important ingredients for pickle and sauces. Its primary use is in tomato ketchup but it is also used in salsas, pizza and pasta sauces, sweet pickle sauces, tomato chutney sauce and other thick sauces. Functionally it provides flavour, colour and consistency. The manufacture and technology of tomato paste and purée has been described in a number of reviews and books including those by Binsted *et al.* [3], Goose and Binsted [28] and Gould [29]. The essential steps of the manufacturing process are shown in sequence in Figure 19.3. These steps are designed to minimize, as far as possible, deterioration of the colour, flavour and consistency provided by the original fruit. Tomato paste plants are run 24 hours a day during the harvesting season, in order to receive and utilize ripe fruit at optimum and consistent quality. Different varieties are blended to provide consistency of final product. Removal of mouldy and damaged tomatoes during the inspection and sorting stage is performed to reduce carryover of moulds and mouldy off-flavour to the finished paste. Seeds are removed at the pulping and screening stages to improve flavour and colour. Controlled heating (breaking) is carried out to inactivate pectolytic enzymes, which would otherwise reduce viscosity, but without significantly affecting colour and flavour. Screening reduces the pulp size and improves consistency. Vacuum concentration is then performed in order to reduce concentration temperatures and thus reduce heat damage. Finally, after pasteurization, paste is nowadays often cooled and aseptically filled into bulk bags rather than hot filled into cans.

Colour is a key attribute of tomato paste, because it gives an indication not only of raw tomato quality but also of the efficiency of the tomato paste manufacturing system. Colour value is therefore one of the primary purchasing criteria for tomato paste. The characteristic red colour of tomato is due to a combination of carotenoid pigments, of which the most abundant is the carmine red lycopene, comprising about 83% of the pigments present. The balance may be composed of α-, β-, γ- and δ-carotenes and xanthophylls. Various workers [29–31] have reported the influence of ripening effects and climatic conditions on pigment formation. Noble [32] has investigated the changes in colour during heat concentration. Lycopene rapidly oxidizes in air and fades in colour, this reaction

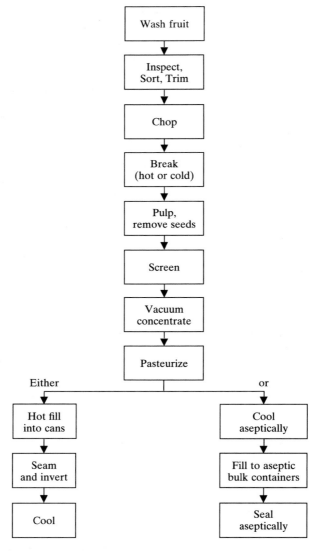

Figure 19.3 Sequence of process operations in the manufacture of tomato paste.

being accelerated by heat. Under prolonged heating, lycopene turns brown.

In view of the importance of colour in relation to tomato paste quality, it is not surprising that numerous methods have been developed for its measurement. These are based either on comparison to physical colour standards, such as in the Munsell or Lovibond systems, or on instrumental techniques. The most popular instrumental technique utilizes tristimulus colorimetry [33].

Another key attribute of tomato paste is its consistency and texture. A number of intrinsic (compositional) and extrinsic (processing) factors contribute to tomato paste and purée consistency. These include pectin content, other cell-wall components such as cellulose and hemicellulose, insoluble solids particularly metal cations, particle size, pH of the final product, break temperature, and concentration and pasteurization conditions. Pectin content is influenced by the maturity and variety of the tomatoes, and also by the method of breaking and pulping. In the hot-break system where the tomatoes are crushed with the minimum inclusion of air and quickly heated to >90°C (>194°F), the high temperatures inactivate the enzymes which would normally catalyse reactions involving degradation of the pectin chains. In the cold-break procedure, tomatoes are crushed at temperatures normally <65°C (<150°F) and then are held for a significant time in a holding tank. During this time, enzymes liberated in the crushing process, such as polygalacturonase and pectinesterase catalyse the break down of the pectin. The amount of breakdown is dependent on the hold time. The insoluble solids content of tomatoes is a function of the variety, the growing conditions and the maturity. Cations, such as calcium, influence consistency because of their ability to form cross-links between the pectin polymer chains. This pectin cross-linked structure can further be affected by the size of the finisher screen, by any milling or homogenization that is carried out and by the temperature and time during concentration and pasteurization. A slightly larger screen size tends to enhance consistency, as does faster pulper speeds or the use of milling. Prolonged heating in the concentrator or the pasteurizer causes a loss in the viscosity. The consistency of tomato pastes and purées can be measured by a number of methods including the Bostwick Consistometer, the Adams Consistometer, various penetrometric techniques and the Brookfield viscometer.

19.5.5 Canning and freezing

Canned or frozen fruits and vegetables are not often used in the manufacture of sauces, pickles and condiments. This is because the significant added cost due to the canning or freezing process in most cases prohibits the use of materials supplied in these forms. Exceptions are canned or frozen items which are added for the specific purpose of providing a certain piquancy, colour or exotic nature to a product. Some examples are canned or frozen chillies added to provide 'hotness', canned fruit juices to provide flavour, canned or frozen mushrooms to provide flavour and colour, and canned beetroots which provide colour and flavour.

The thermal processes involved in canning cause profound changes in the texture, flavour and colour of fruits and vegetables. These changes and their kinetics have been discussed in a number of reviews and books [34–37]. During the initial stages of a heating process, vegetables become firmer due to enzyme activity, but as the process proceeds, softening occurs. Colour and flavour changes also take place. Acidification of the raw material can permit a reduction of the thermal process, whilst still maintaining safety requirements. This means that acidified canned vegetables and fruits have a better texture than unacidified ones. Pre-treatment in solutions containing divalent cations also improves canned vegetable texture, but use of monovalent cations in pre-processes causes tissue softening.

Freezing also causes textural changes in fruits and vegetables [37]. Again these changes can be minimized by pre-blanching in solutions containing divalent cations.

19.5.6 Chilling and controlled atmospheres

It has been mentioned that fresh fruits and vegetables are sometimes utilized in pickle and sauce manufacture. In these cases, the raw material can either be delivered direct from the farm or can be supplied from bulk storage facilities, where they have been held under chilled temperatures in air or under controlled atmospheres.

Bulk storage methods have recently been reviewed by Gorini *et al.* [39]. Storage under controlled atmospheres (CA) has been gaining in popularity over recent years, the advantage being extension of the shelf-life of acceptable quality by some months over that provided by conventional storage methods (see chapter 14). CA storage of produce is a technology relying on the combined effects of gas atmosphere modification and chilled temperatures. Gas atmospheres are usually modified in the direction of reduced CO_2, reduced O_2 and increased N_2. CO_2 is reduced using scrubber systems based on, for example, sodium hydroxide, water, active charcoal or selective membranes. O_2 is reduced by flushing with N_2, or by catalytic combustion methods. Sometimes a high CO_2 pretreatment is used to improve the microbiological state of the fruit and vegetable, and to achieve a rapid reduction in respiration rate and ethylene generation rate. The objective of CA is to extend storage life by decreasing respiration, increasing firmness and turgidity, delaying the appearance of senescence symptoms caused by the effects of ethylene on the metabolism, limiting colour degradation and other physiological changes, reducing nutritional losses and reducing spoilage. CA has been shown to be beneficial for a number of vegetables and fruits used in pickle and sauce production, for example tomato, cauliflower, cucumber, onion, garlic, cabbage and apple. It has little effect on carrot, pepper, beet and rutabaga.

Because reduction of temperature also slows down respiration, ethylene synthesis and various other degredative reactions, CA is usually employed in combination with chilled storage temperatures ($\leq 4°C$, $\leq 40°F$). One of the problems associated with low temperature storage is chilling injury (CI), also known as cold shock. CI has been reviewed by Wang [40] and by Jackman *et al.* [41]. CI may take the form of discoloration, surface changes such as pitting, or textural changes such as limpness or rubberiness. Many vegetable or fruit crops of tropical or subtropical origin, and some from temperate zone origins, are susceptible to cold shock. CI symptoms can be alleviated by application of certain pretreatments prior to chilling, for example chemical treatments such as SO_2, calcium dips and certain CA storage conditions.

19.6 Pickle and sauce processing

Processing methods for pickles and sauces can conveniently be grouped by product type, as follows: methods for sweet pickles, relishes and chutneys, methods for clear pickles, methods for thick sauces, thin sauces and condiments, and methods for mayonnaises and salad dressings. In each case the processing methods are designed to provide products of attractive final appearance, unique often spicy flavour profiles on an acidic base note, and characteristic mouthfeel and texture. When vegetable or fruit particulates are present, methods are designed to preserve, as much as possible, the original texture, flavour and colour properties. Finally, the processing methods must provide products which are shelf stable before opening, and which are either self-preserving or can be kept in a refrigerator with no safety risks, after opening.

19.6.1 Processing of sweet pickles, relishes and chutneys

The common feature of these three sorts of acid preserves is that they all contain vegetables or fruit pieces in a sweet spicy sauce or syrup. The essential processing steps relate therefore to the vegetable or fruit preparation, including debrining where necessary, and to the preparation of the sauce or syrup component. These steps are shown in Figure 19.4.

The first step, particularly for fresh delivered raw produce, is inspection and sorting to remove damaged or blemished material. Large size vegetables and fruits are then cut or diced to the required dimensions. In some installations, vegetable or fruit pieces are colour sorted in automatic machines to separate those pieces which are blemished or outside the colour limits required [42]. Brined vegetables or fruits are then 'freshened', i.e. the salt content is reduced to a level acceptable to the consumer, usually <5%. The freshening is performed by washing the salt out using

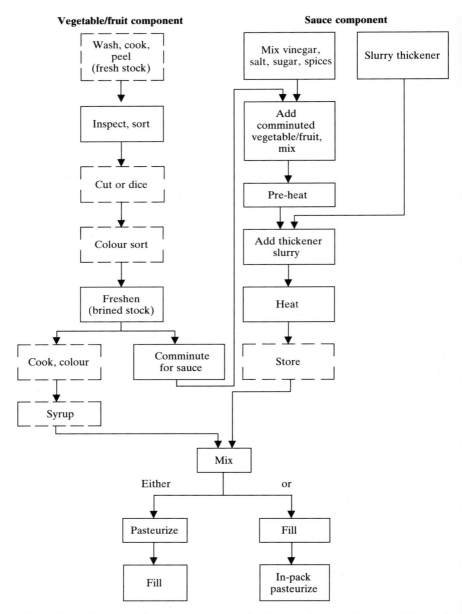

Figure 19.4 Sequence of process operations in the manufacture of sweet pickles and relishes.

potable water or syrup solution. Three types of methods are available: continuous, multi-batchwise or single-batchwise methods. In continuous methods, the freshening solution is circulated through the vegetable mass until the target salt level is reached; in multi-batchwise methods, a number of volumes of the freshening solution are added to and then drained from the freshening vessel; in single-batchwise methods, one volume of freshening solution is utilized.

A number of factors have an impact on the rate of removal of salt during the debrining stage. These factors include the type and size of vegetable, the temperature, and the differences between solute concentrations or water activities when comparing the salt solution in the outer regions of the vegetable pieces with the freshening liquid surrounding these pieces. Concentration gradients can in turn be influenced by the ratio of vegetable to freshening solution, the flow rate of freshening solution and whether or not agitation is being employed. When agitation is absent or slow flow rates are used, stratification of the salt coming from the vegetables can be a problem. The salt sinks to the bottom of the freshening vessel and vegetable pieces in this region therefore debrine more slowly.

After the freshening step, the vegetable or fruit pieces are ready for mixing with the sauce or syrup. Since the method of manufacture of the sauce component is the same as that utilized for thick bottling sauces, this will be discussed in section 19.6.3 below. It is usual to try and arrange for the sauce component and the solids component(s) to be of approximately the same specific gravities, in order to prevent gravitational separation effects. Stokes' law has been utilized to give a theoretical explanation of the factors influencing separation:

$$V = 2/9(D - d)gr^2/\eta \quad (19.3)$$

where V is the velocity of a falling spherical particle of radius r, D is its density, d is the density of the suspending fluid, g is the gravitational constant, and η is the viscosity of the fluid). It follows from this law that if $(D - d)$ approaches zero, or the particle size is very small and the sauce viscosity is very high, then separation is minimized.

After mixing the various product components, the assembled product is usually pasteurized, either in bulk (out-of-pack pasteurization = OPP) or in the final container (in-pack pasteurization = IPP). Pasteurization is a controlled, but limited heat process, performed in order to destroy spoilage micro-organisms, inactivate enzymes of microbiological or ingredient origin and remove entrapped air from the product [5]. Generally, the yeasts and moulds, capable of surviving and growing in the acidic reduced water activity environment associated with pickles and sauces, are destroyed during any process designed to inactivate enzymes. Processes are therefore based on the conditions required to inactivate these enzymes. Inactivation of enzymes, such as peroxidase and polygalacturonase, is advisable in order to prevent deteriorative reactions such as darkening and softening.

For a given heat-processing system, whether this occurs before or after the filling stage, the heating and holding conditions of the slowest heating particle must be known in order to calculate the process value. These conditions depend on the nature of the pasteurizing equipment and the temperature distribution within it, the composition of the product and the sauce to solids ratio, and, for in-pack pasteurization, the size and geometry of the containers. Process values can then be calculated by the method suggested by Shapton *et al.* [43]. For most purposes, a minimum process equivalent to a P_{70} value of 10 (10 minutes at 70°C, 158°F) is sufficient for the slowest heating particle [5]. This type of process will take care of enzymes and spoilage micro-organisms.

19.6.2 Processing of clear pickles

Clear pickles are whole, sliced or cut vegetables or fruits, preserved in a clear acidified liquid. The liquid sometimes contains spices, sugar and colouring. Examples of such products are pickled onions (silverskin, cocktail or brown), mixed pickles (usually cauliflower, gherkins and onions in approximately 50:25:25 proportions), sweet mixed pickles (the same vegetables in a high sugar liquor), pickled gherkins (whole or rings), beetroot (baby or sliced), olives (these are often stuffed with pimento), capers and red cabbage.

The manufacturing process for the vegetable or fruit component is the same as that shown in Figure 19.4, for sweet pickles and relishes. Fresh delivered vegetables or fruits are washed and if necessary cooked and peeled. The vegetable or fruit is cut or diced if applicable and brined stock is freshened. In certain cases, when cooking has not been performed earlier, it may be carried out as part of the freshening stage. Colour, for example yellow turmeric, may be added at this point also. For sweet clear pickles, the vegetable or fruit may then be syruped.

The solid component is filled into jars and covered by the clear vinegar based liquid. In certain cases, such as capers, brown onions and gherkins, caramel can be added to the vinegar to provide brown colour. The vinegar component is often spiced with, for example, ginger, pimento, black pepper, chilli, clove or coriander. If the product is not of the high-acid, self-preserving type, it may then be pasteurized in the jar.

Various types of physico-chemical changes can impact on the quality of clear pickles. These include microbiological growth due to too low an acidity or insufficient heat processing, deterioration of appearance and texture due to enzymatic actions, and deterioration of colour and flavour

19.6.3 Processing of thick and thin sauces

Thick sauces consist essentially of finely divided fruit and vegetable particles suspended in a thickened, spicy, slightly sweet, acidic medium. The most popular thick sauce is tomato ketchup, but others such as brown sauce, mushroom ketchup and mustard sauce are also manufactured in significant volumes. Thick sauces are also used to suspend the vegetable and fruit pieces in sweet pickles.

Tomato ketchup manufacture has been described in detail by Gould [29], Goose and Binsted [28] and Campbell-Platt and Anderson [5]. The process is designed such that the product spends as short a time as possible at high temperatures to minimize heat damage particularly to the red colour, and is deaerated to minimize oxygen induced deterioration reactions. The process steps common to most systems are represented in Figure 19.5. The tomato paste, vinegar, sugar and spices are mixed and pre-heated. The thickening system, if one is needed, is dispersed in vinegar and is then added to the pre-mix. The product mix is heated further, preferably under nitrogen. It may then be homogenized or colloid milled and deaerated. After deaeration, the ketchup is held for a short time and is then filled into pre-heated bottles. The bottles are inverted to sterilize the headspace and the inner surface of the cap. Some modern filling operations are performed under semi-aseptic conditions and bottle inversion is not then necessary. Finally the bottles are cooled rapidly.

Some of the problems encountered with poorly designed manufacturing systems include spoilage, rapid colour and flavour deterioration, browning, plugging in the neck of the bottle, syneresis (water separation by a contraction of the gel structure), particularly at air bubbles and at the product surface, and viscosity loss.

Microbiological spoilage is associated with inadequate temperature–time conditions in the pasteurization process and/or low acidity. It is normal practice to attain a temperature of at least 88°C (190°F) in the hold system prior to filling to prevent this. Colour deterioration is largely due to classic Maillard browning effects, involving reactions between reducing sugars and amino acids in the presence of oxygen. Binstead *et al.* [3] reviewed work on 'blackneck', the formation of a brown ring at the surface of tomato ketchup. It is suggested that this may be due either to Maillard browning or to a reaction between ferric iron and tannins naturally occurring in the product formulation. Oxygen is necessary to oxidize the iron to the ferric state. Deaeration and vacuum filling help to reduce the oxygen component essential to these types of browning effects.

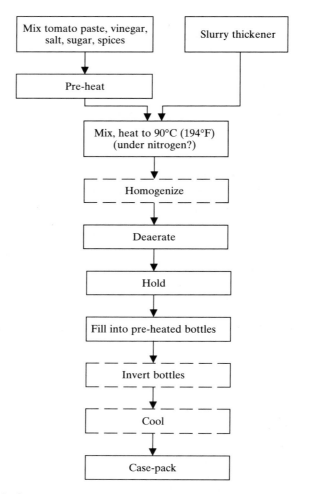

Figure 19.5 Sequence of process operations in the manufacture of tomato ketchup.

Syneresis and plugging are problems associated with time-based reorganizational changes involving the natural pectin structure. Gum or starch thickened products may also be subject to syneresis and plugging. These are due to structural changes in the thickening systems, such as retrogradation, and also to hydrolysis effects because of the high temperature and acid conditions of the process.

The viscosity and consistency of tomato ketchup is provided by the pectin content of the original tomato paste and by the gum or starch thickening system, if this is present. A homogenization or colloid milling stage is sometimes employed in tomato ketchup manufacture to provide additional viscosity, by reducing fibre particle size and 'restructuring' the pectin network. Ketchups and thick sauces are non-Newtonian fluids, that is they have a high yield value and their apparent viscosity is inversely related to the rate of applied shear (see also chapter 5). Particle size reduction, by homogenization or milling, tends to increase the effective viscosity. Viscosity may, however, eventually be reduced with storage time by gradual acid hydrolysis of the thickening system.

Other types of thick sauces such as brown sauce, barbecue sauce or mushroom ketchup, are usually not as susceptible to colour and flavour reactions as tomato ketchup. They are manufactured without the necessity to vacuum deaerate or to heat under nitrogen. The sequence of manufacturing steps is as depicted in Figure 19.4 for the sauce component of sweet pickles and relishes. Condiments can be produced by the same method.

Finely comminuted fruit or vegetable, usually tomato, apple, dates or onion is added to a mix of vinegar, salt, sugar and spices. The mix is heated and a slurry of the thickener in cold water is added. The whole mix is heated further with stirring and stored either for bottling, or for mixing with a vegetable or fruit component to make a sweet pickle or relish.

One of the important constituents of the sauce formulation is the thickener, which can be a starch, flour or gum. Campbell-Platt and Anderson [5] have described the essential characteristics of a thickener for use in bottling sauces or sweet pickle liquors:

- It must provide the required viscosity (as mentioned in section 19.6.1). For sweet pickle liquors this viscosity assists in preventing settlement of the solid component.
- It should be stable to heating in the presence of acetic acid.
- It should have some tolerance to variations in processing conditions.
- It should preferably yield opaque rather than clear dispersions (though this is not the case for tomato ketchup (where it is necessary for the tomato red to be seen through the thickener), salad cream, and some clear relish liquors).
- It should be stable towards storage under acid conditions.
- Either it should not give a gel structure, or if it forms one, then the gel should be capable of being destroyed by mechanical means and should not then reform.

Many natural starches possess some, but not all, of these desirable properties. Cornflower is often used for sweet pickles and bottling sauces; however, it does form a gel structure which has to be mechanically destroyed. Other starches or flours which are sometimes utilized include wheat flour, potato starch, waxy maize starches and rice flour. The last is particularly useful in high acid products because of its good acid stability.

Chemically or physically modified starches are the usual choice, however, since they provide acid tolerance, heat tolerance and storage stability. Traditionally, gums such as tragacanth, guar and carob bean were used in many pickle and sauce formulations, but these are expensive and have largely been superseded by modified starches. Some of the more recently developed thickeners such as CMC (sodium carboxymethyl cellulose), alginate and xanthan gum have found applications in certain formulations.

Thin sauces are very different in composition and manufacture to thick sauces. They are essentially highly flavoured vinegars, with no or very little added stabilizers, containing a certain proportion of finely divided solids. The most popular thin sauce is Worcestershire sauce, but others include mint vinegar and anchovy sauce. Worcestershire sauce usually contains tamarinds, anchovies, onions, fruit pulp, garlic, molasses, vinegar, and hot spices such as pepper, capsicum and ginger. It is manufactured by pre-cooking the tamarinds in water, sieving the softened fruit into the remainder of the ingredients, heating and simmering the product mix for some time, sieving and milling it and finally storing it for some months to mature. The matured product is skimmed to remove floating particles of film, and then filled. The product has to be stirred continuously during filling to ensure an even distribution of the sediment. Very few problems have been reported with thin sauces. Provided the preservation index according to Binstead et al. [3] is >3.6% the products are usually microbiologically stable. Since many spices have antimicrobial properties associated with their essential oil component, the high spice content of thin sauces provides additional microbial stability.

19.6.4 Mayonnaises, salad creams, dressings and spreads

Of all the product groups in the general category of acid preserves, the group encompassing salad creams, salad dressings, salad spreads and mayonnaises has probably been the most widely researched. This is presumably because of the complexity and number of the physico-chemical reactions that have to take place in order to provide the typical quality characteristics of the products. These characteristics include emulsion properties and stability, preservation properties and acidity, typical mustard and acid flavour, colour, viscosity and mouthfeel, and, where applicable, texture, flavour and colour of included vegetable or fruit pieces. Manufacturing methods for this group of products have been reviewed by Binstead et al. [3] and Campbell-Platt and Anderson [5], whilst the various factors influencing quality and shelf-life have been reviewed by Harrison and Cunningham [44] and by Mistry and Min [45]. (Some details concerning the manufacture and rheology of mayonnaise are given in chapter 5, section 5.5.)

Mayonnaises and salad creams are oil-in-water emulsions containing vegetable oil, vinegar and salt, emulsified by egg components (usually egg yolk) and flavoured with spices (usually mustard). Lemon juice can sometimes also be used as flavouring. The water phase in products of lower oil content, below about 50% oil, is thickened using gums or starches. The process steps for manufacture of mayonnaises and salad sauces are shown in Figure 19.6.

The essential spice component of many mayonnaises and salad products is mustard powder obtained by grinding white mustard, *Brassica alba*, black mustard, *Brassica nigra*, or brown mustard, *Brassica juncea*. The first step of the process is to 'activate' the mustard, that is to allow the enzymatic production of the pungent mustard flavour by wetting the powder in warm water at 35–50°C (95–122°F) and maintaining the slurry at this temperature for some minutes. Activation involves the enzyme myrosin (a glucosinolase), which cleaves the glycosides sinalbin and sinigrin present in white mustard and brown or black mustard respectively, to yield either *p*-hydroxy benzyl isothiocyanate or allyl isothiocyanate, by a Lossen rearrangement:

$$R-C\begin{matrix}S-C_6H_{11}O_5\\ \\NOSO_3X\end{matrix} \xrightarrow[\text{glucosinolase}]{H_2O} \underset{\text{isothiocyanate}}{RNCS} + C_6H_{12}O_6 + XHSO_4 \quad (19.4)$$

glycoside

Where R = HO—⟨◯⟩—CH_2 - for white mustard

and R = CH_2=CH=CH_2 - for black mustard

These two compounds have been identified as the key components of the flavours of the two basic types of mustard: *p*-hydroxybenzyl isothiocyanate has sharp taste but virtually no aroma, whereas allyl isothiocyanate is more volatile and has a pungent aroma as well as a sharp taste. Over 95% of the isothiocyanate-flavour compound is released in the first 10 minutes of activation at 40°C (104°F) under neutral pH conditions. If vinegar is added or the temperature is not in the range quoted, then the conditions are no longer optimal and less flavour is developed; activation under non-optimal conditions of reduced pH or elevated temperatures therefore provides a means by which flavour levels can be controlled. Other constituents which contribute to mustard flavour are low molecular weight sulphide compounds generated from hydrolysis of sulphur-containing amino acids. This reaction occurs particularly during later steps in the salad product manufacturing process, when the mustard is heated in the presence of vinegar.

Traditional mayonnaises derive all their viscosity from the oil-in-water emulsion system, but for lower oil-content formulations it is necessary to

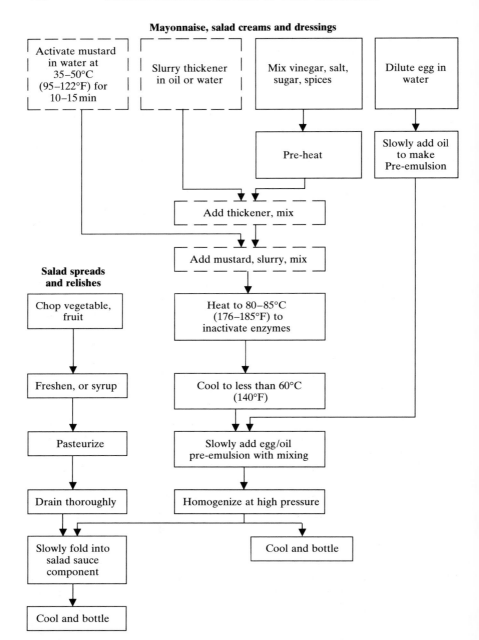

Figure 19.6 Sequence of process operations in the manufacture of mayonnaises, salad creams, dressings and spreads.

have some thickener present to provide additional consistency. In these cases, an early stage in the process is to slurry the thickener in water or oil, and then to add the slurry to a mix of vinegar, salt and sugars. The thickener, which can be a gum, flour or starch, must be adequately stable to heat and to reduced pH. The vinegar–thickener mix is then heated to 80–85°C (176–185°F) and the activated mustard slurry is added. The high temperatures at this stage serve to stop the mustard activation and also to inactivate other enzymes in the mustard and thickener system. If these enzymes are not inactivated they can catalyse the breakdown of the viscoelastic structure provided by the thickener in the continuous water phase. This in turn could permit separation of the emulsion.

After the heating stage, the mix is cooled to <60°C (<140°F) and the egg–oil pre-emulsion is added. The pre-emulsion is prepared firstly by diluting the whole egg or egg yolk with some water and then slowly adding oil with sufficient mixing to provide a coarse oil-in-water emulsion. The pre-emulsion is then added slowly to the cooled vinegar mix, again with sufficient mixing to maintain a coarse emulsion structure. The temperatures must be <60°C (<140°F) when adding the pre-emulsion to the vinegar mix, in order to prevent denaturation of the egg protein. The egg protein is one of the primary factors ensuring stability of the final emulsion. The whole mix is passed through a high-pressure homogenizer to form the final emulsion. It can be cooled further for storage prior to bottling.

If a product containing vegetable or fruit pieces is required, the pieces are added to the mayonnaise or salad cream base after it has been homogenized. The pieces are usually debrined, chopped and then treated with a hot acidified syrup to effect a pasteurization. They must be thoroughly drained before being slowly folded into the salad base. Examples of this type of product are sauce tartare, horseradish sauce and various sandwich spreads.

As has been mentioned, a number of factors are important to the quality and storage stability of mayonnaises and other salad products. These include emulsion properties, flavour and colour, microbial stability, viscosity, and, when relevant, sensory properties of the vegetable or fruit inclusions.

A review of the emulsion properties of dressings and sauces has been given by Holcomb *et al.* [46]. The emulsion properties are, of course, an important feature of the mouthfeel; they also relate to the storage stability of the products. Emulsion properties are dependent on the characteristics of the continuous phase, the size distribution of the dispersed phase and the proportions of continuous phase and dispersed phase. The important characteristics of the continuous phase relate to the viscosity and to the presence of emulsifiers or surface-active agents at the oil–water interface. Increasing the viscosity of the continuous phase enhances the emulsion stability by delaying creaming and coalescence. Creaming, or flocculation, involves the movement of the lighter oil droplets towards the surface of the product (cf. Stokes' law and chapter 3).

Coalescence is the merging of droplets to form larger droplets; it leads ultimately to the complete breakdown of the emulsion to give its original two continuous phases. The presence of emulsifiers in the aqueous phase of the various salad products is essential to the maintenance of a stable oil-in-water emulsion after its formation. Emulsifiers act by gathering at the oil–water interface and reducing the interfacial tension at this surface. They can also exhibit steric forces (forces due to their three-dimensional shape) of repulsion when droplets come into close proximity. The important emulsifier in salad products is egg yolk and it is apparent that its emulsification properties are provided by a complex interaction of various protein, lipoprotein and phospholipid fractions. Other constituents of salad products also contribute some emulsification properties; these include the mustard and the thickener system.

The average size and the size distribution of the dispersed oil phase is another important property. The oil globules should be in the range $1-4\,\mu m$ ($0.0375-0.15 \times 10^{-3}$ in.) with as few globules as possible larger than this. Emulsion stability depends on a balance between the attractive and the repulsive forces between droplets, with the balance being in favour of the repulsive forces. The main force of attraction is the van der Waals force (V_A) described by:

$$V_A = -Aa/6H_0 \qquad (19.5)$$

where A is the Hamaker constant, a is the droplet size and H_0 the distance between droplet surfaces.

The larger the droplet size and the lower the proportion of continuous phase (i.e. the smaller the interdroplet distance), the greater are the attractive forces and the less stable is the emulsion. The type and quantity of the emulsifier, the order of addition of the various ingredients and, most importantly, the amount of work done in mixing and homogenization to form the emulsion, all influence the size and distribution of oil droplets.

Factors influencing the flavour quality of salad products include the extent of activation of the mustard component as already described, the presence of other flavouring ingredients such as lemon juice and spices and the oxidative stability of the system. Oxidative stability is of critical importance to shelf-life, since it profoundly influences the risk of formation of rancid off-flavour and odours. Oxidative stability is influenced by the choice of the type and quality of the oil, the amount of air incorporated in the oil raw material and the product, the holding temperatures of the oil and the final product, the presence of light during storage, the presence of various metal ions which can catalyse autoxidative deterioration reactions, and the presence of lipolytic enzymes derived from mustard or from vegetable pieces. The various types of oxidative reactions, which can occur in salad products to produce off-flavour and rancidity, have been reviewed by Mistry and Min [45].

There are two major pathways leading to rancidity, both of them involving oxidation at some stage. The first, and more important, pathway involves conversion of triglycerides to triglyceride hydroperoxides by action of oxygen at the double bonds of unsaturated fatty acids; the hydroperoxides can then be hydrolysed to free fatty acid peroxides and thence give secondary and tertiary products such as aldehydes, ketones and saturated and unsaturated hydrocarbons. The second pathway involves an initial hydrolysis of the ester linkages in the triglycerides, for example by lipases, to give free fatty acids; these can then be oxidised to give free fatty acid peroxides which can in turn give the secondary and tertiary products. In the initiation step of the first pathway, free radicals are formed from unsaturated fats in the presence of initiators such as heat, light, metal catalysts or lipoxygenases. The free radicals then react with oxygen to form peroxy radicals, which in turn react with the unsaturated fats to form hydroperoxides. The hydroperoxides are odourless and tasteless, but their breakdown products, particularly the short chain volatile monomers, are associated with general rancid off-flavour and odours.

A number of precautions can be taken during manufacture and storage of salad creams and mayonnaises in order to reduce the risk of rancid reactions. The air content of the products can be minimized by allowing oil deliveries to stand at normal ambient temperature to release entrapped air, and by processing the products under nitrogen blanketing. Modified atmosphere packing under nitrogen has also sometimes been practised. Oils of high quality are specified, containing reduced amounts of the minor components, such as free fat acids, which can contribute to reducing oxidative stability. Processing temperatures during the later stages when the oil is present, are kept as low as possible; lipolytic enzymes which are present in mustard or other product constituents, are heat-inactivated at an earlier stage.

The factors influencing microbiological stability of salad sauces and mayonnaises have been reviewed by Smittle [47], de Wit and Rombouts [48] and Brocklehurst *et al.* [8]. In general terms, because of their acidity, low pH and reduced water activity (due to the high solute concentration in the aqueous phase), they are resistant to most spoilage and pathogenic organisms. In addition, raw materials that potentially might be a source of spoilage or food poisoning micro-organisms such as the egg, the mustard and the spices, are pasteurized either before delivery to the factory or during the early stages of the manufacturing process. There is also evidence to suggest that mustard oil, in particular the isothiocyanates associated with the flavour principle, has antimicrobial properties.

Artificial preservatives such as sorbic acid, or benzoic acid, are added to certain formulations in order to reduce spoilage risks. The usual spoilage organisms found in mayonnaises and salad creams are lactobacilli, bacilli and yeasts. Good manufacturing practices must be employed to reduce the risks from these organisms. These practices include microbiological control of raw materials, strict adherence to temperature and time conditions during the process, good hygiene of plant and, in certain cases, 'super-clean' manufacture and filling in high hygiene zones provided with an overpressure of filtered air. With regard to pathogens, mayonnaises and salad creams as traditionally formulated will not support the growth of *Salmonella*, *Staph. aureas*, *Clostridium botulinum* or *Bacillus cereus*. However, some recent developments in the area of low-acid, very low oil content products, and products mixed with high proportions of fresh vegetables, have led to formulations which may be at risk towards the growth of certain pathogens.

19.7 Conclusion

There are a wide variety of different products of different characteristics making up the category of pickles, sauces and salad products. Some of the more important products and the essential process steps in their manufacture have been described in this chapter. It is evident that the pickle and sauce manufacturer has control over most of the physicochemical factors which provide for, or impact on, the unique and desirable characteristics of these products. The common factor across all the products is that their microbiological stability is, in large measure, derived from their organic acid content. Trends in the food market towards milder products have meant that, over recent years, the manufacturer has had to decrease the acidity of many products and lay more emphasis on additional hurdles contributing to microbiological stability, such as the heat process, natural or synthetic preservatives, reduced water activity, modified atmosphere packing, ultra-hygienic manufacture and chilled storage. Other maket trends such as salt-intake reduction, fat reduction, the growth of the ethnic and the chilled food categories, and the growth of snacking products and fast food outlets, have all had their impact on the product mix and on desirable product properties.

It is evident that the food scientist and technologist will continue to be challenged to provide products in this category which, whilst satisfying the traditional prerequisites of attractive appearance and colour, unique texture and mouthfeel, characteristic flavour and good storage stability, will also need to satisfy some new requirements dictated by the market changes.

References

1. Cruess, M.V. Pickles. In *Commercial Fruit and Vegetable Products*, 4th edn., McGraw-Hill, New York, (1958), 708–733.
2. Desrosier, N.W. Principles of food preservation by fermentation and pickling. In *The Technology of Food Preservation*, 3rd edn. (1970), Avi, Westport, CT, 239–267.
3. Binstead, R., Devey, J.D. and Dakin, J.C. *Pickle and Sauce Making*, 3rd edn. (1971), Food Trade Press Ltd., London.
4. Bhasin, V. and Bhatia, B.S. Recent advances in technology of pickles, chutneys and sauces. *Indian Food Packer* (1981), (January–February), 39–58.
5. Campbell-Platt, G. and Anderson, K.G. Pickles, sauces and salad products. In *Food Industries Manual*, 22nd edn. (1988), Blackie, Glasgow, chapter 8, 285–334.
6. Anderson, K.G. Other preservation methods. In *Vegetable Processing*. Eds Arthey, D. and Dennis, C. (1991), Blackie, Glasgow, 154–185.
7. Comité des Industries des Mayonnaise et Sauces Condimentaires de la Communité Economique Européenne. *Code for the Production of Microbiologically Safe and Stable, Emulsified and Non-emulsified Sauces Containing Acetic Acid* (1975).
8. Brockelhurst, T., Parker, M., Gunning, P. and Robins, M. Microbiology of food emulsions: physico-chemical aspects. *Lipid Technol.* (1993) (July/August), 83–88.
9. McClure, P.J., Blackburn, C. de W., Cole, M.B., Curtis, P.S., Jones, J.E., Legan, J.D., Ogden, L.D., Peck, M.W., Roberts, T.A., Sutherland, J.P. and Walker, S.J. Modelling the growth, survival and death of microorganisms in foods: the UK Food Micromodel approach. *Int. J. Food Microbiol.*, **23** (1994), 265–275.
10. Fleming, H.P. Fermented Vegetables. In *Economic Microbiology*. Ed. Rose, A.H. Academic Press, London, (1982), 227–258.
11. Anderson, K.G. The production of brined cauliflower for the pickle industry. *J. Food Technol.*, **3** (1968), 263–267.
12. Etchells, J.L., Bell, J.A., Fleming, H.P., Kelling, R.E. and Thomson, R.L. Suggested procedure for the controlled fermentation of commercially brined cucumbers – the use of starter cultures and reduction of carbon dioxide accumulation. *Pickle Pack Sci.* (1973), 3, 4.
13. Etchells, J.L., Bell, T.A., Fleming, H.P. and Thomson, R.L. Controlled bulk vegetable fermentation, U.S. Patent (1976), 3, 932, 674.
14. Fleming, H.P., Thomson, R.L., Bell, T.A. and Hontz, L.H. Controlled fermentation of sliced cucumbers. *J. Food Sci.*, **43** (1978), 888–891.
15. McFeeters, R.F. and Fleming, H.P. Effect of calcium ions on the thermodynamics of cucumber tissue softening. *J. Food Sci.*, **55** (1990), 446–448.
16. Jewell, G.G. Structural and textural changes in brown onions during pickling. *BFMIRA Res. Rep.*, No. 166 (1971).
17. Edgell, A. and Kilcast, D. *Functions of sodium chloride in pickles – a literature review*. Leatherhead Food RA, Scientific and Technical Surveys No. 153 (1986).
18. Saxton, C.A., Jewell, G.G. and Dakin, J.C. Structural and textural changes in cauliflower during pickling. BFMIRA Res. Rep. No. 142 (1969).
19. Pedersen, C.S. Sauerkraut. In *Advances in Food Research*. Eds Chichester, C.O., Mrak, G.M. and Stewart, G.F. Academic Press, New York, **10** (1960), 233–291.
20. Aurand, L.W., Singleton, J.A., Bell, T.A. and Etchells, J.L. Identification of volatile constituents from pure-culture fermentations of brined cucumbers. *J. Food Sci.*, **30** (1965), 288–295.
21. Vaughn, R.H. Lactic acid fermentation of cucumbers, sauerkraut and olives. In *Industrial Fermentations*. Eds Underkofler, L.A. and Hickey, R.J. Chemical Publishing Company, New York, **2** (1954), 417–478.
22. Fleming, H.P. and Pharr, D.M. Mechanism for bloater formation in brined cucumbers. *J. Food Sci.*, **45** (1980), 1595–1600.
23. Guillou, A.A. and Floros, J.D. Problems associated with the processing of cucumber pickles: softening, bloater formation and environmental pollution. In *Food Science and Human Nutrition*. Ed. Charalambous, G. Elsevier Science Publishers, Amsterdam (1992), 499–514.
24. Scholey, J. Texture of pickles. BFMIRA Scientific and Technical Surveys No. 67 (1971).

25. Brimelow, C.J.B. and Brittain, J.E. Continuous food impregnation. Brit. Patent 1,509,502 (1978).
26. Brimelow, C.J.B. and Brittain, J.E. Continuous food impregnation. Brit. Patent 1,538,686 (1979).
27. Ponting, J.D., Watters, G.G., Forrey, R.R., Jackson, R. and Stanley, W.L. Osmotic dehydration of fruits. *Food Technol.*, **20**(10) (1966), 125–128.
28. Goose, P.G. and Binsted, R. *Tomato Paste, and other Tomato Products*, 2nd edn. (1973), Food Trade Press, London.
29. Gould, W.A. *Tomato Production, Processing and Technology*, 3rd edn. (1992), CTI Publications, Baltimore, MD.
30. Watada, A.E., Norris, K.H., Worthington, J.T. and Massie, D.R. Estimation of chlorophylls and carotenoid contents of whole tomato by light absorbance technique. *J. Food Sci.*, **41** (1976), 329–332.
31. Koskitalo, L.N. and Omrod, D.P. Effects of sub-optimal ripening temperatures on the colour quality and pigment composition of tomato fruit. *J. Food Sci.*, **37** (1972), 56–59.
32. Noble, A.C. Investigation of colour changes in heat concentrated tomato pulp. *Agric. Food Chem.*, **23**(1) (1975), 48–49.
33. Brimelow, C.J.B. Color measurement of tomato paste: some method variables. In *Physical Properties of Foods*, 2. Eds Jowitt, J., Escher, F., Kent, M., McKenna, B. and Roques, M. Elsevier Applied Science, London, (1987), 295–317.
34. Hersom, A.C. and Hulland, E.D. *Canned Foods*, 6th edn., Churchill, London (1969).
35. Bourne, M.C. Applications of chemical kinetic theory to the rate of thermal softening of vegetable tissue. In *Quality Factors of Fruit and Vegetables Chemistry*. Ed. Jen, J.J. (1989), American Chemical Society Symposium Series, Washington, D.C., 98–139.
36. Hersom, A.C. Thermal processing. In *Vegetable Processing*. Eds Arthey, D. and Dennis, C. (1991), Blackie, Glasgow, 69–101.
37. Moreira, L.A., Oliveira, F.A.R., Oliveira, J.C. and Singh, R.P. Textural changes during thermal processing. 1. A descriptive method to segregate effects of process treatments. *J. Food Process. Preserv.*, **18**(6) (1994), 483–496.
38. Brown, M.S. Effects of freezing on fruit and vegetable structure. *Food Technol.*, **30** (1976), 106–114.
39. Gorini, F.L., Eccher Zerbini, P. and Testoni, A. The controlled atmosphere storage of fruit and vegetables. In *Chilled Foods: The State of the Art*. Ed. Gormley, T.R. (1990), Elsevier Applied Science, London, 201–224.
40. Wang, C.Y. Chilling injury of fruits and vegetables. *Food Rev. Int.*, **5**(2) (1989), 209–236.
41. Jackman, R.L., Yada, R.Y., Marangoni, A., Parkin, K.L. and Stanley, D.W. Chilling injury. A review of quality aspects. *J. Food Qual.*, **11** (1988), 253–278.
42. Low, J.M. and Maughan, W.S. Sorting by colour in the food industry. In *Instrumentation and Sensors for the Food Industry*. Ed. Kress-Rogers, E. (1993), Butterworth Heinemann, Oxford, 97–119.
43. Shapton, D.A., Lovelock, D.W. and Laurita-Longo, R. The evaluation of sterilization and pasteurization processes from temperature measurement in degrees Celsius. *J. Appl. Bacteriol.*, **34** (1971), 491–500.
44. Harrison, L.J. and Cunningham, F.E. Factors influencing the quality of mayonnaise: a review. *J. Food Quality*, **8** (1985), 1–20.
45. Mistry, B.S. and Min, D.B. Shelf life of mayonnaise and salad dressings. In *Shelf Life Studies of Foods and Beverages. Chemical, Biological, Physical and Nutritional Aspects*. Ed. Charalambous, G. (1993), Elsevier Science Publishers, Amsterdam, 409–450.
46. Holcomb, D.N., Ford, L.D. and Martin, R.W. Dressings and Sauces. In *Food Emulsions*, 2nd edn. Eds Larsson, K. and Friberg, S.E. (1990), Marcel Dekker, New York, 327–366.
47. Smittle, R.B. Microbiology of mayonnaise and salad dressing: A review. *J. Food Protect.*, **40**(6) (1977), 415–422.
48. de Wit, J.C. and Rombouts, F.M. Chemical preservation of mayonnaise-based salads. In *Food Science: Basic Research for Technological Progress*. Eds Roozen, J.P., Rombouts, F.M. and Voragen, A.G.J. (1989), Pudoc, Wageningen, 125–132.

20 Beer and cider
C.W. BAMFORTH

20.1 Introduction

The production of beer and cider represents two of the oldest biotechnologies practised by mankind. For the vast majority of their histories, brewing and cider-making were not only parochial industries, but owed much more to art than to science. It is only really over the past 125 years that science and technology have increasingly impacted on these processes. Historical treatises on beer [1] and cider-making [2] make compelling reading.

It would be impossible within the scope of this chapter to do justice to the plethora of research pursued on brewing and cider-making matters. The author has attempted to provide a brief and widely-referenced overview of the relevance to brewing and cider-making of the physico-chemical parameters described in the first part of this volume. The topic is discussed in two sections: the influence on product quality and the influence on the production processes.

Features of alcoholic beverage quality relevant to this discussion are foam quality (in beer), mouthfeel, partitioning of flavoursome compounds between the bulk liquid phase and the headspace, and colour.

20.2 Beer foam

Whilst there is a long history of scientific study of beer foams, this has largely been devoted to analysis of the chemical species involved in head formation, retention and adhesion [3]. There is no doubt that such work has led to a far greater appreciation of the changes which take place in foaming species during processing and, to some extent, has explained the interactions which occur between such molecules during the foaming process. As a result, there is an improved understanding of how to ensure adequate foam quality in a product.

However, there is no question that the most dramatic improvements in beer foam quality have been realised by attention to physical issues. Most recently (in the United Kingdom and Ireland) such developments have materialised in the form of the so-called 'widget' in cans, which facilitates the nucleation of gas and promotes foam generation during dispense [4]. Such technology is allied to the use of nitrogen gas which, when present in the range 10–30 parts per million (ppm), enables far more stable foams than can be obtained from carbon dioxide, which is typically present in beers over the range 2000–6000 ppm [5,6].

The principal physical processes pertaining to foaming are bubble formation (which determines 'foamability'), liquid drainage, bubble coalescence and disproportionation (Ostwald ripening). The latter three factors determine 'foam stability' or 'head retention'. Additionally, it is necessary to consider the influence of changes in foam texture, which not only relate to perceived foam quality in terms of appearance, but also lead to the ability of foam to adhere to the glass ('cling' or 'lacing').

20.2.1 Bubble formation

The formation of bubbles not only influences the extent of head production, but also the properties of the foam once formed, for example head retention and appearance. Thus foams with an even distribution of smaller bubbles are more stable and are whiter, creamier and more attractive [7].

Beer is supersaturated with carbon dioxide, but not to the extent that it will spontaneously form bubbles against a thermodynamic energy barrier. Rather, bubble release is catalytically promoted at nucleation sites, which may include cracks in the container surface, insoluble particles or gas pockets introduced during dispense. Considerable effort has been devoted to the design of taps for dispense of beer with excellent foaming properties [8]. Scratched or etched glasses have also been introduced from time to time to ensure a plentiful supply of bubbles to replenish foam as the beer sits in the glass.

Much less desirable is the tendency of molecular nucleation sites to promote overfoaming of beer, so-called 'gushing'. This is the phenomenon whereby beer in can or bottle displays spontaneous 'fobbing' (uncontrolled surging of the contents) when the package is opened. It can be triggered by low-molecular-weight hydrophobic peptides which are in turn derived from fungal infections on barley [9]. Another cause of gushing is oxidised and dimerised derivatives of the hop bitter compounds which can be occasionally found in hop extracts [10].

The size of a bubble emerging from a nucleation site is chiefly a function of the surface tension of the beer, the shape of that nucleation site and the contact angle between the liquid and the nucleation site [11]. The lower the surface tension of a beer, the more readily will it foam.

20.2.2 Foam drainage

Liquid drains from a foam principally through the force of gravity and by suction from plateau borders (regions between bubbles) [11]. Counteracting these influences are bulk beer viscosity and, in particular, surface viscosity in the foam lamellae. Factors promoting high surface viscosity will promote foam stability. Such factors include temperature (reduced temperature improves head retention) and molecular interactions between the bittering iso-α-acids (Figure 20.1; [12]) and the polypeptides of highest amphipathic character [13].

20.2.3 Bubble coalescence

When liquid drains excessively to produce thin interbubble regions, coalescence can occur by rupture of the film and merger of two bubbles (see chapter 3). This process leads to a coarsening of foam when coalescence occurs within the body of the head, but to foam collapse when bubbles coalesce at the surface. Two mechanisms underlying coalescence have been discussed by Ronteltap et al. [11]. It is a process particularly promoted by lipids [14].

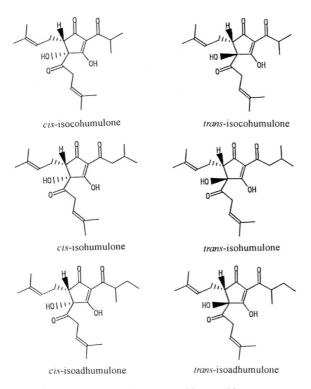

cis-isocohumulone   trans-isocohumulone

cis-isohumulone   trans-isohumulone

cis-isoadhumulone   trans-isoadhumulone

Figure 20.1 Structure of iso-α-acids.

20.2.4 Bubble disproportionation

This is the phenomenon whereby gas passes from a small bubble to a larger one, on account of the higher pressure in the smaller bubble. Eventually the smaller bubble disappears and, once again, there is a coarsening of the foam. It is evident that the initial distribution of bubble sizes is critical: if all bubbles are of a uniformly similar size then disproportionation will be of reduced significance.

The model of De Vires [15] quantitatively describes disproportionation:

$$r_0^2 - r_t^2 = \frac{4RTDS\sigma t}{P_{atm}\theta}$$

where t = time, r_0 = bubble radius at $t = 0$, r_t = bubble radius at $t = t$, R = universal gas constant, T = absolute temperature, D = diffusion coefficient, S = solubility of gas, σ = surface tension, P_{atm} = atmospheric pressure and θ = film thickness between the bubbles.

Disproportionation happens more quickly at elevated temperatures, in liquids of increased surface tension and once substantial liquid drainage has led to a reduced distance between bubbles. It is countered by high atmospheric pressure, as may readily be demonstrated by covering a glassful of freshly poured beer (hence the concept of the traditional covered German stein!). Furthermore, gases of reduced solubility will give a lesser degree of disproportionation. Thus nitrogen affords better foam stability than does carbon dioxide.

In foams of uniform bubble size, surface collapse of bubbles is more important than disproportionation. The speed of foam collapse basically relates to the rate at which successive layers of bubbles come to the surface. For a given depth of foam, smaller bubbles mean more layers and, accordingly, improved foam stability. Furthermore, liquid drains more slowly from the surface of smaller bubbles, as there is a far greater surface area to traverse in a collection of small bubbles than in the same volume of large bubbles.

20.2.5 Surface tension

The surface area of a liquid is greatly increased when bubbles are formed within it. This is in conflict with surface tension. Pure liquids cannot give stable foams, but the presence of surfactants capable of entering into the bubble wall can render foams stable. The principal backbone components of beers are polypeptides, notably those containing high hydrophobic character [16]. At various times there have been claims that specific protein types have a primary role in determining beer foam quality. Most recently, a barley-derived protein of molecular weight 40 000 [17] and a barley lipid-transfer protein [18] have been championed in this context.

However, Onishi and Proudlove [13] have described how both of these species seem to be present in separate fractions recovered from beer and foam by hydrophobic interaction chromatography, fractions which differ greatly in their foam-stabilising properties. This would suggest that it is the degree of hydrophobic character that is important, rather than any other intrinsic feature of a protein or polypeptide. Furthermore, it indicates that both the 40 000 molecular-weight protein and the lipid-transfer protein can adopt conformations which differ in their hydrophobicity in proportion to their surface activity.

20.2.6 Surface viscosity

It has been proposed that the major cause of prolonged survival of protein foams is their high surface viscosity [19]. It is recognised that surface viscosity within a beer foam increases with time (thixotropy) and that this relates to the ability of that foam to lace glass [20,21]. Furthermore, it has been recognised that this increase in surface viscosity is due to the interaction of polypeptides and iso-α-acids.

Maeda et al. [22] made a more recent detailed analysis of surface viscosity, claiming a positive correlation between the increase in surface viscosity over the period 30–50 min after pouring and beer foam stability. It was shown that the beer surface develops plasticity with time. The more rapid the onset of plasticity, the more stable is the foam. Furthermore, polypeptides of greater hydrophobicity displayed the most substantial increase in surface viscosity. Maeda and co-authors used a Donnan–Barker type apparatus to show that the most hydrophobic polypeptides adsorbed to bubble walls to the greatest extent. In turn, the more hydrophobic fractions reduced initial surface tension to the greatest extent, thereby promoting foamability. Yokoi et al. [23] demonstrated that iso-α-acids primarily increase the surface viscosity of the most hydrophobic polypeptides.

Recent work has indicated that the iso-α-acids which have their side-chains chemically reduced are particularly adept at improving beer foam [24]. This probably relates to their increased hydrophobicity and ability to interact with amphipathic polypeptides ('amphipathicity' refers to the presence of both hydrophilic and hydrophobic regions). Such reduced iso-α-acids are sometimes employed to enhance the light stability of beer. Beer is susceptible to the development of 'skunky' character if exposed to light. The precursors of the causative agent, 3-methyl-2-butene-1-thiol, are the iso-α-acid bittering substances (see Figure 20.1). However, if the side chains are reduced they no longer break down to give this substance. The more common (and traditional) way of suppressing light damage in beer is to package it in brown glass. Recent marketing pressures to sell beer in green, or clear, glass bottles are (in the absence of use of reduced side chain iso-α-acids) to the severe detriment of shelf-life. For a review, see [25].

20.3 Flavour

20.3.1 The influence of foam on flavour

Whilst it has been claimed that there is an indirect, psychological influence of foam quality on the perceived flavour of beer [26–28], there is also evidence that foaming does indeed directly affect flavour [29].

Ono and her co-workers poured a lager-type beer such that it either had a coarse foam (a preponderance of large bubbles of 2.2 mm (0.08 in.) diameter, together with finer bubbles of 0.132 mm (0.005 in.) diameter) or a fine foam (preponderance of fine bubbles of 0.115 mm (0.004 in.) diameter and fewer coarse bubbles). Both trained and untrained flavour panellists preferred the beer with the finer foam, discriminating in favour of it on the basis of higher 'and more suitable' carbonation, better flavour balance and less softness/bitterness. The conclusion was that fineness of foam helps determine the mouthfeel characteristics of beer.

It was demonstrated that the production of the finer foam left more carbon dioxide in the liquid beer, which will contribute to the trigeminal sense. (A series of irritants – others include onion and chilli – stimulate the trigeminal nerve ends giving perceptions of heat, burn, prickle, etc.) Several substances are concentrated in foam during dispense, including short-chain fatty acids and their ethyl esters, iron and bitterness compounds. There was no difference, however, in the extent to which concentration occurred in the coarse and fine foams, respectively. Concerning the bitter compounds (iso-α-acids), isocohumulone displayed less tendency to concentrate in the foam than isohumulone or isoadhumulone. A beer containing 25 bitterness units (BU) before dispense was shown to contain 24.2 BU in the liquid portion beneath the foam, irrespective of fineness of the head. The BU level in the foam itself (after collapse) was 30.9 for a coarse foam and 31.7 for a finer foam. Ono *et al.* [29] showed, however, that a more important contributor to the difference in perceived bitterness was the level of residual carbon dioxide in the liquid beer: panellists felt that a beer became more bitter as its carbon dioxide content increased. In contrast to Ono's work, the studies of Langstaff and Lewis [27,28] suggested that, in blind tests, foamed and unfoamed beers could not be distinguished.

20.3.2 Mouthfeel

The organoleptic characteristics of beer are in large part a function of the specific interaction of a range of flavoursome and odoriferous substances with the nose and mouth. In a chapter restricting itself to physico-chemical issues, however, the most relevant attributes warranting attention are mouthfeel and fullness.

Mouthfeel was defined by Jowitt [30] as 'those textural attributes of a food or beverage responsible for producing characteristic tactile sensations on the surfaces of the oral cavity'. The mouthfeel characteristics of beer have been reviewed by Langstaff and Lewis [27,28]. It is difficult to separate mouthfeel and fullness. On the beer-flavour wheel (originally developed by Clapperton [31] and others), mouthfeel can be subdivided into the terms 'alkaline', 'mouthcoating', 'metallic', 'astringent', 'powdery', 'carbonation' and 'warming'. Fullness, meanwhile, has only a single subdescriptor: body. Langstaff et al. [32] proposed a modification to the flavour wheel, indicating that mouthfeel could be subdivided into 'carbonation', 'fullness' and 'afterfeel'. In turn, carbonation is defined by 'sting', 'bubble size', 'foam volume' and 'total carbon dioxide'. Fullness is divided into density and viscosity. Afterfeel comprises 'oily mouthcoat', 'astringency' and 'stickiness'.

The relevance of the foam and carbon dioxide to the mouthfeel characteristics of beer have been referred to above. It is recognised that the inclusion of nitrogen in beer to improve foam also has the impact of 'smoothing' on the palate [5]. This may be a direct (as yet unexplained) phenomenon, but may also relate to the tendency of beers containing nitrogen to have low carbonation, which will cut down palate 'prickle'.

Concerning individual chemical components of beer (other than gases), there is little definitive evidence for the certain involvement of any class as determinants of mouthfeel. Proteins [33], polyphenols [34], chloride [35], dextrins [36], β-glucan [37], ethanol [38] and glycerol [39] have all been championed as contributors to body and mouthfeel. However, it will be noted by the reader that most of these claims have been made in textbooks or reviews. There seems to have been a dearth of major scientific studies of the relationship between beer composition and mouthfeel. Perhaps Langstaff et al. [40] have made the most definitive study. They found the following:

(i) high dissolved carbon dioxide correlated with foam volume and total carbon dioxide;
(ii) polyphenols correlated with density, viscosity, oily mouthfeel and stickiness;
(iii) chloride correlated with density, viscosity, oily mouthcoat, astringency and stickiness;
(iv) lower-molecular-weight dextrins (degree of polymerisation 4–9) correlated with viscosity and stickiness, though not fullness;
(v) there was a better correlation between instrumental viscosity of beers and a selection of mouthfeel terms than there was between

such terms and measurements of the substances generally held to determine viscosity, e.g. β-glucans;
(vi) glycerol correlated with viscosity, oily mouthfeel and stickiness.

Langstaff and her co-workers stressed that, whilst correlations were observed, there may or may not be causal links between any individual beer component and a given mouthfeel sensation.

Szczesnak [41], from studies of a wide range of familiar beverages, concluded that the most significant parameter determining mouthfeel is the perception of viscosity. Whilst the viscosity of a beverage such as beer or cider is relatively low (typically 1.4–2.0 cP), it is known that turbulent flow between the tongue and the roof of the mouth results in an increased perceived viscosity [42,43].

It is well established that a prime contributor to the mouthfeel of cider is tannin [44]. The polyphenols concerned are a series of oligomeric (up to heptamer) procyanidins with a (−)-epicatechin structure (Figure 20.2). The dimers to tetramers afford bitterness, whilst those containing 5–7 monomeric units provide astringency [45]. Lea and Arnold [46] showed that alcohol promotes the perception of bitterness but suppresses astringency.

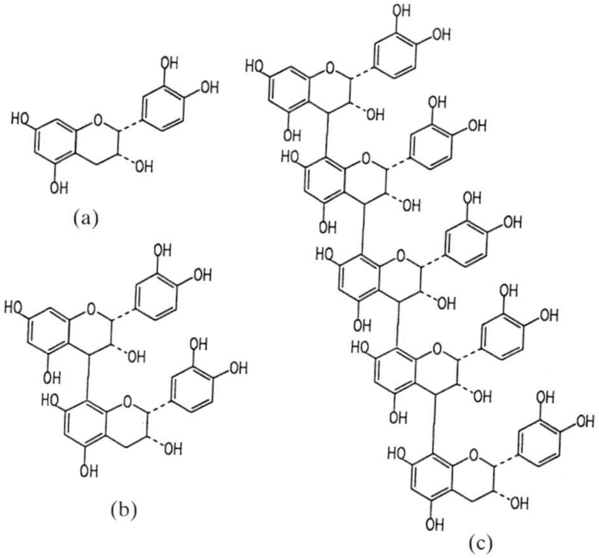

Figure 20.2 Structure of (a) (−)-epicatechin, (b) its dimer and (c) its pentamer.

20.3.3 Flavour partitioning

Relatively little research has been devoted to the physico-chemical aspects of beer or cider flavour delivery. Apart from the work of Ono and co-workers ([29]; see earlier) and the extensive studies by, besides others, Meilgaard on sensory thresholds of substances in beer (e.g. [47]) and similar work by Williams on cider (e.g. [48]), the only significant work in this area is that of Gardner [49]. He suggested that the flavour threshold of a beer component could be correlated with the octanol:water partition coefficient, P, for that substance. Gardner advocated the use of P to predict thresholds for untested compounds according to their structure.

Williams and Prosser [50] demonstrated that the presence of low levels of ethanol (0.5–0.75%) enhances the fruity character of cider, in part due to an effect on the level of esters distributing into the headspace. At higher concentrations, ethanol suppresses the vapour pressure of other volatiles, suggesting that the effect has no significance over the range of alcohol contents encountered in the vast majority of ciders and beers. The effect will be relevant only for very low and non-alcoholic products.

20.4 Colour

The colour of beer derives principally from Maillard interaction of amino acids and sugars. These substances are produced during the germination of barley in the malting process and during the mashing of malt (see later). More extensive enzymolysis at these stages introduces more scope for colour formation, which occurs during the kilning of malt and the boiling of wort. Any adjuncts introduced into the kettle will also potentiate colour formation.

Colour in beer is increasingly determined by the type of malt employed in the grist, although caramels are still employed to an extent [51].

The visible absorption spectrum of beer shows no maxima (Figure 20.3). The relative intensity of red versus yellow absorption varies significantly between beer styles, e.g. ales and lagers. Accordingly, no single wavelength can provide an ideal opportunity for colour measurement. Recently, it has been suggested that existing procedures for specifying colour on the basis of absorption of light at a single wavelength (430 or 530 nm) should be replaced by tristimulus/chromaticity-type techniques [52].

Another contributor to the colour of beer is the oxidation of polyphenols derived from malt and hops. In cider, oxidation of polyphenolic compounds in juice is the principal source of colour [53]. It is limited by the extent to which sulphite is introduced into the system and is reduced during fermentation.

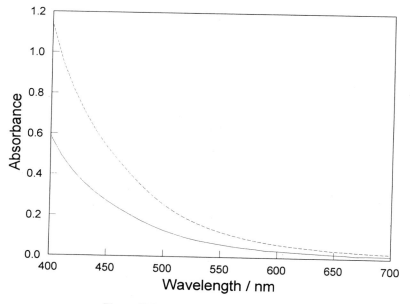

Figure 20.3 Absorption spectra of beers.

The colour of apple-juice concentrate (which is being used increasingly in cider-making on account of its storability and the resultant process flexibility benefits) is largely from Maillard-type reactions occurring during storage.

20.5 The production of beer

Relevant physico-chemical aspects of the malting and brewing processes are depicted in Figure 20.4 and Table 20.1. Readers unfamiliar with malting and brewing can find various overviews in the literature (e.g. see Bamforth and Barclay [54] for a relatively concise account or Hough et al. [55] for a more extensive, if slightly dated, treatise). The discussion which follows is by no means a comprehensive account of either beer brewing or cider-making. Rather it highlights physico-chemical issues at individual process stages.

20.5.1 Malting

The malting of barley is principally carried out for the following reasons:
- To soften the grain and make it more millable.

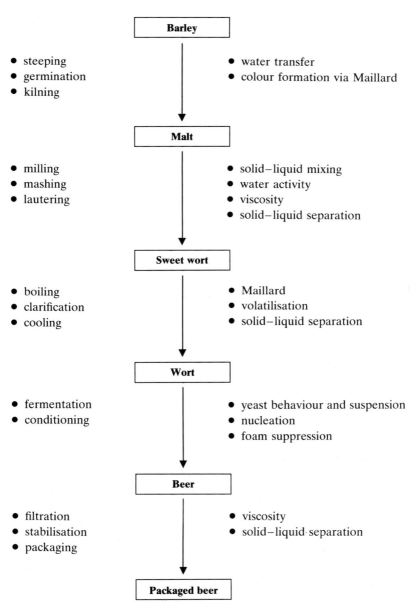

Figure 20.4 Brewing.

Table 20.1 Physico-chemical parameters in relation to alcoholic beverage production

Parameter	Relevance	Reference
Vapour pressure	Barley drying	Marsh [56]
	Malt kilning	Abrahamson [57]
	Wort boiling	Miedaner et al. [58]
	Gas control	Forrest et al. [59]
	De-alcoholisation	Molzahn [60]
Water activity	Susceptibility of grain to infection	Cahagnier et al. [61]
	Osmotolerance of yeasts in high-gravity fermentations	Jakobsen and Piper [62]
	Thermal death of microbes	O'Connor-Cox et al. [63]
Foams	Fermentation	Pierpoint [64]
	Product quality	Bamforth [65]
Maillard	Colour	Kavakus and Scriban [66]
	Flavour	Seaton [67]
	Flavour stability	Moll and Moll [68]
Mixing	Mashing	Moll and Midoux [69]
	Fermenter	Masschelein [70]
Gels and hydrocolloids	Wort separation } Beer filtration	Marchbanks [71]
	Hazes	Dadic [72]
Adsorption	Water purification	Weaver [73]
	Emissions	Schu and Meyer-Pittroff [74]
	Yeast immobilisation	Norton and d'Amore [75]
	Colloidal stabilisation	Dahlstrom and Sfat [76]
	Yeast adherence to glass bottles	Wood et al. [77]
Dissolution	Mashing	Marc et al. [78]

- To develop the proteolytic and amylolytic enzymes which will serve to digest the protein and starch of malt in the brewhouse, thereby generating low-molecular-weight nutrients for yeast fermentation.
- To modify the flavour characteristics of barley to those desired in the beer.

Barley is steeped to 42–46% moisture, which switches on embryo activity and hydrates the endosperm. The embryo produces hormones (principally gibberellins) that stimulate enzyme synthesis. The enzymes migrate into the endosperm which, once hydrated, is rendered susceptible to degradation ('modification'). The malt is dried by kilning to arrest enzymolysis at a stage when the grain is sufficiently friable but before excessive starch is 'lost' to support wasteful embryo growth.

The efficient distribution of water through the barley in the steeping operation is a critical determinant of whether a given barley will malt well. Water distribution is principally influenced by the texture of the endosperm [79]. Mealy endosperms, which have a relatively open structure in which the starch granules are loosely packed in a protein matrix, take up moisture more easily than do the more tightly packed steely endosperms. Such structural features probably also dictate the ease with which the polymers within a fully hydrated endosperm are amenable to enzymic attack [80].

It is important during the kilning of green malt to lower the moisture content to <5% in order to ensure that the malt will remain stable during storage for long periods. Indeed, malt is generally stored for at least one month before brewing in order to optimise performance [81]. It is thought that, during this time, the remaining moisture in the stored grain equilibrates across the bed and makes for more homogeneous handling in the brewhouse.

Bamforth and Barclay [54] provide a detailed account of commercial malting systems. Their account also explains the range of malts available, which principally relates to the severity of kilning which they have received. Kilning dictates (i) the level of enzymes surviving (intense heating destroys enzymes), (ii) the flavour of the malt (intense heat introduces more toffee, caramel, roast and coffee-like characters) and (iii) the colour. This latter parameter derives, of course, from melanoidin reactions between reducing sugars and free amino-nitrogen substances, both of which develop during germination of barley. Therefore, the more extensively a barley is modified during malting, the greater will be the opportunity for colour development on kilning. The colour of a British ale malt may typically be 5 EBC (European Brewery Convention units; for a definition of methods for measuring colour and the units themselves, see [82]). An American lager malt may have a colour of 1.75 EBC. By contrast, crystal malt, which is frequently employed to introduce colour and malty flavour into products, will have a colour in the range 75–300. Chocolate malt has a colour of 500–1200, whilst black malt of the type used in stouts, will have a colour of 1200–1400.

The kilning of green malt (and also the drying of barley) is largely dictated by considerations of vapour pressure. The efficiency with which moisture is removed from the grain bed depends on the water content of the air with which it is in contact. If the relative humidity (RH, the quantity of moisture in air as a percentage of the quantity which saturates air at the same temperature; see chapter 1) in the air flow is below that in the grain, then the grain will dry. The converse applies if air has the higher RH. During evaporation of water, the grain cools to an extent equal to the latent heat of evaporation of the removed water. This is highly significant, as it is essential for most malts to conserve enzyme activity.

In the free drying stage of kilning, an air-on temperature of <60°C (140°F) removes moisture readily. Air flow is adjusted to maintain a RH of 90–95% – total saturation being avoided to prevent condensation of water vapour and resultant 'dripping'. When the moisture in the grain bed has dropped to approximately 23%, the extent of evaporation is restricted by the rate at which moisture migrates to the surface of the grain. Once the moisture content of malt has reached 12%, it is all bound, demanding an increase in temperature to raise the vapour pressure of the grain. Malt 'curing' commences when the moisture content is <5–8%: the air-on temperature is increased (up to a maximum of 110°C (230°F)) to introduce colour and certain flavour characteristics, depending on the type of malt required.

20.5.2 Production of sweet wort

In the brewery, malt is milled before mixing with hot water to commence the extraction operation. Particle size after milling is naturally significant with regard to the solubilisation of materials from the malt: smaller particles enable better access of water and enzymes to the substrates and better leaching. Whilst systems have long been available to facilitate intimate mixing of grist and 'liquor' (for example, the Steel's Masher) there have been no thorough scientific studies of this stage of the process. The particle size distribution is also important from the aspect of wort separation, as it fundamentally influences the rate at which wort can be run-off from a spent grain bed. This procedure, most commonly effected using a lauter tun [83] proceeds according to the equation:

$$\text{Rate of liquid flow from spent grain bed} = \frac{K \Delta P A}{d \eta}$$

where K = constant, ΔP = pressure differential across the bed, A = bed permeability, d = depth of bed and η = viscosity of wort.

The permeability factor depends on the square of the particle diameter and on the shape of the particles.

For lauter tuns it is essential that a compromise is struck between having particles small enough to enable efficient leaching whilst retaining sufficient large husk particles to form a bed through which the wort will 'filter'. Monitoring of milled grist generally involves a plansifter [84], with roller settings being periodically adjusted in response. Clearly the extent of malt modification has as much impact as does mill settings on particle size distribution.

In the 1990s, many new brewhouses have installed mash filters, such as the Meura 2001 [85] as an alternative to the lauter tun. Such equipment demands a finer grist, because the cloths of the filter press reduce the importance of the husks. Therefore hammer mills have tended to replace roller mills for filter press brewhouses.

A further significant physical aspect of the mashing process is the concentration of the mash, generally referred to as the 'liquor-to-grist ratio'. Thinner mashes enable better mixing and contact of enzymes with their substrates. On the other hand, enzymes are protected by the greater number of particles and higher protein contents of thicker mashes.

Muller [86] has made a detailed study of mash thickness with regard to starch gelatinisation and enzyme action. To facilitate efficient starch breakdown within realistically short mashing periods it is essential that the starch is gelatinised [87]. Muller [86] has shown that water activity is a key determinant of gelatinisation and amylolysis. Water is required for gelatinisation, enzymolysis and solvation of the sugars produced during starch degradation. If any of these aspects dominates, then the other processes are inhibited. At the start of a mash, gelatination dominates and enzymolysis is temporarily inhibited. Later, the sugars produced prevent any further gelatinisation as well as inhibiting amylase action. The situation can be relieved by the introduction of more water.

It is essential that, during malting and mashing, the cell-wall polymers of barley are efficiently degraded. Principal amongst these molecules are the β-glucans which, if not solubilised and hydrolysed, elevate wort viscosity and retard the rate at which wort can be separated from the spent grain bed [65]. Wort separation is generally performed at 78°C (172°F), at which elevated temperature viscosity is proportionately lower. It has been claimed that a more important contributor to the permeability of a spent grain bed are the fine particles which collect as a sticky mass ('teig') on the surface of the grains bed. Principal amongst the components of teig are the so-called 'gel proteins' [88].

20.5.3 Wort boiling

The Maillard reaction can occur to a certain extent during the boiling process, leading to substantial flavour effects [89]. A more significant purpose for this stage is the precipitation of unstable protein in association with polyphenols (tannins), forming insoluble trub which is generally separated from wort using a whirlpool [90]. The removal of these colloidal materials ensures that they are not present to 'drop out' of the beer in package.

The boiling stage is also significant for the flavour changes it causes. Hops are introduced at this stage, increasingly in the form of pellets or extracts, thereby introducing bitterness and certain aroma constituents (essential oils). Furthermore, there are certain heat-dependent changes to malt components, principal amongst them being S-methylmethionine, which degrades to dimethyl sulphide (DMS) [91]. DMS is considered to be undesirable in most beers, but does appear to introduce definite benefits to others. It is essential that the level of DMS is controlled to the desired level, and the boil is a critical stage in this regard.

Tressl [92], Kossa [93] and Drawert and Wachter [94] have made detailed analytical studies of the volatile substances to be found in condensates from the boiling stage. However, the most comprehensive study of the physical behaviour of flavour compounds during boiling was made by Buckee et al. [95]. They observed that there are three basic patterns of volatile compound evolution from boiling wort (Figure 20.5). Type A pattern is exhibited by those compounds which are formed and released at the same rate during the boil. Buckee and his co-workers cited Maillard reaction products such as 2-acetylfuran, 5-methyl-2-furfural and 2-methylfuran in this category. Pattern B is displayed by compounds which are, in part, already present before the boil, but which are further produced during the boil, Such compounds include DMS. The third pattern (C) represents those compounds already present in sweet wort and which are entirely, or almost entirely, driven off during boiling. They include dimethyl disulphide and benzaldehyde.

Various approaches to boiling have been considered, primarily with the aim of reducing costs in the most energy-intensive stage of the process [96]. It has also been proposed that mechanical energy might replace thermal energy [97].

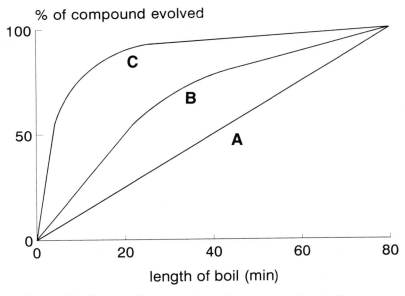

Figure 20.5 Patterns of evolution of volatile compounds from boiling wort.

20.5.4 Beer fermentation

Brewery fermentations have been extensively reviewed by Hammond [98]. With the exception of a substantial production in New Zealand [99], all commerical brewery fermentations worldwide use a batch process. There is, however, considerable interest in the prospects for increased use of immobilised yeast in continuous systems [75].

A prime physical factor influencing fermentations is the behaviour of yeast in suspension. Broadly, yeasts have been classified as 'top' or 'bottom' fermenters, depending on their tendency or otherwise to collect at the surface of the green beer at the end of fermentation (see also chapter 9). Top-fermenting yeasts are typical of traditional British ale fermentations, whilst Continental lager-style fermentations traditionally employ bottom yeasts. The literature on yeast-cell interactions and flocculation has been expertly reviewed by Speers *et al.* [100].

A further important determinant of the rate of fermentation is the extent to which agitation occurs in the fermenter. This will depend on convection currents, the extent to which bubbles of carbon dioxide are formed during fermentation, and whether the contents of the vessel are roused. It has been suggested that solid yeast foods added to fermenters do not act by providing nutrients to the cells, but rather by promoting nucleation sites for carbon dioxide release [101].

Excessive foam formation in fermenters is undesirable from an aspect of reducing fermenter capacity, promoting potential product losses and reduced foam potential in the finished product. In some countries antifoams are employed [102], including silicone-based formulations and those incorporating fatty acids. To avoid jeopardising foam in the finished product, it is essential that any antifoam is removed during filtration. Alternatively, there may be opportunities for the suppression of process foaming by the input of ultrasonic radiation.

The carbon dioxide generated during fermentation carries with it a collection of volatiles, including those already present in pitching wort (such as DMS) and those produced via yeast metabolism.

20.5.5 Beer processing and packaging

Beers are generally clarified in the brewery, after conditioning at temperatures down to $-1°C$ (30°F), using filtration operations. Most frequently such processing involves the use of filter aids, generally kieselguhr [103], although increasing attention is being paid to cross-flow filtration [104]. In either of these processes, both particle and viscosity effects come to bear on beer throughput. As beer is generally filtered at $0°C$ (32°F), viscosity is much more significant here than it is during lautering [105].

Frequently, beer is subjected to stabilisation treatments to remove materials which will lead to colloidal instability in the packaged product. Chief amongst such materials are polypeptides and tannins. The latter can be removed by adsorption on polyvinylpolypyrrolidone (PVPP) [106], the former by adsorption on silica hydrogels [107].

Although there is a progressive shift towards sterile-filtered small pack products [108], most brewery-conditioned beer is pasteurised. A detailed description of beer pasteurisation is given by O'Connor-Cox et al. [63].

Traditional British ale is fined and served from casks. The art and practice of fining, principally using isinglass, is described by Leather [109] and Grimmett [110].

20.6 Cider-making

Relevant physico-chemical aspects of the cider-making process are summarised in Figure 20.6 and Table 20.1. For those seeking a fuller description, this can be found in a review by Lea [44].

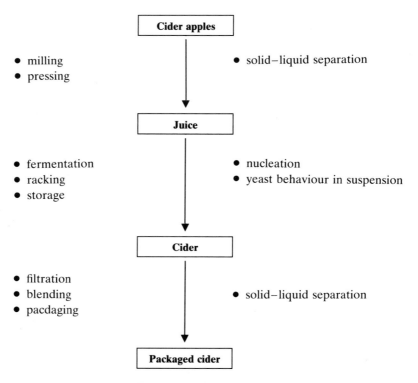

Figure 20.6 Cider-making.

Many of the same principles of solid–liquid separation, fermentation and processing are analogous for cider as for beer. In truth, the literature on cider research is substantially more restricted. The cider-making process is simpler insofar as there is no lengthy pretreatment of the primary raw material (apples).

20.6.1 Milling and pressing of apples

Ripe fruit is stored over a period of weeks to ensure full conversion of starch to sugar, before the apples are washed and (in modern cider manufacture) fragmented via high-speed grater mills, followed by pressing, which in modern facilities involves a high-pressure piston press similar to the Meura described earlier.

The tannins inhibit the breakdown of pectin (Lea [44]). The attendant reduced sliminess makes for an easier pressing operation. Alternatively, pectin hydrolysing enzymes may be used to remove pectin and stimulate pressing and yield [111].

20.6.2 Cider fermentation

Unlike worts, apple juices are generally deficient in assimilable nitrogen, necessitating the introduction of ammonium salts. Also, it is frequently the case that solid materials (e.g. bentonite) are added to bright juice to promote nucleation and a surface which the yeast can colonise [112]. The nucleation prevents inhibition of yeast by accumulating carbon dioxide. Pectolytic enzymes may be added prior to fermentation to further ensure pectin elimination and circumvent filtration problems downstream.

The role of yeasts in cider-making has been comprehensively reviewed by Beech [113]. Unlike brewery fermentations, in cider production, organisms additional to *Saccharomyces cerevisiae* can have a substantial role, e.g. during the malo-lactic fermentation of the traditional process.

20.6.3 Cider processing and packaging

Filtration is performed either just before or after blending (which is the norm in cider-making, but not in brewing). As for beer, filtration may either be powder-based or cross-flow [114]. The latter is more straightforward for cider than for beer as it is performed at higher temperatures (hence, lower viscosity) and cider contains fewer of the gums which foul filters.

20.7 Conclusion

Physico-chemical properties and reactions play an important role in determining the final quality of both beer and cider. The emulsification and surface effects described in chapters 3 and 8 are particularly important with respect to foam formation and retention, whereas Maillard and fermentation reactions (chapter 4 and 9) are vital in the flavour development. Science is therefore enabling us to understand, and to a certain extent control, the production of these two very traditional beverages.

Acknowledgements

The Director-General of BRF International is thanked for permission to publish this chapter. Dr Robert Muller is thanked for this frank observations on the manuscript.

References

1. Anderson, R.G. *J. Inst. Brew.* **98** (1992), 85–109.
2. Beech, F.W. *Ferment* **6** (1993), 259–270.
3. Bamforth, C.W. *J. Inst. Brew.* **91** (1985), 370–383.
4. Bruning, T. *Brew. Guard.* **123**(2) (1994), 22–24.
5. Carroll, T.C.N. *Tech. Quart. Mast. Brew. Assoc. Am.* **16** (1979), 116–119.
6. Archibald, H. *Brew. Guard.* **113**(8) (1984), 30–32.
7. Segel, E., Glenister, P.R. and Koeppl, K.G. *Tech. Quart. Mast. Brew. Assoc. Am.* **4** (1979), 104–113.
8. Little, B. *Brew. Guard.* **121**(11) (1992), 27–32.
9. Amaha, M., Kitabatake, K., Nakagawa, A., Yoshida, J. and Harada, T. *Proc. Eur. Brew. Conv. Congr., Saliburg* (1973), 381–398.
10. Moir, M., Dobson, G. and Seaton, J.C. *Proc. Eur. Brew. Conv. Congr., Lisbon* (1991), 225–232.
11. Ronteltap, A.D., Hollemans, M., Bisperink, C.G.J. and Prins, A. *Tech. Quart. Mast. Brew. Assoc. Am.* **28** (1991), 25–32.
12. Hughes, P.S. and Simpson, W.J. *Eur. Brew. Conv. Monogr. XXII, Symposium on Hops, Zoetwoude* (1994), 128–140.
13. Onishi, A. and Proudlove, M.O. *J. Sci. Food Agric.* **65** (1994), 233–240.
14. Hollemans, M., Tonies, T.R.J.M., Bisperink, C.G.J. and Ronteltap, A.D. *Tech. Quart. Mast. Brew. Assoc. Am.* **28** (1991), 168–173.
15. De Vries, A.J. In *Adsorptive Bubble Separation Techniques*, Ed. Lemlich, R., Academic Press, New York, (1972), 7.
16. Slack, P.T. and Bamforth, C.W. *J. Inst. Brew.* **89** (1983), 397–401.
17. Yokoi, S. and Tsugita, A. *J. Am. Soc. Brew. Chem.* **46** (1988), 99–103.
18. Sorensen, S.B., Bech, L.M., Muldbjerg, M., Beenfeldt, T. and Breddam, K. *Tech. Quart. Mast. Brew. Assoc. Am.* **30** (1993), 136–145.
19. Roberts, R.T. *Brew. Dig.* **52**(6) (1977), 50–58.
20. Klopper, W.J. *J. Inst. Brew.* **60** (1954), 217–222.
21. Klopper, W.J. *Brauwelt* **110** (1970), 1807–1811.
22. Maeda, K., Yokoi, S., Kamada, K. and Kamimura, M. *J. Am. Soc. Brew. Chem.*, **49** (1991), 14–18.

23. Yokoi, S., Yamashita, K., Kunitake, N. and Koshino, S. *J. Am. Soc. Brew. Chem.* **52** (1994), 123–126.
24. Goldstein, H. and Ting, P. *Eur. Brew. Conv. Congr., Monogr. XXII, Symposium on Hops, Zoeterwoude* (1994), 141–164.
25. Templar, J., Arrigan, K. and Simpson, W.J. *Brew. Dig.* (1995), in press.
26. Bamforth, C.W., Butcher, K.N. and Cope, R. *Ferment* **2** (1989), 54–58.
27. Langstaff, S.A. and Lewis, M.J. *J. Inst. Brew.* **99** (1993), 31–37.
28. Langstaff, S.A. and Lewis, M.J. *Tech. Quart. Mast. Brew. Assoc. Am.* **30** (1993), 16–17.
29. Ono, M., Hashimoto, S., Kakudo, Y., Nagami, K. and Kumada, J. *J. Am. Soc. Brew. Chem.* **41** (1983), 19–23.
30. Jowitt, R. *J. Text. Stud.* **5** (1974), 351.
31. Clapperton, J.F. *J. Inst. Brew.* **79** (1973), 495–508.
32. Langstaff, S.A., Guinard, J-X. and Lewis, M.J. *J. Am. Soc. Brew. Chem.*, **49** (1991), 54–59.
33. Narziss, L. *Abriss der Bierbrauerei*, Ferdinand Enke Verlage, Stuttgart (1972).
34. Dadic, M. and van Gheluwe, G.E.A. *Tech. Quart. Mast. Brew. Assoc. Am.* **10** (1973), 69–73.
35. Comrie, A.A.D. *J. Inst. Brew.* **73** (1967), 335–341.
36. Charalambous, G. In *Brewing Science*, Vol. 2, Ed. Pollock, J.R.A. Academic Press, London, (1991), 167–254.
37. Luchsinger, W.W. *Brew. Dig.* **42**(2) (1967), 56–63.
38. Meilgaard, M.C. and Peppard, T.L. In *Food Flavours, Part B. The flavour of beverages*. Ed. Morton, I.D. and Macleod, A.J. (1986), Elsevier, New York.
39. Parker, W.E. and Richardson, P.J. *J. Inst. Brew.* **76** (1970), 191–198.
40. Langstaff, S.A., Guinard, J-X. and Lewis, M.J. *J. Inst. Brew.* **97** (1991), 427–433.
41. Szczesnak, A.S. In *Food Texture and Rheology, Proceedings of IUFOST Symposium, London*, Academic Press, London (1977).
42. Parkinson, C. and Sherma, P. *J. Text. Studies*, **2** (1971), 451.
43. Christensen, C.M. *J. Text. Studies* **10** (1979), 153.
44. Lea, A.G.H. In *Fermented Beverage Production*. Eds Lea, A.G.H. and Piggott, J.R. Blackie Academic and Professional, Glasgow (1995), 66–96.
45. Lea, A.G.H. In *Bitterness in Foods and Beverages (Dev. Food. Sci. 25)*. Ed. Rouseff, R. Elsevier, Amsterdam (1990) 123–143.
46. Lea, A.G.H. and Arnold, G.M. In *Sensory Quality in Foods and Beverages*. Eds Williams, A.A. and Atkin, R.K. Ellis Horwood, Chichester (1983), 203–211.
47. Meilgaard, M.C. *Food Qual. Prefer.* **4** (1993), 153–167.
48. Williams, A.A. and May, H.V. *J. Inst. Brew.* **87** (1981), 372–375.
49. Gardner, R.J. *Tech. Quart. Mast. Brew. Assoc. Am.* **4** (1979), 153–167.
50. Williams, A.A. and Prosser, P.R. *Chemical Senses*, **6** (1981), 149–153.
51. Jupp, D.H. *Brewer* **80** (1994), 466–469.
52. Smedley, S.M. *J. Inst. Brew.* **98** (1992), 497–504.
53. Goodenough, P.W., Kessell, S., Lea, A.G.H. and Loeffler, T. *Phytochemistry* **22** (1979), 359–363.
54. Bamforth, C.W. and Barclay, A.H.P. In *Barley Chemistry and Technology*. Eds MacGregor, A.W. and Bhatty, R.S. American Association of Cereal Chemists, St. Paul, Minnesota (1993), 297–354.
55. Hough, J.S., Briggs, D.E., Stevens, R. and Young, T.W. *Malting and Brewing Science* (2 vols), Chapman and Hall, London (1984).
56. Marsh, J.B. *Brew. Guard.* **104** (1975), 21–37.
57. Abrahamson, M. *Proc. Conv. Inst. Brew. (Aust. NZ section)* (1978), 247–259.
58. Miedaner, H., Narziss, L. and Schneider, F.P. *Eur. Brew. Conv. Mongr. XVIII, Symposium Wort Biol. Clarif., Strasbourg* (1991), 37–44.
59. Forrest, I.S., Cuthbertson, R.C. and Dickson, J.E. *Proc. Eur. Brew. Conv. Congr., Lisbon* (1991), 505–512.
60. Molzahn, S.W. *Brew. Rev.* **20** (1990), 7–9.
61. Cahagnier, B., Lesage, L. and Richard-Molard, D. *Lett. Appl. Microbiol.* **17** (1993), 7–13.

62. Jakobsen, M. and Piper, J.U. *Tech. Quart. Mast. Brew. Assoc. Am.* **26** (1989), 56–61.
63. O'Connor-Cox, E.S., Yiu, P.M. and Ingledew, W.M. *Tech. Quart. Mast. Brew. Assoc. Am.* **28** (1991), 67–77.
64. Pierpoint, D.J. *Brewer* **74** (1988), 280–294.
65. Bamforth, C.W. *Brew. Dig.* **69**(5) (1994), 12–21.
66. Karakus, M. and Scriban, R. *Bios* **5** (1974), 498–519.
67. Seaton, J.C. *Proc. Eur. Brew. Conv. Congr. Madrid* (1987), 177–188.
68. Moll, M. and Moll, N. *Brau. Rundsch.* **101** (1990), 2–10.
69. Moll, M. and Midoux, N. *Tech. Quart. Mast. Brew. Assoc. Am.* **22** (1985), 67–72.
70. Masschelein, C.A. *Brauwelt* **115** (1975), 608–648.
71. Marchbanks, C.J. *Brew. Dist. Int.* **20** (1989), 16–21.
72. Dadic, M. *Tech. Quart. Mast. Brew. Assoc. Am.* **21** (1984), 9–26.
73. Weaver, F.B. *Tech. Quart. Mast. Brew. Assoc. Am.* **21** (1984), 140–150.
74. Schu, G. and Meyer-Pittroff, R. *Brauwelt Int.* **1** (1986), 79–89.
75. Norton, S. and d'Amore, T. *Enzyme Microb. Technol.* **16** (1994), 365–375.
76. Dahlstrom, R.V. and Sfat, M.R. *Brew. Dig.* **47** (1972), 75–80.
77. Wood, K.A., Quain, D.E. and Hinchliffe, E. *J. Inst. Brew.* **98** (1992), 325–327.
78. Marc, A., Engasser, J.M., Moll, M. and Flayeux, R. *Util. Enzymes Technol. Aliment. Symp. Int.* (1982), 115–119.
79. Palmer, G.H. and Harvey, A.E. *J. Inst. Brew.* **83** (1977), 295–299.
80. Palmer, G.H. *Proc. Am. Soc. Brew. Chem.* **33** (1975), 174–180.
81. Rennie, H. and Ball, K. *J. Inst. Brew.* **85** (1979), 247–249.
82. European Brewery Convention Analytica, 4th edn. *Brauerei und Getranke Rundschau*, Zurich (1987).
83. Herrmann, H. *Brewer* **81** (1995), 12–14.
84. Anger, H-M. *Brauwelt* **133** (1993), 146–148.
85. Hermia, J. and Rahier, G. *Ferment* **5** (1992), 280–286.
86. Muller, R.E. *Proc. Eur. Brew. Conv. Congr. Zurich* (1989), 283–290.
87. MacGregor, A.W. and Fincher, G.B. In *Barley Chemistry and Technology*. Eds MacGregor, A.W. and Bhatty, R.S. American Association of Cereal Chemists, St. Paul, Minnesota (1993), 73–130.
88. Moonen, J.H.E., Graveland, A. and Muts, G.C.J. *J. Inst. Brew.* **93** (1987), 125–130.
89. Moll, M. *Beers and Coolers*, Intercept, Andover (1991).
90. Denk, V. *Ferment* **7** (1994), 299–312.
91. Dickenson, C.J. *J. Inst. Brew.* **85** (1979), 329–333.
92. Tressl, R., Koss, T. and Renner, R. *Proc. Eur. Brew. Conv. Congr., Nice* (1975), 737–756.
93. Kossa, T. Thesis. Technical University of Berlin (1976).
94. Drawert, F. and Wachter, H. *Monat. Brauwiss.* **37** (1984), 304–313.
95. Buckee, G.K., Malcolm, P.T. and Peppard, T.L. *J. Inst. Brew.* **88** (1982), 175–181.
96. Pensel, S. *Brauwelt* **129** (1989), 1242–1248.
97. Reed, R.J.R. and Jordan, G. *Proc. Eur. Brew. Conv. Congr., Lisbon* (1991), 673–680.
98. Hammond, J.R.M. In *The Yeasts*, Volume 5, 2nd edn. Ed. Rose, A.H. Academic Press, London (1993), 7–67.
99. Stratton, M.K., Campbell, S.J. and Banks, D.J. *Proc. Conv. Inst. Brew. (Asia Pacific section)* (1994), 96–100.
100. Speers, R.A., Tung, M.A., Durance, T.D. and Stewart, G.G. *J. Inst. Brew.* **98** (1992), 525–531.
101. Axcell, B., Kruger, L. and Allan, G. *Proc. Conv. Inst. Brew. (Aust. NZ section)*, Brisbane (1988), 201–209.
102. Evans, J.I. *Process Biochem.* **7** (1972), 29–32.
103. Reed, R.J.R. *J. Inst. Brew.* **92** (1986), 413–419.
104. Burrell, K.J. *Tech. Quart. Mast. Brew. Assoc. Am.* **31** (1994), 42–50.
105. Freeman, G.J. and McKechnie, M.T. In *Fermented Beverage Production*. Eds Lea, A.G.H. and Piggott, J.R. Blackie Academic and Professional, Glasgow (1995), 325–350.
106. Marchbanks, C.J. *Brew. Dist. Int.* **16** (1986), 30–37.
107. Fernyhough, R. and Ryder, D.S. *Tech. Quart. Mast. Brew. Assoc. Am.* **27** (1990), 94–102.

108. Ryder, D.S., Barney, M.C., Daniels, D.H., Borsh, S.A. and Christiansen, K.J. *Eur. Brew. Conv. Monogr. XXI, Symposium on Process Hygiene, Nutfield* (1994), 64–80.
109. Leather, R.V. *Brewer* **80** (1994), 429–433.
110. Grimmett, C. *Brewer* **80** (1994), 522–524.
111. Lea, A.G.H. In *Production and Packaging of Non-carbonated Fruit Juices and Fruit Beverages*, 2nd edn. Ed. Ashurst, P.R. Blackie Academic and Professional, Glasgow (1994), 182–225.
112. Ough, C.S. and Groat, M.L. *Appl. Environ. Microbiol.* **35** (1978), 881–885.
113. Beech, F.W. In *The Yeasts*. Ed. Rose, A.H., Academic Press, London (1993), 169–213.
114. Scott, J.A. *Process Biochem.* **23** (1988), 146–148.

21 Multi-component foods
C.A. STREET

21.1 Introduction

The chapters in the second part of this book have discussed some of the physico-chemical factors which explain the changes occurring during the processing and storage of various classes of food. Although this knowledge is not complete and further research is needed to enhance understanding and control of the manufacturer's available options, the product development worker has a considerable wealth of literature on which to base his/her deliberations and experimental programmes. However, with the increasing use of convenience foods, many manufacturers are processing more than one of these food classes together in order to make a single product. Some typical examples are listed in the Table 21.1.

In all cases, the different product types interact. Sometimes the interaction involves significant chemical and physical changes during manufacture, as with the baking of a meat pie. Alternatively, it may only entail physical migration of fat or water during storage and distribution. Between these extremes are frozen meals, which can undergo chemical change during thawing and reheating by the consumer. Of these interactions, some are highly desirable and produce characteristic qualities in the product, while others may have detrimental effects.

It is clear from patent and trade literature, that many manufacturers are familiar with these changes and are seeking solutions which overcome the difficulties they introduce. For example, a German patent [1] claims retardation of pigment migration between stewed fruit and a cream topping. However, despite the obvious interest, the published information directly relating to these issues is very sparse. A qualitative picture of the changes to be expected can be predicted from knowledge of the individual product types involved, but quantitative information is required, if the effects are to be adequately controlled.

The physico-chemical changes of the individual classes of food discussed in previous chapters depend on the local environment and the time they are exposed to those conditions. In multi-component products, the situation is much more complex. Two types of difference are specific to these products:

1. The compositions of the components are different and hence, there is scope for diffusion of fat, water, salt, etc. from one to the other

Table 21.1 Multi-component products

Product	Components	Comments
Meat pie	Meat Gravy or sauce Pastry	Partially cooked meat is enclosed in a pastry case before baking.
Lemon meringue pie	Pastry Fruit/custard Meringue	Fruit, usually in a custard type paste, is poured into a pre-baked pastry base and covered with meringue before baking.
Choc ice	Ice cream Chocolate	Ice cream is enrobed in chocolate.
Quiche	Pastry Fruit/vegetable	An open pastry case containing a sweet or savoury filling. Often stored and sold frozen.
Pizza	Pastry or similar Cheese Meat/vegetables	A pre-baked pasta base is covered with a mixture of cheese and meat or vegetables and cooked to melt the cheese.
Torte/gateaux	Cake/pastry Fruit Cream	Layers of flour confection are separated by layers of fruit and/or cream. The resultant product is often frozen.
Oxtail soup	Meat Vegetables	Meat and vegetables are suspended in liquid thickened with flour.
Chocolate chip cookies	Chocolate Biscuit	Pieces of chocolate are mixed into a biscuit type batter before baking.
Frozen meals	Meat Potato or pasta Vegetables Gravy or sauce	The components are cooked before being combined but may interact during reheating.

leading to chemical reactions and physical changes, which do not occur when the materials are considered alone.

2. One component, such as the pastry in a meat pie, can shelter the other from the external environment, thus affecting the rate at which changes occur, relative to those occurring when directly subjected to the prevailing conditions.

Hence, heat and mass transfer within the product have a major role to play in defining the interactions which occur and the extent to which they develop within the product.

21.2 Mathematical modelling of heat and mass transfer

Mathematical simulation can greatly aid the food technologist in understanding and optimizing his or her processes, for example it can determine the minimum time for removing moisture (this is not always obvious, as just raising the temperature might cause case hardening, which will reduce the drying rate, whilst each component will have its own diffusion rate which will probably vary with moisture content). Alternatively it may be possible to determine the relationship between temperature and cooking time, so as to establish the most cost-effective conditions to use, whilst retaining the required eating qualities.

Many multi-component, prepared products have a well defined shape, and any changes brought about during cooking, cooling or storage are due to factors outside of it. Hence, heat within the product, in a conventional oven, can be completely defined in an equation of the form:

$$\partial \psi(x,y,z,t)/\partial t = D[k(x,y,z,t) \, D\psi(x,y,z,t)] \qquad (21.1)$$

where $\psi(x,y,z,t)$ is a function which defines the temperature (t) at any point within the sample at any time, and $k(x,y,z,t)$ is the coefficient (dependent on the thermal conductivity, density and specific heat) defining the rate of change at any point within the material at any time. D is the differential operator $[\partial/\partial x + \partial/\partial y + \partial/\partial z]$.

The same equation describes the mass transfer within the product, with the function $\psi(x,y,z,t)$ defining the value of the variable (for example, the activity of the relevant molecular species) at any position and any time. In this case, the coefficient $k(x,y,z,t)$ is the appropriate diffusion coefficient, which will change with temperature and composition. Many texts refer to the concentration of a molecular species as being related to the motive force, with molecules in high concentration regions moving to low-density ones at a rate dependent upon the concentration difference. While this may work well for mixtures of non-reacting gases and when considering fat migration (in this case several, closely related, molecular species can sometimes be treated as one), it can cause major problems when dealing with water transport through solid and liquid materials. Here it is the activity difference which provides the driving force for movement, and the constant relating water content to water activity is composition dependent, and it is even possible for moisture to move to regions of higher water content, if this has a lower water activity (chapter 1). Hence, this constant is different for each product component, even if it does not change during processing. Furthermore, the assumption that the activity of water within a single component is proportional to the water content of that component, is an acceptable approximation only when changes are relatively small, which is seldom the case in multi-component products. Here, for example, the amount of one component may be small relative to another, and some of the molecules from the major section will migrate into the small one and vice versa. Certain of these molecules, in particular salts and surfactants can dramatically change the water activity of the area where they are present.

In order to mathematically model these situations many workers have used this second-order partial differential equation, which is a generalized mathematical statement of Fick's second law. For example, Palmia [2] showed that the migration of salt into ham during curing obeyed this law.

In most cases, the equation can be simplified to a single space dimension form:

$$\partial\psi(x,t)/\partial t = \partial[k(x,t) \cdot \partial\psi(x,t)/\partial x]/\partial x \qquad (21.2)$$

which considerably reduces the computation required to solve the equation.

In simple cases, the expressions $k(x,y,z,t)$ and $k(x,t)$ are constants allowing analytical solution of the equation, but for these products this is clearly not the case. For example, the thermal conductivity of pastry dough is about half that of the pork it surrounds in a pork pie. It would also be surprising if the value for the pastry did not change as the water activity (related to the water content) changes. In addition, the diffusion coefficient of water within both pork and pastry will change with temperature. Similarly, we would expect the diffusion coefficient for fat molecules in chocolate to differ from that in ice cream.

Fortunately, provided the initial condition (initial state of the sample with respect to the variable of interest) and the boundary condition (value of the variable at the sample surface, e.g. oven temperature, at all times) are known, these equations can be readily solved numerically using an implicit Crank–Nicholson or similar procedure, in one dimension using a standard desk-top computer. The numerical procedure involves dividing the space dimension into N intervals and the time axis into a number of small, equal steps (say M in number). The value of the variable $u_{i,j+1}$ (the value of u at position i along the x-axis after $j + 1$ time steps) can be defined in terms of u_{ij} and its neighbours by rewriting equation (21.2) in the form:

$$-au_{i-1,j+1} + bu_{i,j+1} - cu_{i+1,j+1} = au_{i-1,j} + du_{i,j} + cu_{i+1,j}. \qquad (21.3)$$

a, b, c and d depend on the size of the space/time steps and the current value of the coefficient $k_{i,j}$. The scheme is shown diagrammatically in Figure 21.1. The boundary value of $u_{0,j}$ is defined by the external applied conditions and thus is known at all times. Similarly, the value of $u_{N,j}$ may be defined by the external conditions, or in some cases, by invoking symmetry considerations. The values $u_{i,0}$ are known, for all i, from the initial state of the material, which can be readily measured. Hence, the system becomes an iterative process based on a set of $N - 1$ simultaneous equations in $N - 1$ unknowns, to be solved for each time step. Several methods of solving these equations are described by Smith [3].

By using the same space and time steps for each set of equations changes in temperature, water activity, fat distribution, etc. can be followed at the same time, by computing the changes in each factor for a single time step, then updating the values of $k_{i,j}$ for the effects of change

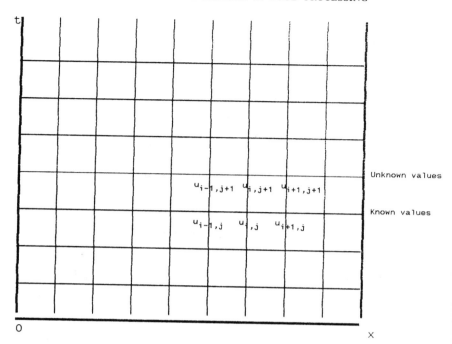

Figure 21.1 Diagramatic representation of the solution of the diffusion equation.

in the relevant variables (including any chemical change occurring in the time interval under the prevailing conditions), prior to proceeding with the next time step. This updating between time steps is essential as diffusion coefficients change with both temperature and concentration in a system, which is being heated or cooled.

Figures 21.2 and 21.3 illustrate the pattern of change in temperature and water content in a pie, respectively, with time during baking.

In conventional ovens the heat is transferred directly to the surface of the product being heated by radiation, convection or conduction, but in microwave ovens part of the energy input penetrates the product, where it generates heat. However, heating and drying in microwave ovens can still be modelled using the same equations, if an additional term is added. The additional term describes this internal heat generation, which results from electrical coupling. The subject is discussed in detail by Mudgett [4].

21.3 Application of models to control and development of processes

The models described above can be very effective for developing an understanding of processes in the controlled environment of the research

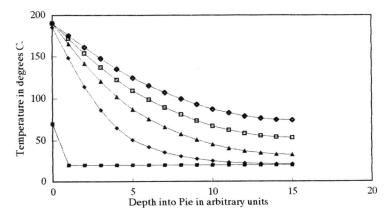

Figure 21.2 Change in temperature distribution with time during baking.

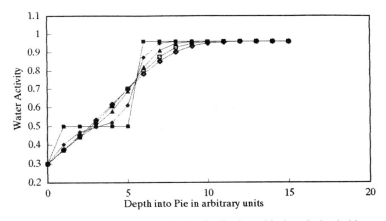

Figure 21.3 Change in water activity distribution with time during baking.

laboratory. They can be used effectively to define new process routes to a given product. However, when used for refining and controlling process operations, four main difficulties limit their value:

1. The computing time may be long relative to the reaction time of the process operation.
2. Material non-uniformity, random variation of controlled variables (e.g. temperature hot spots) and imperfections in the model structure lead to uncertainty in the predicted distribution functions, which define the state of the product at any given time.
3. Product quality, the variable of interest to the consumer, is a qualitative experience rather than numeric quantity. Hence, the relation between

the state of the product (as defined by the set of distribution functions or any other set of quantitative measurements related to it) and the quality of the product is necessarily ill-defined (fuzzy).
4. Different inputs to a model will produce different outputs. In many cases more than one such output will relate to a satisfactory product. Few, if any, models incorporate a mechanism for identifying acceptable outputs.

Simpler, traditional, often empirical, models can overcome the first problem as they involve simple, rapid calculations but only at the expense of making the second difficulty worse. For example, the Casson equation, used in chocolate manufacture (chapter 17), uses a small number of measurements to produce values for two dependent variables (i.e. yield value and plastic viscosity). Since each of these variables records the combined effect of at least four factors (i.e. temperature, particle size distribution, water content, volume fraction of the solid material), each of which changes during the process operation (although none of them is included in the equation), the advantage of simplicity is overshadowed by the dubious nature of assumptions required to make use of the results. The third and fourth problems are still present.

Because of these limitations the value of many mathematical models is restricted to setting the limits within which a closely defined process can operate. To increase the flexibility of process control, make full use of available knowledge and give both engineering and purchasing departments the maximum freedom of choice, a new methodology is required. Fuzzy logic and neural networks, usually in combination, offer a possible approach.

A craftsman confectioner can look at, smell and feel a sample and deduce that it requires an additional 5 minutes in the oven, with reasonable precision. It may not be known what the senses are measuring, but the observations can be compared with those from past experience. Confectioner experience suggests the action required to obtain an adequately correct conclusion. Having observed the result of that action, the confectioner updates the experience for future use.

A 'set' can be defined as a collection of objects, quantities or ideas which have specified qualities in common. Logic is a branch of mathematics which deals with the membership of and relationships between sets. In traditional logic, the contents of sets are clearly defined and the relationships between them are quantitative, usually one to one. Human thought, however, tends to be expressed qualitatively in words, the deductions involve approximate rather than exact reasoning and the transition from membership to non-membership of sets is gradual rather than abrupt. Fuzzy logic was developed as a generalization of traditional logic (which is included as a limiting case) to enable ideas of an approximate nature, characteristic of common-sense reasoning, to be manipulated by computer. Fuzzy logic, therefore, may be considered as a language with the structural form required to describe past experience. Dohnal et al. [5] have applied this approach to meat chilling, malt modification and sugar refining and refer their readers to Lee [6] for more detail of the approach.

Neural networks are structures, designed for computers, based on understanding of brain function. They can take in large arrays of data, reduce them to smaller arrays, using a suitable calculating procedure, and compare the resultant patterns with a library of patterns entered during a learning stage. The library patterns can be recorded as sequences generated during carefully designed experiments and hence, the changes in pattern can be related to the settings of the controlled variables of the process. Human 'expert' assessment of the material being processed may also be recorded. The library data may be in 'fuzzy' form, when appropriate. Having found the best match (or matches) between pattern and library data, the network system contains the necessary information to suggest possible courses of action, which can be activated via feedback loops. Since the data patterns generated during use of the network and machine settings can be recorded and 'expert' judgements are easily input as required, the system also has the capacity to refine its library in a manner similar to that used by updating personal craft experience.

Many modern techniques, such as NMR (nuclear magnetic resonance) or infrared spectroscopy, can generate complex spectra almost instantaneously. These spectra, in some complex but barely understood way, relate to the composition and structure of a sample and are ideally suited to neural network applications. Each spectrum can be recorded as a large number of data points, which can be combined using suitable weightings to produce a smaller number of variables; a number of about 15 has been found optimal for some applications [7]. More than one source of spectral data can be used and other measurement signals included. In fact, a combination of fuzzy logic and neural network can utilize all available information and simulate craft experience and judgement, thus suggesting an appropriate course of action. The measurements and selected action may be related directly to the quality of the resultant product, defined in terms of good, satisfactory, below par, etc. and does not require correlation with some quantitative measure which only partially defines quality.

There are many possible uses of these techniques in industry. One interesting example is in the production of iced cakes, where the manufacturer wants each one to look slightly different, so as to give them a 'home-made' appearance rather than a mass-produced one. At the same time, it is necessary to ensure that each cake has at least an ice coating and the correct type of pattern. Thus, unlike many plastics or metal plants, each item will be different. It is possible, however, to use a TV camera and image-analysis system to produce the data. The problem is that many cakes are being produced per minute, so the analysis must be fast. Neural networks, however, can be used to handle the data with fuzzy logic to decide if the cake is within the desired limits. It is possible to get an automated quality control system. This is, therefore, an example where more than one model output satisfies the product quality criteria, and the system needs information as to which outcomes are acceptable and which are not.

Material flow can also be modelled numerically and computer fluid dynamics (CFD) software is becoming increasingly available and used in the food industry. This can be used to predict how materials will flow through a die at the outlet of an extruder, or how the air will pass through a drier or a smoking system, thus allowing processing conditions to be modified to ensure a uniform product.

21.4 Meat pies

The cooking of a meat pie provides a good example of the problems encountered when making a multi-component product. When baked alone, pastry dough undergoes a series of changes as the surface temperature rises from ambient to that prevailing in the oven. Water is lost from the outer surface, which at the end of the bake has very low water activity. The conditions of high temperature and low water activity are ideal for the interaction between amino acids and sugars, both produced from the flour, leading to browning and development of caramel flavours (chapter 4). These changes occur throughout the pie, but are more noticeable at the surface, where the temperature is higher and the water activity lower. Similarly, meat cooked alone or in gravy heats up during the cooking period and changes, such as degradation and hydrolysis of proteins, occur. The extent of the change depends on time, temperature and water activity.

In a meat pie, the part-processed meat is enclosed in pastry dough before baking. Hence, as indicated in Figure 21.2, the temperature of the meat rises much more slowly than when cooked alone. The temperature at the centre may not reach 70°C (158°F) (the minimum required to ensure pasteurization) during the time required to bake the pie. Hence, the extent to which the meat is pre-cooked and its subsequent handling is of great importance.

The outer surface of the pastry changes in much the same way as when cooked alone. However, the changes at the boundary between filling and pastry are different and of prime significance to the eating quality.

Pastry dough has a lower water activity than the meat. Hence, during cooking, despite the water loss from the exposed surface, the overall water content of the pastry crust can increase. As indicated in Figure

21.3, the water activity (and water content) increases significantly at the interface with the inner layers of the pastry crust becoming wetter than in the original dough. The temperature of the dough–meat interface may reach temperatures around 100°C (212°F) allowing the starch to swell and begin to gelatinize, yielding a soft custard-like texture. The inner surface of the pastry will not brown readily, despite the free amino acids from the meat layer, due to the increase in water activity and the lower temperature resulting from the presence of the meat; the Maillard reaction develops most readily at temperatures above 100°C (212°F) and relatively low water activity. In the presence of gravy, starch may leach into the gravy from the pastry; this starch may also swell or gelatinize causing a thickening of the fluid surrounding the meat. The meat layer will tend to dry as water moves into the pastry, although this is seldom noticeable organoleptically.

The changes will always occur during the baking of a pie, but the manufacturer has to control the extent, in each case, both to achieve consistent quality and to obtain the optimum balance for best organoleptic character.

The migration of water from filling to pastry will continue during storage, after the baking and cooling processes are complete. This will be exacerbated by absorption of water from the surrounding atmosphere into the low-moisture pastry surface. These increases in water content can lead to a toughening of the pastry surface and an increased rate of staling.

21.5 Other examples of changes during the processing of multi-component foods

21.5.1 Oxtail soup

Oxtail soup is essentially pieces of meat and vegetable suspended in water thickened with gelatinized starch. The cooking conditions required to ensure the product is sterile are sufficiently severe to release amino acids from the meat components. Similarly, there will be some release of reducing sugar from the starch thickener and within the vegetable pieces. The amino acids will diffuse throughout the liquid bulk and into the vegetable matter, where it can react with free aldehyde groups, leading to change in colour, flavour and texture of these pieces.

21.5.2 Pizza

Most pizzas consist of a dough layer on which cheese, with or without other ingredients, is spread before cooking. The pasta and cheese are often cooked and frozen, allowing a topping to be added at the time of reheating, immediately before consumption. Changes occur in the cheese during cooking, which include separation of fat. This fat can migrate into the pasta layer, changing its texture. Piquet and Kalab [8] report a study of changes in cheese during the cooking of pizzas. However, they separated the dough base from the cheese with a cotton cloth, thus the interaction between the cheese and the base was not observed.

The fat components of pizzas are particularly prone to oxidation during storage.

21.5.3 Chocolate chip cookies

Chocolate chip cookies are cake and biscuit type products with pieces of chocolate distributed within the mass. The process relies on a short baking time and the fact that the chocolate pieces have a significantly lower thermal conductivity than the dough or batter. The chocolate fat does not completely liquefy during the baking, leaving sufficient crystal of the desired form to maintain temper (correct fat crystal type) so as to set correctly on cooling. When the dough has a high fat content, fat migrates from the dough into the chocolate during the baking period. This change in fat composition can affect the crystal structure causing bloom (droplets of surface fat), which detracts in both appearance and texture. Even if the crystal structure is maintained, the increased polydispersity of the resultant fat can lead to a lower melting point (chapter 7) and greater risk of melting in the hand of the consumer.

Cooling the chocolate pieces before incorporating them reduces the risk of them melting completely, but leads to the possibility of condensation on the chip surfaces. As the condensate warms up, it will dissolve sugar and subsequently begin to evaporate. In the later stages of baking the sugar from this syrup layer can caramelize, producing a blotchy coarse surface coating on the chips exposed to view.

21.5.4 Choc ices

These are made by enrobing ice cream with chocolate. For this purpose the chocolate, necessarily, has a temperature of $>30°C$ ($>86°F$), while the ice cream will have a surface temperature below the freezing point of water. Hence, a thin layer of the ice cream will melt and subsequently refreeze as the chocolate cools and solidifies. During this period, when both ice cream and chocolate are semi-solid, there is opportunity for exchange of fats between the two components. The resultant change in composition of the chocolate fat will alter the solidification/melting temperature of the chocolate and could lead to fat bloom (see also chapter 17).

21.6 Frozen meals

Frozen meals usually consist of meat, potato or pasta, other vegetables and a gravy or sauce. The component foodstuffs are cooked or partially cooked and possibly frozen before being combined. The overall combination is packed and frozen for storage. Hence, there is little scope for interaction between components up to this point. Also, little change will occur during storage, since most rate constants are very low, at the recommended freezer temperatures.

However, the meal has to be thawed and heated prior to consumption. During this phase, the various components will heat at different rates. There is a risk that some components (particularly vegetables with short cooking times) become overcooked, while others have not reached the desired temperature. The nature of this problem could be very different according to the type of heating used.

When reheated in conventional ovens, the recommended heating time can be $\geqslant 40$ minutes. Drying of the meal surface is prevented by heating in the sealed packet, as is usually recommended on the wrapper. However, many vegetables will be overcooked during this reheating period, even when the absolute minimum of blanching is used prior to freezing. Defrosting at room temperature prior to heating could reduce overcooking of vegetables, but would allow more scope for interaction between components.

Most frozen meals are reheated using microwave ovens. In this case defrosting occurs almost as rapidly in the centre as it does at the surface. However, the components of the meal which defrost fastest will be subjected to additional cooking, while others are still warming up. If the meal is perfect when heated in a microwave oven, it will be less than ideal when heated in a conventional oven. Hence, the manufacturer has to compromise in formulating his product.

21.7 Conclusion

Although there are very many multi-component food products on the market, there is very little in the scientific literature about the physico-chemical changes which take place during their manufacture and use. Changing lifestyles, like the increased use of microwaves in the home, are making the situation more complicated for the food manufacturer, who often has to formulate a product to enable it to be used in several different ways. Fortunately, rapid advances in techniques like modelling, fuzzy logic and neural networks are providing help in aiding the design of new products, determining optimum processing conditions and greatly improving quality control.

References

1. German Federal Republic Patent DE3240068C2 (1987).
2. Palmia, F. *Revista Española de Cienca y Tecnología de Alimentos*, **32**(1) (1992), 71–83.
3. Smith, G.D. *Numerical Solution of Partial Differential Equations*. Oxford University Press, Oxford (1965, reprinted with corrections 1974).
4. Mudgett, R.E. *Microwaves in Food Processing*. Ed. Decareau, R.V. (1985), 38–56.
5. Dohnal, M. *et al. Journal of Food Engineering*, **19**(2) (1993), 171–201.
6. Lee, C.C. *IEEE Transactions on Systems, Man, and Cybernetics*, **20** (1990), 404–428.
7. Black, J.D. and Kerr, N.C. *Application of Neural Networks* (1992).
8. Piquet, A. and Kalab, M. *Food Microstructure*, **7**(1) (1988), 93–103.

Index

abrasive peeling 297
acetaldehyde 206
acetic acid 207, 208, 356, 388, 389–392, 407
acetolactate 204, 205
acetylation 317
2-acetylfuran 432
acid hydrolysis 407
acid preserves 387
acidification 298, 400
actin 222, 224
actomyosin 222
adenosine diphosphate (ADP) 193
adenosine triphosphate (ATP) 193, 198, 223–229
Admul-WOL 360
aeration 264
aerobic conditions 350
agarose 132, 133
ageing 279
agglomeration 191, 238
aggregation 134, 265
air vacuole 251, 252
albumin 222
alcohol 194, 200, 202, 392, 395, 424
aldehyde 395, 413, 449
aldose 67
ale 203, 425
alginate 120, 135–137
alkalizing 352
allyl isothiocyanate 409
Amadori rearrangement 68–70, 76
amino acids 65, 68, 69, 72, 75, 196, 273, 274, 340, 356, 384, 385, 405, 409, 425, 448, 449
amino group 14
ammonium salts 196, 435
amorphous form 18, 226, 338, 370, 372, 378
amorphous glass 332, 335
amphipathicity 421
amylolysis 431
amylopectin 36–38, 135, 218, 222, 268, 370, 379
amylose 134, 135, 215, 218, 219, 268, 370, 379
anaerobic glycolysis 228
anaerobic respiration 295
anchovies 408
anhydride bridge 133
animal tissues 222–230

antibiotic 201, 207
anti-oxidants 76
antiviral 207
apples 214, 294, 295, 298, 317, 388, 397, 400, 407, 435
apricots 216, 293
arabinans 312
arabinose 317
aroma 203
Arrhenius 27, 32, 34, 41, 51, 102
aseptic packaging 302, 311, 312, 321, 325, 328, 397, 405
asepton *see* aseptic packaging
asparagus 73, 293
Aspergillus 194
associative adsorption 59
atmosphere modification 400
atomisation system 254
Australia 388

bacon 282
bacteria 2, 194, 236, 237, 245, 246, 250, 283, 302, 391, 396
  acetic acid 205
  Acetobacter 205, 209
  halophilic 396
  lactic acid 206, 209, 210, 245, 249
  mucogenic 249
  propionic acid 246
bakery products 65, 192, 218, 236, 242, 258, 259, 268, 270, 272
baking powders 265
banana 295
barbecue sauce 407
barley 203, 278, 425–429
'bath tub' curve 300
batter 258, 264
bean, faba 218
beef 73, 76, 281, 287
beefburgers 280
beer 202–204, 417–436
beer flavour wheel 423
beet 400
beetroot 399, 404
bentonite 435
benzaldehyde 432
benzoic acid 390, 414
beta-glucan 423, 424, 431
Bifidobacteria 250
Bingham model 89
biomass 200

biopolymers 258, 262, 368
biosynthesis 193
birefringent 262
biscuits 258, 259, 263
bitterness units (BU) 422
black pepper 404
'blackneck' 405
blanching 75, 216, 292, 298–301, 309, 400, 451
blast freezer 252, 280
bottling 301
Botulinum 414, 302
'botulinum cook' 302
bran 380, 382
bread 65, 69, 194, 197, 258, 259, 268–272
breadcrumb 270
breakfast cereals 368–385
brewing 196, 202, 203, 417–436
brine 210, 217, 227–229, 391, 392, 394
brining 391, 394–396
Brix 310, 311
broilers 280
brown onions 404
brown sauce 405
Brownian motion 53, 54, 86, 102, 118
browning reactions 259, 268, 273, 274, 304, 340, 396, 405 *see also* Maillard
Brunauer–Emmett–Teller theory 10
bubble
  coalescence 419
  disproportionation 420
  formation 418
  pressure tensiometer 175
butter 199, 234–239
  creams 268
  oil 354
butterfat *see* milk fat
buttermilk 207, 238, 239

cabbage 295, 400
Cadbury's Dairy Milk 355
cake crumb 268, 270
cakes 258, 259, 263–273
calcification 298
calcium 196, 246, 249, 255
  calcium chloride treatment 298
  ion 135, 136, 216, 217, 223, 229, 246, 247, 318, 321, 399
  salts 245
Camembert 246
candied fruits 305
candy 248
candying 305
canned foods 75
canning 210, 271, 296, 301, 328, 399
capers 389, 404
Capparis spinosa L. 389

caramelisation 259, 272, 334, 340, 341, 384
caramels 13, 332, 338, 404, 425, 429
carbohydrate 195, 234, 252, 256, 272, 273, 300, 392
carbon dioxide 197, 200–203, 206, 208, 210, 224, 246, 295, 298, 350, 392, 394, 400, 418, 420–423, 433, 435
carbonation 422, 423
carbonyl-amine reaction *see* Maillard
carborundum 297
carboxymethyl cellulose 138
carcomeres 226
carmine red lycopene 397
carob beans 408
carotenes 397
carrageenan 120, 132, 133
carrots 73, 214, 298, 301, 400
casein 35, 61, 246, 249, 251, 339
  micelle 246, 249
  protein 339
  ratio 255
Casson equation 89, 446
Casson yield value 357–360
catalyst 143, 279
cauliflower 389, 395, 400, 404
caustic soda 297
cells 193–211, 270, 298
  fish 228
  hypodermal 297
  membranes 303
  structure 212–230, 292, 297, 312
  vacuole 212, 251, 252
  wall 394
cellulose 9, 212, 214
centrifugation 235, 237, 311
cereal 248, 258–274
charcoal filter 311
cheddaring 246
cheese 70, 194, 199, 207, 211, 234, 244, 245, 247, 449
  blue 208
  Brie 208
  Camembert 208
  Cheddar 246
  Emmental 208
  processed 245
  Roquefort 246
  Stilton 246
chemical peeling 297
cherries 305, 308
chicken 226, 283, 286
chill factor 363
chill storage 271
chilli 399, 404
chilling 279, 400, 447
chilling injury (CI) 401
chilling rates 279

chirality   *see* optical isomerism
chlorination   267
chlorine   267
chlorophyll   221
choc ices   450
chocolate   72, 88, 90, 94, 107, 150–153, 158, 191, 208, 347, 360, 450
  chip   450
  crumb   354
  dark   355–358
  fat bloom   152, 154, 450
  heat resistant   364
  milk   358, 361
  sugar bloom   364
  temper   153
  white   355
cholesterol   62, 249, 250
chromatography   421
chrymosin   246
chutney   387, 389, 401
cider   417–436
cis-formation (fat)   142, 146, 157
citrate   324
citric acid   390
citrus fruit   294, 295, 310, 315, 317, 388, 397
citrus juices   311
cleaning (vegetables)   295, 296
Clostridium botulinum   302, 307
clotted cream   207
cloves   404
coalescence   50, 51, 411, 412
coating   105, 173
cobalt   196
Coberine   159
cocktail onions   404
cocoa   167, 208
  beans   65, 147, 208, 347–350
  butter   144, 147–153, 158–163, 348, 351–355, 360, 364, 365
  butter equivalent   155, 158, 159
  liquor   351–355, 360
  nib   347, 351, 352
  pods   208, 347, 349
  powder   159, 351, 352
  shell   348, 350, 351
coconut oil   144, 156–158
cod   228
coffee   65, 69, 70, 243
coffee whitener   253–255
cold-shortening   225
collagen   134, 222, 227, 229
colloid mill   405, 406
colloids   312
colour formation   417, 425, 449
Comité des Industries des Mayonnaises et Sauces Condimentaire de la CEE (CIMSCEE)   390

competitive adsorption   58, 59
complex modulus   127
Computer Fluid Dynamics (CFD)   448
concentration   311, 312, 397, 399
conche   350, 355–357, 360, 361
condiments   387–414
conduction   363, 444
Consistometer
  Adams   399
  Bostwick   399
contact angle   167–192
continuous phase   411
controlled atmospheres (CA)   295, 400, 401
convection   287, 363, 444
convective air cooking   287
convenience foods   280, 440
cook value (Co)   307
cooked meat   283
cookies   46, 364, 450
cooking   276, 292, 301, 306, 307, 315, 333, 373
cooling tunnels   363
coriander   404
corn chips   374
corn flakes   369, 373
corn syrup   13, 196
cornflour   407
cottonseed oil   161
cotyledons   347
Couchman–Karasz (C–K) equation   36, 37
courgettes   309
covalent bond   131
cowpeas   217
Cox–Merz rule   129, 138
crack testing   341
crackers   3
Crank–Nicholson procedure   443
cream   234, 235, 240–244
cream of tartar (potassium hydrogen citrate)   335
creaming   49, 50, 411
crisps   382, 384
critical micelle concentration (CMC)   168–170, 172, 175
croissants   242
cross flow filtration   433
crumb structure   382
crust   272–274, 376
cryogenic freezing   304
cryogenic tunnels   280
crystallisation   3, 26, 27, 41, 105, 142–166, 221, 237–244, 251, 268, 270, 337, 342, 343, 361, 362
cucumber   295, 389, 395, 400
cultured foods   248
curd   207, 246

curd syneresis 246
cuticular waxes 297
cysteine amino acid 196
cytoplasma 212, 215, 219

dairy products 234–256
Dalton unit 317, 328
dates 407
De Vrie model 420
deaeration 405, 407
Deborah number 17, 30
debrining 401
defrosting 451
degree of methoxylation (DM) 317
dehydration 30, 46, 221, 222, 262, 273, 293, 304, 309, 313
dehydroascorbic acid 221
demersal 228, 277
Dent maize 372
deoxyribonucleic acid (DNA) 193, 196
desserts 117
dew point 286, 364
dextran 61
dextrins 423
dextrose 224, 335, 340, 344
diacetyl 70
diatomaceous earth 312
dielectric relaxation 33
dietetic 255
differential scanning calorimetry (DSC) 21, 25–27, 107, 150, 152, 154, 378
differential thermal analysis (DTA) 1, 37
diffusion 30–32, 41, 268, 442
diffusion coefficient 3, 443
dimethyl disulphide 432
dimethyl sulphide (DMS) 206, 431, 433
disaccharide 204, 390
dispersability 167, 253
dispersed phase 411
dispersion 191, 237, 240
disproportionation 420
dissociation 389
'doctoring' 335
Donnan-Barker type apparatus 421
doughs 205, 206, 218, 242, 258, 260, 270, 368–378, 383, 443, 448, 449
  biscuit 262
  masa 374
dough mixer 370
doughnuts 273
dressings 408
drip 224, 225, 279
drop volume technique 178
drying 272, 349, 350, 373, 376, 378
Du Nouy ring 170, 176
Dupré equation 182
Dutching 352

dynamic mechanical thermal analysis (DMTA) 23, 29, 37
dynamic surface tension 174

Easter eggs 362
egg 258, 263, 265, 271, 411, 413
  albumen 265
  frappe 341
  liquid 263
  protein 265, 266, 411
  yolk 411, 412
'egg-box' structure 135–137
elastic forces 380
elastic modulus 24, 126, 127
elasticity 84, 250, 260, 381, 382, 395
elastin 222
electric field sterilisation/pasteurisation 330
electrolyte 120
electron beams 305
electron spin resonance (ESR) 24, 27, 29, 306
electrostatic forces 86
electrostatic repulsion 55, 56, 134
elongational flow 80, 86
Embden–Meyerhof pathway see glycolytic pathway
emulsification 107, 168, 247, 339, 340
emulsifiers 49–62, 168, 244, 251, 255, 359, 360, 366, 412
emulsifying salts 247
emulsions 49–62, 109, 110, 167, 237, 240, 247, 251, 254, 339, 364, 409, 411
endomysial-sarcolemmal sheath 224
endomysium 222, 224
endosperm 218, 271, 370–372, 375–378, 428, 429
enrobing 362, 364, 450
Enterococcus faecium 250
enthalpy 8, 17, 19, 130
entropy 36, 139
enzymatic browning 305
enzymatic proteolysis 246
enzymes 1, 40, 195, 216, 220–225, 229, 245, 247, 258, 259, 274, 283, 293–303, 307, 309–312, 316, 327, 328, 344, 350, 397–404
  activity 344
  amylase 195, 274, 431
  amylolytic 428
  cellulase 195
  degradation 307
  glucosinolase 409
  hydrolysis 256
  lipolytic 245, 413
  myrosin 409
  pectin see pectin
  pectinase 216

pectinesterase (PE)   399
pectolytic   312, 327, 397, 435
polygalacturonase   216
protease   245, 274
proteolytic   428
rennet   245, 246
enzymic   258, 259, 283
enzymic browning   296
enzymolysis   425, 431
epicuticular   297
epidermal   297
epimysium   222
equilibrium relative humidity (ERH)   1–15, 268, 270, 335, 337, 343–345, 365   *see also* water activity
essential oils   431
ester linkages   413
ethanol   204–206, 209, 350, 395, 423, 425   *see also* alcohol
ethyl esters   422
ethylene   294, 295, 400, 401
eukaryotes   194
eutectic temperature   17
eutectics (fat)   25, 142–166, 361, 365
extruder   104, 271, 370, 374, 375, 378, 379, 383, 448
extrusion   42, 271, 382–384

fat   142–166, 263, 365   *see also* *cis*-formation, illipe, lauric, polyunsaturated, shea butter, soya bean oil, *trans*-formation, triglyceride, vegetable
  bloom   360
  globules   251
  migration   157, 162
  unsaturated   228
fatty acids (free, saturated and unsaturated)   142–146, 153, 278, 279, 282, 309, 413, 433
  short chain   422
fermentation   193–211, 293, 344, 348–350, 356, 391–395, 433, 435
  aerobic   193, 194, 228
  anaerobic   193, 194, 209
fermentation brining   391
fermenters   433
Fick's laws   30, 32, 443
ficoll   46
fillets   280
filter press   430
filtration   312, 433
fining   312, 434
fish   276–289, 387
fish oil   145
flaked   371
flaking rolls   373

Flint maize   372
floating (jam)   396
flocculation   50, 53, 55, 411, 433
Flory–Fox equation   119
flour   196, 218, 258–269, 274, 370, 384, 388, 407
flour confectionery   263
flowpack   343
fluctuations in temperature   309
FMBRA   262
foam   237, 306, 370, 376, 377, 379, 383, 417–423, 433
foamability   418
fobbing   418
fondant   332, 341, 342, 344
foreign bodies   296
formic acid   207
Fowke's theory   182
fractal flocs   55
fractionation   142, 155
freeze burn   221
freeze concentration   311, 312
freeze drying   2, 75, 304, 310
freezing   210, 212, 219–222, 225, 228, 229, 276, 277, 289, 293, 296, 297, 302, 307, 312, 328, 399, 400
freezing tunnels   304
French fries   *see* potato chips
freshening   403
Fritz butter making process   239, 242
frozen meals   451
frozen storage life   279
fructose   13, 37, 38, 274, 318, 335, 340, 344, 385
fruit   65, 204, 212, 216, 248, 292, 315–330, 387–414
  fining   312
  flotation (jam)   328
  gum   345
  juice   292, 308, 310, 399
  pulp   408
frying   219, 309, 383–385
fudge   332, 334, 338, 340–344
fugacity ratio   5
full cream   354
fungi   194
furans   69
furfurals   311
fuzzy logic   446, 447, 451

Gaio   *see* yoghurt
galactomannan   137, 138
galactose   317
galacturonate   137, 138
galacturonic acid   213, 316, 317
gamma rays   305
garlic   400, 408

gel 41, 117–140, 221, 228, 246, 249, 250, 265, 269, 315, 317, 323, 324, 332, 366, 372, 407
  formation 324
  proteins 431
gelatin 132, 134, 248, 312, 344
gelatinisation 217–221, 264, 266, 306, 431, 449
gelatinised starch 306, 309, 370
gelling agent 317
Ghana 348
gherkins 404
gibberellins (hormone) 428
Gibbs energy 4
Gibbs function 27
ginger 404
glass transition 3, 17–46, 338, 366, 372–382
glass transition temperature (T_g) 14, 17–46, 260, 268, 269, 274, 304, 338, 343, 345, 371–375, 379, 381, 385
glass transition theory 255
glassy
  forms 262
  state 272, 292, 304, 353, 375, 382, 383
  structure 271
  textures 368
glazing 273
gliadin 218
gloss 360, 362
glucose 13, 14, 67, 72, 134, 195–199, 273, 274, 318, 332, 335, 340, 341, 384, 385
glucosylamine 67
gluten 35, 206, 218, 260, 262, 269, 372
glutenin 218, 269
glycerin 224
glycerol 7, 13, 144, 423, 424
glycerol monostearate 339
glycogen 224, 228, 278, 279
glycol 240
glycolytic pathway 197
glycoside 409
good manufacturing practices (GMP) 414
Gordon–Taylor equation 36
grain endosperm 369
graining 337
granulation 243
grapes 202, 313
grape juice 312
gravy 448, 451
grinding 353, 366
grist 430
grits 369, 370, 378, 380
guluronate 135, 136
gum 332, 334, 344, 345, 388, 407–411
  arabic 61

guar 99, 408
  xanthan 99, 121, 138–140, 222

haem pigments 279
half-products 383
ham 227, 228, 287, 443
ham tumbling 228
Hamaker constant 412
hammer mills 430
hard-to-cook (HTC) legumes 217
headspace 417
heat capacity 27
heat resistance 216
heat transfer 219, 441
heat treatment 253
helical structure 132–135, 138, 268, 270
hemi-cellulases 274
hemicellulose 212, 214, 394
herring 278
Herschel Bulkley model 89
hexose 390, 391
Heyns rearrangement 68
high boiled sweets 334, 344
high boilings 332, 341
high pressure treatment 306, 329, 330
high temperature short time (HTST) 302
high temperature steam 296
HLB value 61
homogenisation 234, 235, 243, 251, 253, 356, 405–407, 411, 412
honey 204
hops 203, 418, 425
hormones 428
horseradish sauce 411
hot spices 408
HPLC 328
humectant 13
hydration 3, 258, 268, 271, 371, 372, 383
hydration force 55, 57
hydrocarbons 413
hydrocolloids 117–140, 269, 332, 344, 345
hydrodynamic volume 118
hydrogels 372
hydrogen 198
hydrogen peroxide 312
hydrogenated fats 157
hydrogenation 143, 158
hydrolysis 193, 344, 409, 413
hydroperoxides 413
hydrophobic attraction 55
hydrophylic 372
p-hydroxy benzyl isothiocyanate 409
hydroxyl group 216
hygrometer 12
hypertonic solution 197
hypochlorite 296

hysteresis 9

ice cream 234, 250–252, 365, 450
iced cakes 447
icings 268
illipe (fat) 158
image analysis 297
immersion cooking 287
immunoglobulins 255
in-pack pasteurisation (IPP) 403
India 383
inert atmosphere packaging 255
infant formula 253–256
infra-red 24
infra-red radiation 297
infra-red spectroscopy 447
instant milk powder 253
instant powders 253
interfacial tension 167–192
International Institute of Refigeration (IIR) 276, 277, 281
International Organisation of Cocoa, Chocolate and Confectionery (IOCCC) 357
inversion (sugar) 334, 335
invert sugar 332
invertase 344
iodine numbers 278
ionic strength 121, 131
iron 196, 422
irradiation 301, 305, 306
isinglass 434
iso-alpha-acids 419–422
iso-solids phase diagram 159
isoadhumulone 422
isocohumulone 422
isohumulone 422
isomer 143
isomerisation 143
isopiestic technique 11
isotherm 11, 13, 15
isothiocyanate 409
Ivory Coast 348

jam 301, 306, 308, 315–330
  extra 315
  reduced sugar 315, 320
Java cracker 383
jellies 315–330, 332

ketchup 110, 113, 405, 407
ketones 413
kieselguhr 433
kilning 429
Kloeckera spp. 202
Kraemer plot 119
Kraft wood pulp 312
Krieger–Dougherty equation 92

Kyoto University 306

lactalbumin 61, 255
lactic acid 203, 207–209, 224, 245–250, 279, 390, 392
Lactobacillus 199, 206–210, 414
  acidophilus 250
  brevis 392
  bulgaricus 249
  plantarum 392
lactoferrin 255
lactoglobulin 61, 249, 255
lactoperoxidase 255
lactose 41, 234, 246, 249, 255, 340, 384, 385
  amorphous 255
lactose intolerance 249
lager 203, 204, 422, 425, 429, 433
lamb 278–281
lamella 213–217, 298
Laplace pressure 50
Laplace theory 190
lard 156
laser light scattering instrument 358
latent heat 270, 307
lauric fats 157–165
lauter tun 430
lautering 433
layer adsorption 59
lecithin 62, 145, 255, 339, 359, 360, 366
lemon 317
lemon juice 409, 412
lettuce 73, 213, 295
Leucanostoc mesenteroides 392
leucine 72
Leuconostoc mesenteroides 210
light sterilisation/pasteurisation 330
lime 317
Lincoln biscuits 262
linoleic acid 155
lipase 245
lipid degradations 384
lipid oxidation 304, 305
lipids 195, 219, 234, 273, 279, 421 *see also* fat
lipolysis 246
lipoprotein 412
lipoxygenases 413
Listeria 305
live steam 301
loaf 270
longlife 244
loss modulus 24, 84
Lossen rearrangement 409
Lovibond systems 398
lozenge 334
lycopene 397, 398
lye peeling 297

lysine (amino acid) 75, 76, 196

Macara and Morpeth index 390
mackerel 228, 278
Madeira cakes 268
magic angle spinning (MAS) 23
magnesium 196, 217
magnetic resonance imaging (MRI) 30
Maillard 65–77, 206, 254–259, 272, 273, 292, 304, 307, 332–341, 345, 351, 356, 366, 384, 385, 405, 425, 426, 431, 432, 449
maize 372, 374, 377, 380, 407
Malaysia 348
Malaysian Kerapok 383
malic acid 203, 388, 390
malt 425, 428–431, 447
  black 429
  chocolate 429
  'curing' 430
  green 429
malting 425, 426, 431
maltodextrin 39, 273
maltose 27, 204, 224, 385
manganese 196
mango 389, 397
mannose 138
manufacture 333
margarine 51, 157, 237
Mark–Houwink relationship 120
marmalade 315, 325, 328
marshmallow 13
mashing 430, 431
mass transfer 219, 441
mathematical modelling 441
mathematical simulation 441
maturation 246
mayonnaise 88, 108, 387, 389, 408–414
measurement of set 322
meat 65, 76, 117, 213, 224, 276–289, 448, 451
meat pies 440, 441, 448
mechanical freezers 304
Meidi-Ya 330
melanoidin 429
melt transition 372
melting points T_m 375–379
meringues 263
metabolic respiration 293
metabolite 193, 200, 201
methode champenoise 203
methoxyl group 317
3-methyl-2-butene-1-thiol 421
methyl ester 325
methyl esterase 312
5-methyl-2-furfural 432
methylation 318
2-methylfurfuran 432

methylmethionine 431
Meura 435
Meura 2001 430
micelles 249, 339
micro-organism 1, 10, 14, 193–211, 245, 246, 294, 299–302, 312, 327–330, 393–396, 404, 413
microbial growth 395
microbial spoilage 296
microbiology 194, 262, 283
micronizing 351
microscopy 214
microwave heating 219, 273, 302, 304, 308, 309, 313, 444, 451
milk 70, 167, 206, 207, 234–256, 263, 278, 368
  cultured products 248
  fat 154, 156, 158, 340
  pasteurised 236
  powder 10, 74, 95, 167, 249, 253–256, 339, 354
  proteins 332, 339, 340
  solids 274, 385
  standardization 235
milling 218, 269, 350, 354, 374–377, 399, 407, 435
minimal processing 293, 300
mitochondria 212, 225
MIVAC process 313
modified atmosphere packaging (MAP) 301, 413, 414
modulus of elasticity 214
moisture diffusion 272
moisture migration 365
molasses 408
molecular mobility 28
molecular packing 147
molecular weight 317, 325–328, 409, 418–421
Monilliella acetoabutans 391
monosaccharides 118, 204, 334
morning goods 268
mould 295, 350, 359, 392, 397, 403
moulding 362, 364
mousse 88
mouthfeel 417, 422, 423
mucopolysaccharide 222
multi-component foods 440–451
Munsell 398
muscles 222, 228
mushroom ketchup 407
mushrooms 399
mustard 389, 408–414
  black mustard *Brassica nigra* 409
  brown mustard *Brassica juncea* 409
  white mustard *Brassica alba* 409
Mycobacterium leprae 77
myocommata 228

myofibres 222-227
myofibril 222, 224, 228, 229
myosin 222, 224, 229
myotomes 228

navy beans 217
near infrared (NIR) spectroscopy 12, 107
neural network 446-448, 451
Newtonian flow 83, 87, 124-127, 138
Nigeria 348
nitrogen 195, 196, 400, 417, 423
nitrogen blanketing 413
non-Newtonian flow 81, 86, 357, 407
non-enzymatic browning 41, 65-77, 272, 274 *see also* Maillard
nuclear magnetic resonance (NMR) 1, 17, 21-24, 29, 447
nucleation 240, 435
nucleation site 418
nuts 65

octanol 425
off-flavours 309
ohmic heating 308
oleic acid 153
oleine 155
olive oil 144
olives 389, 395, 404
onions 294, 400, 404, 407, 408
  pickled 389, 404
  silverskin 404
open boiling process (jam) 318
optical isomerism 147
orange 317, 325
  dummy 325, 326
  juice 310
  Seville 325, 328
orgelles 212
orthokinetic flocculation 53
oscillatory rheometry 250
osmosis 193, 330, 372
osmotic dehydration 305, 396
osmotic pressure 51, 57, 197, 213
Ostwald ripening 49, 50, 418
oven 258, 270, 284, 286
overrun 250, 252
oxidation 279
oxidative stability 412
oxido-reduction reaction 198
oxtail soup 449
oxygen 197, 400

packaging 259, 264, 281-283, 289, 292, 312, 338, 382, 433, 435
palm kernel oil 144, 156-159
palm mid-fraction 156
palm oil 154-158, 161
pan boiling process (jam) 319

panning 337
papain injections 224
particle size distribution 358, 359, 430
pasta 258, 259, 271, 272, 383, 451
Pasteur, Louis 193, 205
pasteurisation 75, 205, 238, 244, 245, 251, 253, 298, 306, 307, 311, 316, 320, 321, 397, 399, 403, 405, 411, 434
pastille 345
pastry 237, 242, 258-262, 272, 273, 441, 443, 448, 449
pastry crust 448, 449
pasty 448
peaches 216, 294
pearling 377
pears 294, 295
peas 217, 302
pectin 137, 213, 216, 217, 297, 312, 316-318, 321-328, 394, 399, 406, 407, 435
  lyase 327
  methoxyl 216
  methylesterase 327
Pediococcus cerevisiae 392
peeling 297
pelagic fish 228, 277, 280
Penicillium 194, 208
pentosan 269
peppermint 345
peppers 295, 400, 408
peptides 273, 274, 356, 384, 385, 418
pericarp 376, 378
pericarp surface 297
pericellular 229
perikinetic flocculation 53
perimysial connective tissue 227
perimysium 222, 227
perinysial 227
peroxidase 404
peroxidase test 299
peroxidation 309
peroxide values 279
pesticide residues 296
petiole 292
pH 3, 117, 122, 130, 131, 194, 197, 200, 207, 210, 224, 229, 246, 249, 279, 301, 307, 317, 324, 344, 390, 391, 394, 399, 409, 411, 413
phase inversion 237, 238
Phaseolus vulgaris *see* navy beans
phenols 356
phosphate 224, 227, 247
phosphoglyceride 145
phospholipid 57, 244, 279, 412
phosphorus 246, 255
phosphovitin 61
phosvitin 61
photosynthesis 195

pickle 210, 387–414, 404, 407
pickling 30
pies 280
pigs 278
pimento 404
pineapple 308
pizza 3, 397, 449, 450
pK_a value 389
plansifter 430
plant tissue 212–222
plasmalemma 212, 214, 222, 229
plasmolysis 394
plastic 383
plastic viscosity (PV) 357–360, 446
plasticiser 34, 213, 259, 260, 368, 370, 372, 382
plasticity 421
plate evaporator 320
plate freezers 280
plate heat exchanger 321
plugging 406
pneumatic fingers 297
Poiseuille's formula 191
polygalacturonase (PG) 312, 327, 399, 404
polygalacturonic acid 327
polyglyceril polyricinoleate (PGPR) 360
polymerisation 309, 335
polymers 49–62, 99, 117, 178, 259, 265, 268–271, 306, 312, 368–383, 399, 429
polymorphic forms 107
polymorphism 143, 147–153, 165, 166
polyols 340
polypeptides 419–421, 434
polyphenols 312, 356, 423–425, 431
polyphosphate 228
polysaccharides 58, 108, 118, 135–139, 195, 209, 249, 252, 297, 310
polyunsaturated fat 277
polyvinylpolypyrrolidone (PVPP) 434
popcorn 3, 370, 376–378, 383
pork 69, 277–282
potassium carbonate 352
potato 213, 215, 217, 218, 221, 294, 297, 308, 380, 451
  chips 3, 219, 307
  crisps 219
  hoops 376
  starch 407
  sticks 384
poultry 213, 280, 286
practical storage life (PSL) 276, 278, 281
pre-freezing treatment 277
pregelatinised flour 374
preservation index 390, 408
preserves 315–330
preset of jams/preserves 324
pressing 435

pressure 287, 288
pressure cooking 287
pressure vessels 373
Prestine 154
procyanidins 424
prokaryotes 194
propionate 246
Propionibacterium 208
propionic acid 208, 390
protein 14, 58, 69, 75, 95, 108, 109, 134, 195, 209, 217–238, 246–256, 262–272, 278, 292, 306, 310, 356, 368, 372, 376, 420–423, 429, 448
  barley 420
  denaturation 258
  emulsifier 255
  foams 421
  gel 259, 269
  myofibrillar 224, 227, 228
  phase gel 259
  sarcoplasmic 224
proteolytic action 193, 207, 246
proton 389
protoplast 212
proximity equilibrium cell 12
pseudoplastic 124, 139, 140
puff pastries 242
puffed cereals 370
puffed rice 376
puffed wheat 370, 376, 377
puffing process 370, 376–378
pullulan 26, 27, 325
pulp 311
purée 308, 397
purines 196
pyrazines 69, 72, 73
pyrimidine 196
pyruvate 198, 199
pyruvic acid 70, 75, 198

radiation 363, 444
radio frequency (RF) 308
rainbow trout 229
raisin 313
rancidity 277
Raoult's law 4
rapeseed oil 158
reconstitution 309
red cabbage 404
redox potential 3
reducing sugars 274, 340, 384, 385, 429, 449
refractrometry 310
refrigeration 271
rehydration 30, 293, 304, 310
relative humidity (RH) 268, 295, 429
relishes 389, 401
rennet 245, 246

INDEX

retortable pouches 302
retorting 301
retrogradation 373
reverse osmosis 312
rhamnose unit 137, 316
rhamsan 138
rheology 80–114, 247–250, 253, 258, 269–272, 292, 308, 322, 330, 345, 353, 354, 366, 387, 408
rheoplexy 99, 100
rhubarb 292
riboflavin vitamin 196
rice 46, 407
rigidity 323
rigor mortis 224, 225, 229, 279, 280
ripening 246
roasting 350, 351, 356, 366
roll refiner 353
roller drying 254
roller mills 430
rotational viscometry 250
rubbery state 17–46, 262, 269, 337
rutabaga 400

Saccharomyces 194, 195, 202–205, 209
Saccharomyces cerevisiae 435
salad 412
    cream 408, 413, 414
    crops 295
    dressing 387, 408
Salmonella 305, 414
salsas 397
salt 51, 225, 227, 237, 239, 274, 377, 388–392, 401, 440, 443
salting 246
sandwich spreads 411
saponification 352
sarcolemma 222, 227, 228
sarcomere 223, 226
sarcoplasmic fluid 225
sarcoplasmic reticulum 223, 229
sauces 3, 308, 387–414, 451.
    tartare 411
    thick 388
    thin 388
    Worcestershire 389, 408
sauerkraut 210, 395
scraped surface evaporator 320
scraped surface heat exchanger 240, 308, 321, 341
scraped surface process 240, 242, 251
sedimentation 53
semolina 267, 271, 272
sensor 12
sensory analysis 80
sessile drop method 186
shea butter (fat) 158
shear 117, 139, 269, 271, 407

flow, stress and rate 80–114, 124, 126, 250
    modulus 214
    -thinning 124, 125
sheeting 271
shelf life 15, 62, 256, 259, 267–270, 274, 276, 292–295, 300, 301, 318, 384, 400, 412
shredded wheat 369, 373
silverside 287
sinalbin 409
sinigrin 409
skimmed milk 235, 236, 273, 354
    powder 247, 340
    solids 251, 273
snackfoods 368
sodium 247
    carboxy methyl cellulose (CMC) 408
    caseinate 255
    chloride 393, 394
    hydroxide 297
sol state 130, 132
sol–gel transformation 239
solubility 253
sorbic acid 390, 414
sorbitan ester 61
sorbitol 340
soups 75, 307, 308
soy sauce 194
soya 274
soya bean oil 145, 157, 158, 161, 242
specific heat 21, 442
specific mechanical energy 381
specific surface area 358
spices 388, 405, 407, 412, 413
spinning drop method 177
sponge 206, 263, 365
sports powders 255
spray drying 253, 254
spraying 105, 173
spreads 117, 408
squash 397
stabilisers 117–140, 157, 248, 251
Stable Microsystems TA–XT 2 Texture Analyser 322
starch 14, 61, 91, 132, 135, 195, 213, 217–219, 248, 259–273, 300, 302, 345, 368–383, 388, 406–411, 435
    gelatinisation 258, 431
    granules 262, 371, 375, 377, 429
    moulding 345
    polymer 381
state diagram 25
steaks 280, 286
steam 297
stearine 155
Steel's masher 430
steeping 427

steric
  clashes 120
  forces 412
  hindrance 75
  repulsion 55, 56
sterilant 312
sterilisation 205, 302, 306, 309, 405
sterilisation value (Fo) 308
Stevens LFRA Texture Analyser 322
stickiness 423
sticky 255
stochastic process 51, 55
Stoke's law 50, 264, 265, 403, 412
storage 221, 222, 259, 271, 279, 292–296, 303, 304, 392
storage life 276, 278
storage modulus 84, 127
stoves 345
strawberries 292, 295
Strecker degradation 72, 73
Streptococcus 199, 207
Streptococcus thermophilus 249, 250
subliming 304
sucrose 7, 9, 13, 25–27, 29, 33, 41, 51, 264, 270, 273, 274, 318, 337, 340
sucrose inversion 324
sugar 21, 39, 65, 69, 95, 202–204, 213, 251, 258, 262–267, 273, 274, 292, 302, 305, 317–324, 328, 350, 353, 388, 404, 405, 425, 431, 435, 448 *see also* arabinose, dextrose, fructose, glucose, hexose, lactose, maltose, monosaccharides, polysaccharides, sucrose
  amorphous 353, 354
  confectionery 332–345
  glass 262
  refining 447
  syrup 264
sugaring 391, 396
sulphur 195, 196
sulphur dioxide 305, 311, 325, 328, 335, 401
sunflower oil 145, 358
surface
  active agent 167–192, 359, 411, 420, 442
  properties 254
  tension 50, 236–238, 380, 418, 420 *see also* interfacial tension
  wetting 345
surfactant *see* surface active agent
sweetened condensed milk 340
syneresis 221, 246, 405, 406
syruping 396

tableting 345
tacos 374, 375
tallow 156
tamarinds 397, 408
Tamarindus indica 397
tannins 203, 312, 352, 405, 431, 434, 435
tartaric acid 390
teig 431
temperature fluctuation 281, 282
tempering 105, 280, 360–363
texture 80–114, 247, 251, 271, 277, 296
thaw-shortening 225
thawing 222, 225, 280, 309, 311, 328, 440, 451
Theobroma cacao 208, 347
thermal conductivity 442, 443
thermal death time (TDT) 302
thermal runaway 309
thermally accelerated short time evaporator (TASTE) 311
thermoluminescence 306
thiamin 196, 307, 308
thickeners 117–140
thixotropy 99, 100, 237, 239, 250, 421
time temperature tolerances (TTT) 303
toasting 373
toffees 332, 334, 338, 340, 343, 365
  continuous cooker 341
tomato ketchup 397, 405, 406, 407
tomatoes 216, 292–295, 298, 307, 309, 389, 397–400, 407
toppings 242
tortilla 370, 371, 374–376
  chips 219, 385
  snacks 369
tragacanth 408
trans-formation (fat) 144, 157
transmission electron microscopy 226
transport 279
treacles 332, 340
trehalose 310
tricarboxylic acid (TCA) 198
trichloroethane 358
trigeminal nerve 422
triglyceride 50, 142–166, 251, 255, 413
trihydric acid 144
trilaurin 156
tropical fruits 302
troponin 222
trub 431
tubes 384
tubules 214
turgor pressure 213–217, 220, 394, 395
turkey 279, 287
turmeric 404
Tuynenberg-Muys index 390
twistwrap 343

UHT 236, 243, 249
UHT coffee 243

INDEX

ultra filtration 311, 312
ultra high temperature (UHT) 236
ultrasonic radiation 433
UNIFAC 27
Uniplast 335
UNIQUAC 27
uronic acid 137

vacuole *see* cells vacuole
vacuum 271, 287, 288, 297, 301, 307, 333, 341
  boiling process 320
  concentration 397
  cookers 307
  cooling 287
  drying 305, 313
  evaporation 312
  filling 405
  freezing 309
  packing 276, 287
  pan 318
van der Waals force 55, 56, 86, 412
vapour pressure 1–15, 263, 366, 382, 395
veal 73
vegetable fat 237, 251, 255, 338
vegetables 65, 73, 210, 212, 292, 387–414, 449, 451
vinegar 204, 205, 388, 389, 405–411
  cider 204
  malt 204
  mint 387
  wine 204
viscoelasticity 84, 89, 100, 126, 127, 268, 269, 368, 372, 378, 411
viscometer 82, 107
  Brookfield 399
viscosity 17, 32, 39, 80–114, 118, 236, 243, 250, 264–270, 339, 343, 361, 373, 379, 381, 388, 403, 406–411, 419, 423, 424, 431 *see also* rheology
  apparent 81, 87, 221, 407
  batter 265
  dynamic 127, 129, 138
  intrinsic 95–99, 118–120, 126
  viscosity 421
vitamins 196, 256, 300, 329 *see also* thiamin
  biotin 196
  vitamin C 221, 301, 311

Vogel, Tamman and Fulcher (VTF) equation 33
Vromann series 60

wafer 258, 270, 271, 365
Warner–Bratzler shear press 227

Washburn equation/technique 188–191
water activity 1–15, 17, 30, 34, 41, 75, 122, 197, 218, 247, 254, 259, 267, 268, 274, 292, 305, 337, 366, 382, 395, 396, 403, 431, 442, 448, 449 *see also* equilibrium relative humidity
water adsorption isotherms 305
water immersion cooking 287
water vapour 379
wax coating 295
weight losses 287
Weissenberg number 112
welan 138
wetting 167–192, 253, 293
  adhesional 182
  immersional 183
wheat 258, 267, 272, 407
  biscuits 369
  durum 271, 274
  flour 206, 378, 381
  proteins 260, 268
  starch 378
whey 207, 236, 246, 248, 251, 255, 273, 385
whips 242
widget 417
Wilhelmy Pt Plate technique 170, 172, 187
wine 202, 203, 211
wine gum 345
winnowing 351
WLF equation 33, 39–43
wood pulp 312
wort 204, 425, 430–435

X-ray 135, 147, 148, 156
X-ray diffraction 45
xanthophylls 397
xylose 317

yeasts 193–211, 268, 274, 350, 392, 396, 403, 414, 433, 435
  osmophilic 396
  top fermenting 433
yield value 81, 90, 362, 364, 407, 446 *see also* Casson yield value
YN 359, 360
yoghurt 206, 207, 248–250, 308
  bio 250
  set 249
Yorkie 355
Young's equation 182

zinc 196
Zisman plot 184